T0140140

Emergence, Complexity and Computation

Volume 28

The Emergence, Complexity and Computation (ECC) series publishes new developments, advancements and selected topics in the fields of complexity, computation and emergence. The series focuses on all aspects of reality-based computation approaches from an interdisciplinary point of view especially from applied sciences, biology, physics, or chemistry. It presents new ideas and interdisciplinary insight on the mutual intersection of subareas of computation, complexity and emergence and its impact and limits to any computing based on physical limits (thermodynamic and quantum limits, Bremermann's limit, Seth Lloyd limits...) as well as algorithmic limits (Gödel's proof and its impact on calculation, algorithmic complexity, the Chaitin's Omega number and Kolmogorov complexity, non-traditional calculations like Turing machine process and its consequences,...) and limitations arising in artificial intelligence field. The topics are (but not limited to) membrane computing, DNA computing, immune computing, quantum computing, swarm computing, analogic computing, chaos computing and computing on the edge of chaos, computational aspects of dynamics of complex systems (systems with self-organization, multiagent systems, cellular automata, artificial life,...), emergence of complex systems and its computational aspects, and agent based computation. The main aim of this series it to discuss the above mentioned topics from an interdisciplinary point of view and present new ideas coming from mutual intersection of classical as well as modern methods of computation. Within the scope of the series are monographs, lecture notes, selected contributions from specialized conferences and workshops, special contribution from international experts.

More information about this series at http://www.springer.com/series/10624

Susan Stepney · Andrew Adamatzky
Editors

Inspired by Nature

Essays Presented to Julian F. Miller
on the Occasion of his 60th Birthday

 Springer

Editors
Susan Stepney
Department of Computer Science
University of York
York
UK

Andrew Adamatzky
Unconventional Computing Centre
University of the West of England
Bristol
UK

ISSN 2194-7287   ISSN 2194-7295   (electronic)
Emergence, Complexity and Computation
ISBN 978-3-319-88528-5   ISBN 978-3-319-67997-6   (eBook)
https://doi.org/10.1007/978-3-319-67997-6

Printed on acid-free paper

This Springer imprint is published by Springer Nature
The registered company is Springer International Publishing AG
The registered company address is: Gewerbestrasse 11, 6330 Cham, Switzerland

Preface

Julian Francis Miller

This book is a tribute to Julian Francis Miller's breadth of ideas and achievements in computer science, evolutionary algorithms and genetic programming, electronics, unconventional computing, artificial chemistry, and theoretical biology. Well-known for both Cartesian Genetic Programming and evolution *in materio*, Julian has further interests from quantum computing to artificial chemistries. He has over 200 refereed publications (http://www.cartesiangp.co.uk/jfm-publications.html); here, we highlight just a few of his major accomplishments.

Julian started his life in science as mathematical physicist working on the interaction of solitons in various nonlinear partial differential equations such as the sine-Gordon equation [3, 5], and the modified Korteweg-de Vries equation [4]. He entered classical computer science with his paper on synthesis and optimisation of networks implemented with universal logic modules [1, 16, 30]. Julian's interest in optimisation led him to genetic algorithms, which he employed for optimisation of field-programmable arrays [26], Reed-Muller logical functions [15], finite-state machines [2], and evolving combinatorial logic circuits [18, 22, 23] and non-uniform cellular automata [27, 28].

Julian combined his interests in physics and computer science in work on constant complexity algorithm for solving Boolean satisfiability problems on quantum computers, and quantum algorithm for finding multiple matches [31]. Julian's ideas in optimisation of circuits and quantum computing are reflected in Younes' Chapter "Using Reed-Muller Expansions in the Synthesis and Optimization of Boolean Quantum Circuits".

Julian's interest in combining natural processes and computation expanded from physics to include the exciting world of biological processes, such as evolution and morphogenesis. He used principles of morphogenesis to evolve computing circuits and programs [14, 17, 19]. These aspects of Julian's work are reflected in Chapters "Evolvable Hardware Challenges: Past, Present and the Path to a Promising Future" by Haddow and Tyrell, "Artificial Development" by Kuyucu et al., and Banzhaf's "Some Remarks on Code Evolution with Genetic Programming".

In 2000, Julian, together with Peter Thomson, presented a fully developed concept of Cartesian Genetic Programming (CGP) [24]. There, a program is genetically represented as a directed graph, including automatically defined functions [29] and self-modifying operators [10]. This approach has become very popular, because it allows the discovery of efficient solutions across a wide range of mathematical problems and algorithms. Several chapters of the book manifest the success of CGP in diverse application areas: "Designing Digital Systems Using Cartesian Genetic Programming and VHDL" by Henson et al.; "Breaking the Stereotypical Dogma of Artificial Neural Networks with Cartesian Genetic Programming" by Khan and Ahmad; "Approximate Computing: An Old Job for Cartesian Genetic Programming?" by Sekanina; "Medical Applications of Cartesian Genetic Programming" by Smith and Lones; "Multi-step Ahead Forecasting Using Cartesian Genetic Programming" by Dzalbs and Kalganova; "Cartesian Genetic Programming for Control Engineering" by Clarke; "Bridging the Gap Between Evolvable Hardware and Industry Using Cartesian Genetic Programming" by Vasicek; "Combining Local and Global Search: A Multi-objective Evolutionary Algorithm for Cartesian Genetic Programming" by Kaufmann and Platzner.

In 2001, Miller and Hartman published "Untidy evolution: Evolving messy gates for fault tolerance" [21]. Their ideas of exploiting of "messiness" to achieve "optimality"—"natural evolution is, par excellence, an algorithm that exploits the physical properties of materials"—gave birth to a new field of unconventional computing: evolution *in materio* [7, 12, 20]. The evolution *in materio* approach has proved very successful in discovering logical circuits in liquid crystals [11–13], disordered ensembles of carbon nanotubes [6, 7, 25] (and Chapter "Evolution in Nanomaterio: The NASCENCE Project" by Broersma), slime mould (Chapter "Discovering Boolean Gates in Slime Mould" by Harding et al.), living plants (Chapter "Computers from Plants We Never Made: Speculations" by Adamatzky et al.), and reaction-diffusion chemical systems ("Chemical Computing Through Simulated Evolution" by Bull et al.).

Julian's inspiration from nature has not neglected the realm of chemistry: he has exploited chemical ideas in the development of a novel form of artificial chemistry, used to explore emergent complexity [8, 9]. Chapter "Sub-Symbolic Artificial Chemistries" by Faulkner et al. formalises this approach.

The book will be a pleasure to explore for readers from all walks of life, from undergraduate students to university professors, from mathematicians, computers scientists, and engineers to chemists and biologists.

York, UK   Susan Stepney
Bristol, UK   Andrew Adamatzky
June 2017

References

1. Almaini, A.E.A., Miller, J.F., Xu, L.: Automated synthesis of digital multiplexer networks. IEE Proc. E (Comput. Dig. Tech.) **139**(4), 329–334 (1992)
2. Almaini, A.E.A., Miller, J.F., Thomson, P., Billina, S.: State assignment of finite state machines using a genetic algorithm. IEE Proc. Comp. Dig. Tech. **142**(4), 279–286 (1995)
3. Bryan, A.C., Miller, J.F., Stuart, A.E.G.: A linear superposition formula for the sine-Gordon multisoliton solutions. J. Phys. Soc. Japan **56**(3), 905–911 (1987)
4. Bryan, A.C., Miller, J.F., Stuart, A.E.G.: Superposition formulae for multisolitons. II.—The modified Korteweg-de Vries equation. Il Nuovo Cimento B **101**(6), 715–720 (1988)
5. Bryan, A.C., Miller, J.F., Stuart, A.E.G.: Superposition formulae for sine-Gordon multisolitons. Il Nuovo Cimento B **101**(6), 637–652 (1988)
6. Dale, M., Miller, J.F., Stepney, S.: Reservoir computing as a model for *in materio* computing. In: Advances in Unconventional Computing, pp. 533–571. Springer, Berlin (2017)
7. Dale, M., Miller, J.F., Stepney, S., Trefzer, M.A.: Evolving carbon nanotube reservoir computers. In: International Conference on Unconventional Computation and Natural Computation, pp. 49–61. Springer, Berlin (2016)
8. Faulconbridge, A., Stepney, S., Miller, J.F., Caves, L.: RBN-world: The hunt for a rich AChem. In: ALife XII, Odense, Denmark, August 2010, pp. 261–268. MIT Press, Cambridge (2010)
9. Faulconbridge, A., Stepney, S., Miller, J.F., Caves, L.S.D.: RBN-World: A sub-symbolic artificial chemistry. In: ECAL 2009, Budapest, Hungary, September 2009, vol. 5777 of LNCS, pp. 377–384. Springer, Berlin (2011)
10. Harding, S., Miller, J.F., Banzhaf, W.: Developments in *Cartesian Genetic Programming*: Self-modifying CGP. Genet. Program. Evolvable Mach. **11**(3/4), 397–439 (2010)
11. Harding, S., Miller, J.: Evolution in materio: Initial experiments with liquid crystal. In: Proceedings of NASA/DoD Conference on Evolvable Hardware, 2004, pp. 298–305. IEEE (2004)
12. Harding, S., Miller, J.F.: A scalable platform for intrinsic hardware and in materio evolution. In: Proceedings of NASA/DoD Conference on Evolvable Hardware, 2003, pp. 221–224. IEEE (2003)
13. Harding, S., Miller, J.F.: Evolution in materio: Evolving logic gates in liquid crystal. In: Proc. Eur. Conf. Artif. Life (ECAL 2005), Workshop on Unconventional Computing: From Cellular Automata to Wetware, pp. 133–149. Beckington, UK (2005)
14. Liu, H., Miller, J.F., Tyrrell, A.M.: A biological development model for the design of robust multiplier. In: Workshops on Applications of Evolutionary Computation, pp. 195–204. Springer, Berlin, Heidelberg (2005)
15. Miller, J., Thomson, P., Bradbeer, P.: Ternary decision diagram optimisation of reed-muller logic functions using a genetic algorithm for variable and simplification rule ordering. In: Evolutionary Computing, pp. 181–190 (1995)

16. Miller, J.F., Thomson, P.: Highly efficient exhaustive search algorithm for optimizing canonical Reed-Muller expansions of boolean functions. Int. J. Electr. **76**(1), 37–56 (1994)
17. Miller, J.: Evolving a self-repairing, self-regulating, french flag organism. In: Genetic and Evolutionary Computation—GECCO 2004, pp. 129–139. Springer, Berlin, Heidelberg (2004)
18. Miller, J., Thomson, P.: Restricted evaluation genetic algorithms with tabu search for optimising boolean functions as multi-level and-exor networks. In: Evolutionary Computing, pp. 85–101 (1996)
19. Miller, J.F.: Evolving developmental programs for adaptation, morphogenesis, and self-repair. In: European Conference on Artificial Life, pp. 256–265. Springer, Berlin, Heidelberg (2003)
20. Miller, J.F., Downing, K.: Evolution in materio: looking beyond the silicon box. In: Proceedings of NASA/DoD Conference on Evolvable Hardware, pp. 167–176. IEEE (2002)
21. Miller, J.F., Hartmann, M.: Untidy evolution: evolving messy gates for fault tolerance. In: International Conference on Evolvable Systems, pp. 14–25. Springer, Berlin, Heidelberg (2001)
22. Miller, J.F., Job, D., Vassilev, V.: Principles in the evolutionary design of digital circuits—part I. Genetic Prog. Evolvable Mach. **1**(1–2), 7–35 (2000)
23. Miller, J.F., Thomson, P.: Combinational and sequential logic optimisation using genetic algorithms. In: GALESIA. First International Conference on Genetic Algorithms in Engineering Systems: Innovations and Applications, 1995 (Conf. Publ. No. 414), pp. 34–38. IET (1995)
24. Miller, J.F., Thomson, P.: Cartesian genetic programming. In: European Conference on Genetic Programming, pp. 121–132. Springer, Berlin Heidelberg (2000)
25. Mohid, M., Miller, J.F., Harding, S.L., Tufte, G., Massey, M.K., Petty, M.C.: Evolution-in-materio: solving computational problems using carbon nanotube–polymer composites. Soft Comput. **20**(8), 3007–3022 (2016)
26. Thomson, P., Miller, J.F.: Optimisation techniques based on the use of genetic algorithms (gas) for logic implementation on fpgas. In: IEE Colloquium on Software Support and CAD Techniques for FPGAs, pp. 4–1. IET (1994)
27. Vassilev, V., Miller, J., Fogarty, T.: The evolution of computation in co-evolving demes of non-uniform cellular automata for global synchronisation. In: Advances in Artificial Life, pp. 159–169 (1999)
28. Vassilev, V.K., Miller, J.F., Fogarty, T.C.: Co-evolving demes of non-uniform cellular automata for synchronisation. In: Proceedings of the First NASA/DoD Workshop on Evolvable Hardware, 1999, pp. 111–119. IEEE (1999)
29. Walker, J.A., Miller, J.F.: The automatic acquisition, evolution and re-use of modules in Cartesian genetic programming. IEEE Trans. Evol. Comput. **12**, 397–417 (2008)
30. Xu, L., Almaini, A.E.A., Miller, J.F., McKenzie, L.: Reed-Muller universal logic module networks. IEE Proc. E-Computers Dig. Tech. **140**(2), 105–108 (1993)
31. Younes, A., Rowe, J., Miller, J.: A hybrid quantum search engine: A fast quantum algorithm for multiple matches. arXiv preprint quant-ph/0311171 (2003)

Contents

Part I   Evolution and Hardware

Evolvable Hardware Challenges: Past, Present and the Path to a Promising Future   3
Pauline C. Haddow and Andy M. Tyrrell

Bridging the Gap Between Evolvable Hardware and Industry Using Cartesian Genetic Programming 39
Zdenek Vasicek

Designing Digital Systems Using Cartesian Genetic Programming and VHDL 57
Benjamin Henson, James Alfred Walker, Martin A. Trefzer
and Andy M. Tyrrell

Evolution in Nanomaterio: The NASCENCE Project 87
Hajo Broersma

Using Reed-Muller Expansions in the Synthesis and Optimization of Boolean Quantum Circuits 113
Ahmed Younes

Part II   Cartesian Genetic Programming Applications

Some Remarks on Code Evolution with Genetic Programming 145
Wolfgang Banzhaf

Cartesian Genetic Programming for Control Engineering 157
Tim Clarke

Combining Local and Global Search: A Multi-objective Evolutionary Algorithm for Cartesian Genetic Programming 175
Paul Kaufmann and Marco Platzner

Approximate Computing: An Old Job for Cartesian Genetic Programming? 195
Lukas Sekanina

Breaking the Stereotypical Dogma of Artificial Neural Networks with Cartesian Genetic Programming 213
Gul Muhammad Khan and Arbab Masood Ahmad

Multi-step Ahead Forecasting Using Cartesian Genetic Programming 235
Ivars Dzalbs and Tatiana Kalganova

Medical Applications of Cartesian Genetic Programming 247
Stephen L. Smith and Michael A. Lones

Part III Chemistry and Development

Chemical Computing Through Simulated Evolution 269
Larry Bull, Rita Toth, Chris Stone, Ben De Lacy Costello
and Andrew Adamatzky

Sub-Symbolic Artificial Chemistries 287
Penelope Faulkner, Mihail Krastev, Angelika Sebald and Susan Stepney

Discovering Boolean Gates in Slime Mould 323
Simon Harding, Jan Koutník, Júrgen Schmidhuber
and Andrew Adamatzky

Artificial Development 339
Tüze Kuyucu, Martin A. Trefzer and Andy M. Tyrrell

Computers from Plants We Never Made: Speculations 357
Andrew Adamatzky, Simon Harding, Victor Erokhin, Richard Mayne,
Nina Gizzie, Frantisek Baluška, Stefano Mancuso
and Georgios Ch. Sirakoulis

Part I
Evolution and Hardware

Evolvable Hardware Challenges: Past, Present and the Path to a Promising Future

Pauline C. Haddow and Andy M. Tyrrell

Abstract The ability of the processes in Nature to achieve remarkable examples of complexity, resilience, inventive solutions and beauty is phenomenal. This ability has promoted engineers and scientists to look to Nature for inspiration. Evolvable Hardware (EH) is one such form of inspiration. It is a field of evolutionary computation (EC) that focuses on the embodiment of evolution in a physical media. If EH could achieve even a small step in natural evolution's achievements, it would be a significant step for hardware designers. Before the field of EH began, EC had already shown artificial evolution to be a highly competitive problem solver. EH thus started off as a new and exciting field with much promise. It seemed only a matter of time before researchers would find ways to convert such techniques into hardware problem solvers and further refine the techniques to achieve systems that were competitive (better) than human designs. However, almost 20 years on, it appears that problems solved by EH are only of the size and complexity of that achievable in EC 20 years ago and seldom compete with traditional designs. A critical review of the field is presented. Whilst highlighting some of the successes, it also considers why the field is far from reaching these goals. The chapter further redefines the field and speculates where the field should go in the next 10 years.

P.C. Haddow (✉)
CRAB Lab, Department of Computing and Information Science,
Norwegian University of Science and Technology, Trondheim, Norway
e-mail: pauline@ntnu.no

A.M. Tyrrell
Intelligent Systems & Nano-Science Group, Department of Electronic Engineering,
University of York, York, UK
e-mail: andy.tyrrell@york.ac.uk

© Springer International Publishing AG 2018
S. Stepney and A. Adamatzky (eds.), *Inspired by Nature*, Emergence,
Complexity and Computation 28, https://doi.org/10.1007/978-3-319-67997-6_1

1 Introduction

Yao and Higuchi published a paper in 1999 entitled "Promises and Challenges of Evolvable Hardware" which reviewed the progress and possible future direction of what was then a new field of research, Evolvable Hardware [1]. A little more than ten years on, this chapter[1] considers the progress of this research field, both in terms of the successes achieved to date and also the failure to fulfil the promises of the field highlighted in the 1999 paper. Through a critical review of the field, the authors' intention is to provide a realistic status of the field today and highlight the fact that the challenges remain, to redefine the field (what should be considered as Evolvable Hardware) and to propose a revised future path.

In the mid 1990s, researchers began applying Evolutionary Algorithms (EAs) to a computer chip that could dynamically alter the hardware functionality and physical connections of its circuits [2–10]. This combination of EAs with programmable electronics—e.g. Field Programmable Gate Arrays (FPGAs) and Field Programmable Analogue Arrays (FPAAs)—spawned a new field of EC called Evolvable Hardware.

The EH field has since expanded beyond the use of EAs on simple electronic devices to encompass many different combinations of EAs and biologically inspired algorithms (BIAs) with various physical devices or simulations thereof. Further, the challenges inherent in EH have led researchers to explore new BIAs that may be more suitable as techniques for EH.

In this chapter we define the field of EH and split the field into the two related but different sub-fields: Evolvable Hardware Design (EHD) and Adaptive Hardware (AH).

Evolvable Hardware, as the name suggests, should have a connection to embodiment in a real device. However, in a number of EH papers, results produced are interpreted as hardware components e.g. logic gates, illustrated as a circuit diagram and justified as a hardware implementation, despite the lack of a realistic hardware simulator. In this chapter such work is not considered as Evolvable Hardware. The lack of grounding in real hardware, either physical or through realistic simulators, defies their inclusion in the field of EH. In other cases, authors attempt to speed up their EC process by implementing part or all of the BIA in hardware. In other words, hardware accelerators are certainly not, as defined in this chapter, Evolvable Hardware.

In this chapter we define **the field of Evolvable Hardware (EH) as the design or application of EAs and BIAs for the specific purpose of *creating*[2] physical devices and novel or optimised physical designs.**

---

[1]This chapter is a revised and updated version of: Pauline C. Haddow, Andy M. Tyrrell (2011) Challenges of evolvable hardware: past, present and the path to a promising future. Genetic Programming and Evolvable Machines 12(3):183–215.

[2]"creating" refers to the creation of a physical entity.

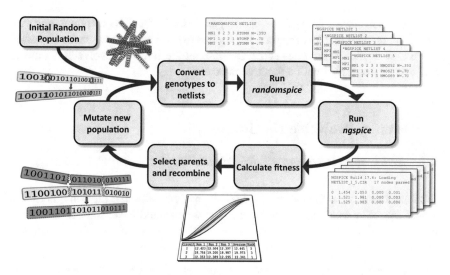

Fig. 1 An evolvable hardware cycle, for circuit design where high-fidelity simulation is used [13]

With this in mind, there are some successes in the field including analogue and digital electronics, antennas, MEMS chips, optical systems as well as quantum circuits. Section 3 presents an overview of some of these successes.

However, let us step back for a minute and consider what evolving hardware should consist of. Figure 1 illustrates an example of EH where an accurate model of the device physics is applied to produce the fitness function used to evaluate the efficacy of the circuits. So while the evolutionary loop is using no hardware, the simulation models used are very accurate with respect to the final physical system (in this case transistors with variable characteristics). Other forms of realistic hardware designs may include device simulators, as in [11], or actual physical devices, as in [12].

Such work can be classed under the subfield of Evolvable Hardware Design. The goal is novel or optimised non-adaptive hardware designs, either on physical hardware or as solutions generated from realistic simulators.

The sub-field of Adaptive Hardware can be defined as the application of BIAs to endow real hardware, or simulations thereof, with some *adaptive characteristics*. These adaptive characteristics enable changes within the EH design/device, as required, so as to enable them to continue operating correctly in a changing environment. Such environmental pressure may consist of changes in the operational requirements i.e. functionality change requirements, or maintenance of functionality e.g. robustness, in the presence of faults. Examples of such Adaptive Hardware include an FPAA that can change its function as operational requirements change or a circuit on an FPGA that can "evolve" to heal from radiation damage.

The remainder of the chapter is structured as follows: Sect. 2 provides a definition of the field together with the inherent advantages one can expect from such a field. Some actual success stories are reviewed in a little more detail in Sect. 3.

Section 4 considers many of the challenges that still face the community. Some newer approaches that are, or might be, applied to evolvable hardware are discussed in Sect. 5. Section 6 provides some thoughts on the future of evolvable hardware and Sect. 7 concludes with a short summary.

2 Defining Evolvable Hardware

2.1 *Evolvable Hardware Characteristics*

In the literature, it is difficult to find a set of characteristics that the community has agreed upon for what characterises an EH system should have. Indeed most work does not address this issue at all. However, there are a number of fundamental characteristics that should be included in such a definition:

- The system will be evolved not designed (an obvious one, but worth pointing out) N.B.: The term "evolved" is used to encompass any BIA and does not assume that an EA is used.
- It is often not an optimal design, but will be fit for the purpose (typical of biological systems in general—due to the evolution process).
- Typically, evolved systems do show levels of fault tolerance not seen in designed systems (again typical of biological systems).
- They may be designed to be adaptable to environment change.
  Such adaptivity is dependent upon when the evolutionary cycle stops. This is not necessarily true for all evolved systems, for example those that stop once the goal has been reached (again this is a characteristic of all biological systems).
- They produce systems that are most often unverifiable formally; indeed in many cases it is difficult to analyze the final system to see how it performs the task.

What is EH?

From considering the characteristics involved in EH one can start to be more precise as to what is and is not EH. Evolvable hardware will include a form of evolutionary loop, a population (even very small ones), variation operators and selection mechanisms. These will vary from one implementation to another but will be present. In addition to this, there are a number of other characteristics that are important to consider:

- The final embodiment of the evolutionary process (the final phenotype) should be hardware, or at least have the potential to be implemented in hardware. An obvious example would be evaluating the evolving design on a given FPGA device and for the evolved design to be implemented on that FPGA. A less obvious example might be the evolution of part or all of a VLSI device where the fabrication of such a device may not be feasible at that point in time i.e. in terms of cost, but could later be fabricated.

- The evolutionary process may be applied to the investigation of some future technology medium (non electronics).

- The evolution can be either intrinsic ("on-chip") or extrinsic ("off-chip", applying a realistic simulator).

- The evolved design should solve or illustrate a viable approach to solving a "useful" problem—application or device—not one that could be solved by traditional means e.g. n-bit multipliers.

- The final embodied implementation includes features that traditional designs could not produce e.g. inherent fault-tolerance.

What is not EH?

The following are not considered to be evolvable hardware in this chapter:

- Simple optimisation processes (where "simple" refers to nothing more than when the optimization process itself is trivial).

- Where there is never any intension or reason ever to embed the final solution into hardware (whatever the platform).

- Where there are no benefits from evolving a solution.

- Where hardware is applied purely as an accelerator.

Taking a few examples from the literature and placing these into this context, we can summaries these in Table 1.

2.2 Possible Advantages of Evolvable Hardware

Evolvable Hardware is a method for circuit or device design (within some media) that uses inspiration from biology to create techniques that enable hardware designs to emerge rather than be designed. At first sight, this would seem very appealing. Consider two different systems, one from nature and one human engineered: the human immune system (nature design) and the computer that we are writing this chapter on (human design). Which is more complex? Which is most reliable? Which is optimised the most? Which is most adaptable to change?

The answer to almost all such questions is the human immune system. This is usually the case when one considers most natural systems. Possibly the only winner from the engineered side might be the question, "Which is optimised the most?" However, this will depend on how you define optimised. While this is appealing, anyone who has used any type of evolutionary system knows that it is not quite that straightforward. Nature has a number of advantages over current engineered systems, not least of which are time (most biological systems have been around for

Table 1 Comparison of different proposed evolvable hardware systems

	Analogue/Digital	Intrinsic/Extrinsic	Real hardware	Realistic simulator	Hardware accelerator?	EH?
Higuchi's team e.g. (2, 11, 52)	Digital	Intrinsic	Yes	No	No	Yes
JPL team e.g. (16, 32, 47)	Analogue	Intrinsic	Yes	No	No	Yes
N-bit digital circuit evolution e.g. (39, 48, 71)	Digital	Extrinsic	No	No	No	No
Lohn's team e.g. (34, 83, 91)	Analogue	Extrinsic	Yes	Yes	No	Yes
NanoCMOS project e.g. (13)	Analogue (for digital)	Extrinsic	Potentially	Yes	No	Yes
EvoDevo work e.g. (59, 69, 73)	Digital	Both	Yes	No	No	Yes
Koza's team e.g. (23, 26, 28)	Analogue	Extrinsic	No	Yes	No	Yes
Sekanina's team e.g. (101)	Digital	Both	Yes	No	No	Yes
Neural network (NN) projects e.g. (5, 6, 61)	Both	Both	Sometimes	Sometimes	Yes	No
Liquid crystal evolution e.g. (84, 85, 86)	Analogue	Intrinsic	Yes	No	No	Yes

If the structure and connections of an NN project are actually evolved and may be developed then it could be considered as EH

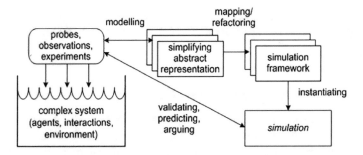

Fig. 2 Conceptual framework [14]

quite a long time) and resources (most biological systems can produce new resources as they need them, e.g. new cells). In addition, we need to be very careful as practitioners of bio-inspired techniques that we pay due regard to the actual biological systems and processes. All too often researchers read a little of the biology and charge in applying this to an engineering system without really understanding the biology or the consequences of the simplifications they make. On the other hand, we are not creating biological systems but rather artificial systems. An exact replication of a complex biological process may be neither needed nor appropriate for the hardware design in question. Improvements in BIAs for EH cannot be justified purely by their "improvements in biological realism", as may be seen in the literature. Such improvements should be reflected in improvements in the efficiency of the BIA or/and the hardware solutions achievable.

It is important to note that biological-like characteristics will be observed in an engineering system that is referred to as bio-inspired but that has in fact little or no real resemblance to the biology that "inspired" it. It is thus important to have a proper framework when using bio-inspired techniques. A framework for such systems is suggested in [14] and illustrated in Fig. 2. We are, however, not suggesting that every EH researcher needs to turn to biological experimentation. However, the biological inspiration and validation does need to come from real biological knowledge.

To summarise, use evolvable hardware where appropriate, associate evolution with something that has some real connection with the hardware and if using a simulator, make sure the simulator is an accurate one for the technology of choice and ensure that your biological inspiration is not only biologically inspired but also beneficial to the evolved hardware.

3 Platforms for Intrinsic EH

Evolvable hardware has, on the whole, been driven by the hardware/technology available at a particular time, and usually this has been Electronic hardware. While this chapter does not go into great detail on the different hardware platforms, it is at least an important enough driver for the subject to mention a few of the main platforms/devices. These are presented under Analogue and Digital Devices.

(A) Analogue Devices

Field Programmable Analogue Arrays (FPAA)

The Lattice Semiconductor ispPAC devices are typical of what is currently available from manufacturers that allow reconfiguration of "standard" analogue blocks (in this case based around OpAmps). The isp family of programmable devices provide three levels of programmability: the functionality of each cell; the performance characteristics for each cell and the interconnect at the device architectural level. Programming, erasing, and reprogramming are achieved quickly and easily through standard serial interfaces. The device is depicted in Fig. 3. The basic active functional element of the ispPAC devices is the PACell, which, depending on the specific device architecture, may be an instrumentation amplifier, a summing amplifier or some other elemental active stage. Analogue function modules, called PACblocks, are constructed from multiple PACells to replace traditional analogue components such as amplifiers and active filters, eliminating the need for most external resistors and capacitors. Requiring no external components, ispPAC devices flexibly implement basic analogue functions such as precision filtering,

Fig. 3 PACell structure [15]

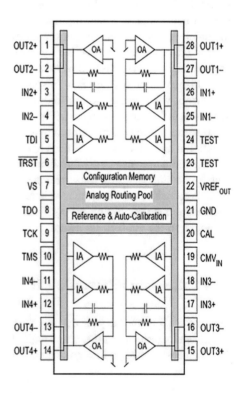

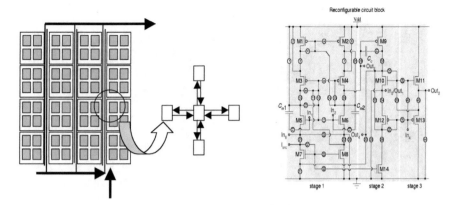

Fig. 4 The FPTA2 architecture. Each cell contains additional capacitors and programmable resistors (not shown) [16]

summing/differencing, gain/attenuation and conversion. An issue for someone wishing to undertake evolution on these devices is that the changes that can be made are at a relatively high functional level i.e. at the OpAmp level. This, together with the overhead for changing functionality, have limited their use in the EH field.

JPL Field Programmable Transistor Array

The Field Programmable Transistor array, designed by the group at NASA, was the first such analogue device specifically designed with evolvable hardware in mind. The FPTA has transistor level reconfigurability and supports any arrangement of programming bits without danger of damage to the chip (as is the case with some commercial devices). Three generations of FPTA chips have been built and used in evolutionary experiments. The latest chip, the FPTA-2, consists of an 8 × 8 array of reconfigurable cells (see Fig. 4). The chip can receive 96 analogue/digital inputs and provide 64 analogue/digital outputs. Each cell is programmed through a 16-bit data bus/9-bit address bus control logic, which provides an addressing mechanism to download the bit-string of each cell. Each cell has a transistor array (reconfigurable circuitry shown in Fig. 4), as well as a set of other programmable resources (including programmable resistors and static capacitors). The reconfigurable circuitry consists of 14 transistors connected through 44 switches and is able to implement different building blocks for analogue processing, such as two- and three-stage Operational Amplifiers, logarithmic photo detectors and Gaussian computational circuits. It includes three capacitors, Cm1, Cm2 and Cc, of 100 fF, 100 fF and 5 pF respectively.

Heidelberg Field Programmable Transistor Array (FPTA)

The FPTA consists of 16 × 16 programmable transistor cells. As CMOS transistors come in two types, namely N- and P-MOS, half of the transistor cells are designed as programmable NMOS transistors and half as programmable PMOS

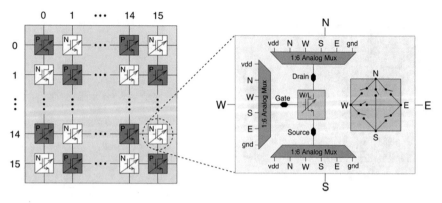

Fig. 5 Schematic diagram of the FPTA [17]

transistors. P- and N-MOS transistor cells are arranged in a checkerboard pattern, as depicted in Fig. 5. Each cell contains the programmable transistor itself, three decoders that allow the three transistor terminals to be connected to one of the four cell boundaries, Vdd or Gnd and six routing switches. Width W and Length L of the programmable transistor can be chosen to be 1, 2, ... 15 μm and 0.6, 1, 2, 4, or 8 μm respectively. The three terminals—Drain, Gate and Source—of the programmable transistor can be connected to either of the four cell edges (N, S, E, W), as well as to Vdd or Gnd. The only means of routing signals through the chip is via the six routing switches that connect the four cell borders with each other. Thus, in some cases, it is not possible to use a transistor cell for routing and as a transistor.

(B) Digital Devices

Field Programmable Gate Arrays (FPGA)

The FPGA is a standard digital device and is organised as an array of logic blocks, as shown in Fig. 6. Devices from both Xilinx and Alter a have been applied to EH. Programming an FPGA requires programming three tasks: (1) the functions implemented in logic blocks, (2) the signal routing between logic blocks and (3) the characteristics of the input/output blocks i.e. a tri-state output or a latched input. All of these, and hence the complete functionality of the device (including interconnect), are defined by a bit pattern. Specific bits will specify the function of each logic block, other bits will define what is connected to what. To date FPGAs have been the workhorse of much of EH that has been achieved in the digital domain.

The CellMatrix MOD 88

As with analogue devices, the other path to implement evolvable hardware is to design and build your own device, tuned to the needs of Evolvable Hardware. The Cell Matrix, illustrated in Fig. 7, is a fine-grained reconfigurable device with an 8 × 8 array of cells, where larger arrays may be achieved by connecting multiple

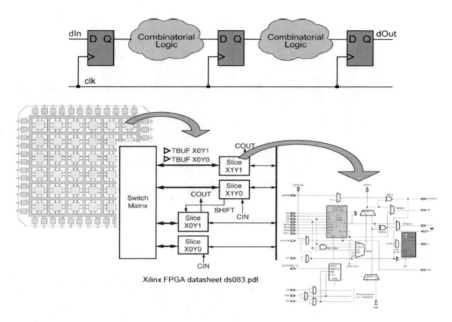

Fig. 6 Generic FPGA block diagram [18]

Fig. 7 The cellmatrix MOD 88, 8 × 8 Array [19]

8 × 8 arrays. Unlike most other reconfigurable devices, there is no built-in configuration mechanism. Instead, each cell is configured by its nearest neighbour, thus providing a mechanism not only for configuration but also for self-reconfiguration and the potential for dynamic configuration. Further, the Cell matrix cannot be accidently damaged by incorrect configurations, as is the case for FPGAs if the standard design rules are not followed.

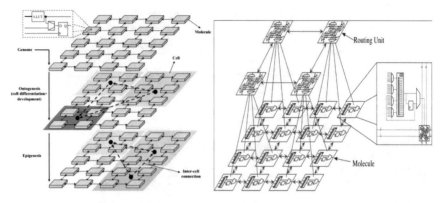

Fig. 8 Organic subsystem, schematic and organic views [21]

The POEtic Device

The goal of the "Reconfigurable POEtic Tissue" ("POEtic") [20], completed under the aegis of the European Community, was the development of a flexible computational substrate inspired by the evolutionary, developmental and learning phases in biological systems. POEtic applications are designed around molecules, which are the smallest functional blocks (similar to FPGA CLBs). Groups of molecules are put together to form larger functional blocks called cells. Cells can range from basic logic gates to complete logic circuits, such as full-adders. Finally, a number of cells can be combined to make an organism, which is the fully functional application. Figure 8 shows a schematic view of the organic subsystem.

The organic subsystem is constituted from a regular array of basic building blocks (molecules) that allow for the implementation of any logic function. This array constitutes the basic substrate of the system. On top of this molecular layer there is an additional layer that implements dynamic routing mechanisms between the cells that are constructed from combining the functionality of a number of molecules. Therefore, the physical structure of the organic subsystem can be considered as a two- layer organization (at least in the abstract), as depicted in Fig. 8.

The RISA Device

The structure of the RISA cell enables a number of different system configurations to be implemented. Each cell's FPGA fabric may be combined into a single area and used for traditional FPGA applications. Similarly, the separate microcontrollers can be used in combination for multi-processor array systems, such as systolic arrays [10]. However, the intended operation is to use the cell parts more locally, allowing the microcontroller to control its adjoining FPGA configuration. This cell structure is inspired by that of biological cells. As illustrated in Fig. 9, the microcontroller provides functionality similar to a cell nucleus. Each cell contains a microcontroller and a section of FPGA fabric. Input/output (IO) Blocks provide interfaces between FPGA sections at device boundaries. Inter-cell communication is provided by dedicated links between microcontrollers and FPGA fabrics.

Fig. 9 The RISA
architecture comprising an
array of RISA Cells [22]

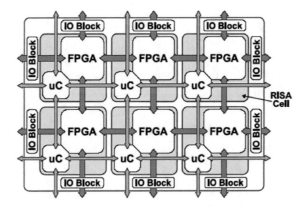

4   Success Stories

There have been some real successes within the community over the past 15 years.
This section gives a short review of some of these.

Extrinsic Analogue

Some of the earliest work on using evolutionary algorithms to produce electronic
circuits was performed by Koza and his team [23–27]. The majority of this work
focuses on using Genetic Programming (GP) to evolve passive, and later active,
electronic circuits for "common" functions such as filters and amplifiers. All of this
work is extrinsic, that is, performed in software with the results generally presented
as the final fit individual. Much of the early work considers the design of passive
filters with considerable success [25]. Figure 10 illustrates such results. The
interesting aspect in that particular paper (which is carried on and developed in
subsequent work) is the use of input variables (termed *free variables*) and conditional statements to allow circuit descriptions to be produced that are solutions to
multiple instances of a generic problem, in this case filter design. In Fig. 10 this is
illustrated by specifying, with such input variables, whether a low-pass or high-pass
filter is the required final solution of the GP.

   A more sophisticated example of evolving electronic circuits is shown in Fig. 11
[28]. Here the requirement is to create a circuit that mimics a commercial amplifier
circuit (and even improve on it). There are a number of both subtle and complex
points that come out of that paper that are of interest and use to others trying to use
evolvable hardware. The paper illustrates how the use of domain knowledge, both
general and problem-specific, is important as the complexity of the problem
increases. New techniques, or improvements in actual GP techniques, are developed
to solve this problem. Finally, and potentially most important, the use of
multi-objective fitness measures are shown to be critical for the success of
evolution.

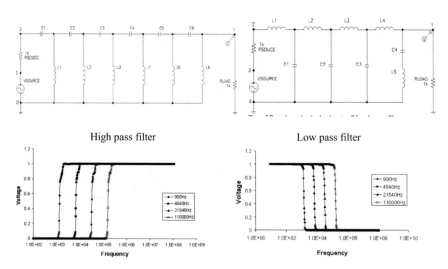

Fig. 10 Examples of filter circuits and frequency plots obtained by evolutionary processes [25]

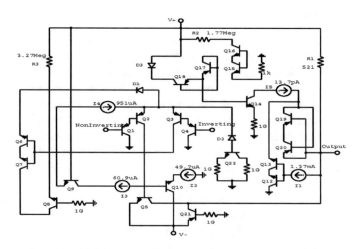

Fig. 11 Best run of an amplifier design [28]

In that case, 16 objective measures are incorporated into the evolutionary process, including: gain, supply current, offset voltage, phase margin and bias current. When evolving for real systems in real environments the use of multi-objective fitness criteria is shown to be extremely important.

Koza and his group continue to evolve electronic circuits in an extrinsic manner, and have many results where the evolutionary process has replicated results of previously patented human designs.

Intrinsic Digital

One of the very first research teams to apply evolvable hardware in an intrinsic manner was Higuchi and his group in Japan [1, 2, 29–31]. Their work includes applying evolvable hardware (almost always intrinsically) to myoelectric hands, image compression, analogue IF filters, femto-second lasers, high-speed FPGA I/O and clock-timing adjustment. Here we give a little more detail on the application of clock-timing adjustment as an illustration of the work conducted by Higuchi and his group.

As silicon technology passed the 90 nm process size new problems in the fabrication of VLSI devices started to appear. One of these is the fact that, due to process variations cause by the fabrication process, the clock signals that appear around the chip are not always synchronised with each other. This clock skew can be a major issue in large complex systems. This is a difficult issue for chip designers since it is a post-fabrication problem where there is a significant stochastic nature to the issues. One solution would, of course, be to slow the clocks down, but that is generally an unacceptable solution. Another possibility is to treat this as a post-fabrication problem, where actual differences in paths and actual speeds can be measured.

Figure 12 illustrates the basic philosophy behind the idea of using evolution to solve the problem. Typically, clock signals are transmitted around complex VLSI devices, in a hierarchical form, often known as clock-trees, as illustrated in the left-hand picture in Fig. 12. At the block level (middle picture), sequential circuits should receive clock signals at the same time for the circuit to perform correctly. The idea behind this work is to introduce programmable delay circuits (right-hand picture) that are able to fine-tune the delay on all the critical path lengths that the clock propagates. The issue is, what delay should be programmed into each individual path on each individual VLSI device? Each device is likely to be different due to the fabrication process and we will not know what this is until after fabrication. However, the delay can be controlled (programmed) by a short bit pattern

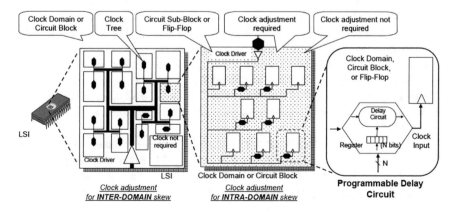

Fig. 12 Basic process of clock skew adjustment using evolvable hardware [29]

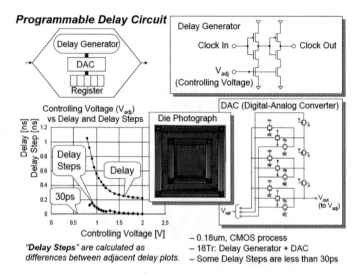

Fig. 13 Realisation of clock skew adjustment on VLSI device [29]

stored in a register, on-chip. This bit pattern can form part of a genome that is used within an evolutionary loop that is using a Genetic Algorithm (GA) to optimise the delay in each of the critical paths in the circuit. Figure 13 illustrates in more detail the actual circuit elements required, and the final chip design, to achieve the required functionality for this evolvable chip. The results suggest that not only can the clock skew be optimised, and consequently the frequency that the device can run at be increased (in the paper by 25%) but that also the supply voltage can be reduced while maintaining a functioning chip (in some cases by more than 50%).

Intrinsic Analogue

Section 3 has given a brief outline of the work performed by the team at NASA JPL on the design and manufacture of a number of analogue programmable devices aimed at assisting with intrinsic analogue evolution, Field Programmable Transistor Arrays (FPTAs) [16]. Here, one of these devices is presented to illustrate, not only the evolution of useful circuits (in real environments), but that through the continued use of evolution throughout the lifetime of the system, fault tolerance is increased. In this case, the functionality considered is that of a 4-bit digital-to-analogue converter (4-bit DAC). The basic evolvable block used in these experiments is shown in Fig. 4, and the results obtained in the work are summarised in Fig. 14.

Hardware experiments use the FPTA-2 chip, with the structure illustrated in Fig. 4. The process starts by evolving first a 2-bit DAC—a relatively simple circuit. Using this circuit as a building block, a 3-bit DAC is then evolved and, again reusing this, a 4-bit DAC is evolved. The total number of FPTA cells used is 20. Four cells map a previously evolved 3-bit DAC (evolved from a 2-bit DAC), four

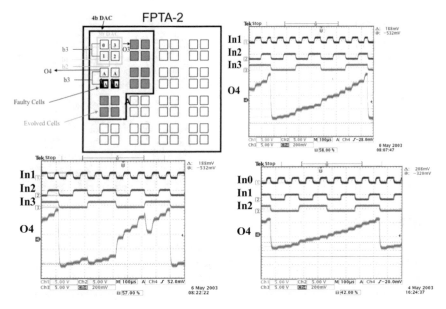

Fig. 14 FPTA topology (top-left), evolved 4-bit DAC response (top-right), evolved 4-bit DAC response with two faulty cells (bottom-left) and recovered evolved 4-bit DAC response with two faulty cells (bottom-right) [32]

cells map a human designed Operational Amplifier (buffering and amplification) and 12 cells have their switches' states controlled by evolution. When the fault is injected into the operating circuit, the system opens all the switches of the 2 cells of the evolved circuit.

Figure 14 (top-left) shows the faulty cells in black. In these cells all switches are opened (stuck-at-0 fault). The 3-bit DAC, cells '0', '1', '2' and '3' map the previously evolved 3-bit DAC, whose output is O3. The Operational Amplifier cell (Label 'A') is constrained to OpAmp functionality only. The evolved cell (in grey) switches' states are controlled by evolution. O4 is the final output of the 4-bit DAC.

The top-right plot in Fig. 14 illustrates the initial evolved 4-bit DAC with inputs 1, 2 and 3 (4 is missing due to number of oscilloscope channels) and output O4 which can be seen to be functioning correctly. The bottom left plot in Fig. 14 shows the response of the 4-bit DAC when two cells are made faulty. Finally, the plot at the bottom-right of Fig. 14 illustrates the response of the DAC after further evolution (after fault injection) has taken place. This response is achieved after only 30 generations of evolution. Again, the only cells involved in the evolutionary process are those in grey in Fig. 14. This example shows very well that recovery by evolution, in this case for a 4-bit DAC, is possible. The system makes available 12 cells for evolution. Two cells are constrained for the implementation of the OpAmp. Four cells are constrained to implement the simpler 3-bit DAC. Two faulty cells are implemented by placing all their switches to the open position.

Fig. 15 Prototype evolved
antenna, which on March 22,
2006 was successfully
launched into space [34]

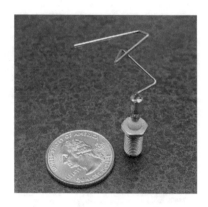

Extrinsic and Intrinsic Analogue

As a final example in this section, the domain of evolved antenna design [33] is presented. In the work presented in [34], a GA is used in conjunction with the Numerical Electromagnetic Code, Version 4 (NEC, standard code within this area) as the simulator to create and optimise wire antenna designs. These designs not only produce impressive characteristics, in many cases similar or better to those produced by traditional methods, but their structure is often very different from traditional designs (see Fig. 15). The 'crooked-wire' antennas consist of a series of wires joined at various locations and with various lengths (both these parameters determined by the GA). Such unusual shapes, unlikely to be designed using conventional design techniques, demonstrate excellent performance both in simulation and physical implementation [34].

Apart from the difference in domain, an interesting feature of this application of EH is the relatively large number of constraints (requirements) that such systems impose on a design. Requirements of a typical antenna include [34]: Transmit Frequency; Receive Frequency; Antenna RF Input; VSWR; Antenna Gain Pattern; Antenna Input Impedance, and Grounding. From this list, which is not exhaustive, it can be seen that this is certainly a non-trivial problem requiring a complex objective function to achieve appropriate results.

Given the nature of the problem and the solution (wire antenna), a novel representation and construction method is chosen. The genetic representation is a small "programming language" with 3 instructions: place wire, place support, and branch. The genotype specifies the design of one arm in a 3D-space. The genotype is a tree-structured program that builds a wire form. Commands to undertake this process are: Forward (length radius); rotate_x (angle); rotate_y (angle) and rotate_z (angle). Branching in genotype equates to branching in wire structure. The results are very successful (see Fig. 15) and the characteristics (electrical and mechanical) of the final antenna more than meet the requirements list.

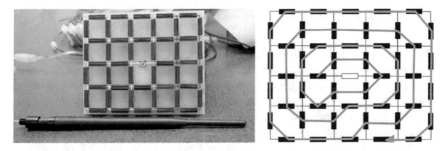

Fig. 16 Reconfigurable antenna platform (left) and mapping of chromosome to structure (right) [33]

The previous example is situated within the extrinsic world, that is, evolution runs within a simulation environment and only when a successful individual is found, is a physical entity built. Those authors then went on to consider how you might do evolution intrinsically. The goal now is to explore the in situ optimisation of a reconfigurable antenna. One motivation behind this work is to consider how a system might adapt to changes in spacecraft configuration and orientation as well as to cope with damage. A reconfigurable antenna system (see Fig. 16) is constructed. The software evolves relay configurations. In the initial stages of this work, the user subjectively ranks designs based on signal quality, but the plan is to automate this process in future work. In the system shown in Fig. 16-left, the system consists of 30-relays that are used to create the antenna designs. It has been shown that the system is able to optimise effectively for frequencies in the upper portion of the VHF broadcast TV band (177–213 MHz). A simple binary chromosome is used, with 1s and 0s corresponding to closed and open switches respectively. The connections between the Pads and the switches define the chromosome mapping (Fig. 16-right). The antenna is able to automatically adapt to different barriers, orientations, frequencies, and loss of control, and in a number of cases shows superior performance over conventional antennas.

5 Challenges

The previous section has considered a range of successes for evolvable hardware. While these are not the only successes, they do represent a fair range of achievements: from analogue to digital electronics and from extrinsic to intrinsic evolution. However, these successes have been few relative to the efforts put into the field so far. So where, if anywhere, does evolvable hardware go from here? The rest of this chapter gives some speculative suggestions to the answer to this question.

5.1   From Birth to Maturity

There are many fundamental challenges in the field but one major challenge, which is perhaps rarely mentioned, is the challenge of any field: the need to publish. In the early stages of a new field, there is much uncertainty but at the same time there are many ways to achieve simple exciting results, raising the field's prominence quickly. Although publishing is a challenge in a new area, there is also the excitement of a new venture. As the newness wears off, without realistic results matching the more traditional field, you need to rely on specialised workshops or conferences. Such specialised workshops and conferences do provide a valuable venue for development of the field. Specialised researchers are able to recognise more subtle, but important, contributions and their value to the field. Unfortunately, this can often lead to over-acceptance of small advancements whilst the main goal is forgotten. That is, EH needs to one day compete with traditional electronic design techniques, either in general or for specific sub-areas of electronics or other mediums. As the field becomes more mature, special sessions/tracks become easier to establish. However, if the results are not there, such sessions begin to disappear and the numbers at conferences begin to wane.

Such a process is quite typical of a new field, and EH is no exception. The main challenges have either not been solved, or perhaps even addressed, so as to enable the field to take the bigger steps forward.

Neural networks provide us with a sub-field of AI that almost died out in the late 60s—in fact for a decade. Here it was the work of Minsky and Papert [35] that highlighted a major non-solvable challenge to the field. Although in the early 70s, the work of Stephen Grossberg [36] showed that multi-layered networks could address the challenge, and the work of Bryson and Ho [37] solved the challenge through the introduction of back-propagation, it was not until the mid 80s, and the work of Rumelhart et al. [38], before back-propagation gained acceptance and the field was revived. However, in the mean-time, a significant decline in interest and funding resulted from the publishing of the Minsky and Papert's 1969 book. A further relevant example is that of the field of AI itself. The field paid a high price due to the over hype in the mid 80s to mid 90s.

We do not consider that EH is dying out, but the field is in a critical stage. There is certainly no argument to suggest that the field cannot go forward. However, it is also not clear that there is a path forward that will allow the field to progress in the way that neural networks have progressed. Also, is the field suffering from over-hype, similar to that of AI?

Section 3, highlights a number of success stories. However, as stated, this far from represents the vast research that has gone on in the field. In the early stages of a new field, results are simply that—results, enabling the field to take some small steps forward. In theory, small steps should enable researchers to build on these results and take bigger and bigger steps. However, on the whole, this has not been the case with EH. Bigger steps have proven to be hard to achieve, and fewer researchers have been willing to push the boundaries and use the time to achieve such steps. The reality is that small steps give results and publications, bigger steps are much more uncertain, and research is becoming more and more publishing driven.

In our view, this focus on smaller steps has not only hampered progress in the field, but also caused industry and traditional electronics to turn away from EH. In general, one can say that industry is fed-up of seeing results from "toy problems". They want results that will make a difference to them. Further, funding challenges in many countries means that research fields have to rely more and more on industrial funding, providing more pressure to take the bigger steps towards EH becoming a viable alternative to traditional electronics so as to secure funding for further development of the field.

5.2 Scalability

One of the goals of the early pioneers of the field was to evolve complex circuits, to push the complexity limits of traditional design and, as such, find ways to exploit the vast computational resources available on today's computation mediums. However, the scalability challenge for Evolvable Hardware continues to be out of reach.

Scalability is a challenge that many researchers acknowledge [49, 53, 59] but it is also a term widely and wrongly used in the rush to be the one to have solved it. Evolving an $(n + 1)$-bit adder instead of a previously evolved n-bit adder is not solving the scalability challenge (it is simply iteration), especially when one considers the fact that traditional electronics can already design such circuits. Scaling up structures alone, i.e. the number of gates, without a corresponding scaling of functionality, is also not solving the scalability challenge.

The scalability challenge for Evolvable Hardware may be defined as finding a refinement of bio-inspired techniques that may be applied in such a way to achieve complex functions without a corresponding scaling up of resource requirements (number of physical components, design time and/or computational resources required).

Examples of one-off complex designs may be seen, such as Koza's analogue circuits [28]. However, here functionality is scaled up at the expense of huge computational resources—clusters of computers (a brute force solution). Although such work provides a proof of concept that complex designs are possible, much refinement is needed to the techniques to provide a truly scalable technique for designs of such complexity.

So if incrementally increasing complexity, i.e. from an n-bit adder to an $(n + 1)$-bit adder, is not the way forward, how should we approach solving this challenge that has evaded EH researchers for 15 years?

The goal of much EH is to evolve digital and analogue designs (at least if we stick to electronic engineering applications). Little is known about the phenotypic search space of digital and analogue designs. Since any search algorithm provides implicit assumptions about the structures of the search space, our lack of knowledge of the phenotypic space suggests these assumptions and how they match the problem in hand, i.e. electronic circuits, and whether the problem is effectively represented gives us little chance of efficiently solving the problem.

Evolvable Hardware, has suffered from the lack of theoretical work on the evolutionary techniques applied to this particular problem to provide stronger guidelines for how the field might approach the achievement of a scalable technique. Hartmann et al. [39] studied the Kolmogorov complexity (complexity of a bit string) of evolved fault tolerant circuits through the application of Lempel-Ziv complexity [40] to these circuits. The complexity analysis involves analysing the compressibility of strings (representing circuits), and comparison of such analysis to the complexity spectrum (simple to highly complex strings). The complexity of the circuits analysed has a significantly smaller region of complexities than that covered by the evolutionary search, indicating that it would be beneficial to bias the search towards such areas of the complexity search space.

It should be borne in mind that Kolmogorov complexity may or may not be the most suitable theoretical measure for the analysis of EH complexity and that this is just one example of a possible approach. However, that work is presented to highlight how theoretical approaches to EH can help us to understand our application area and help us refine our techniques to such applications.

5.3 Measurements

Measurement refers to the metric used to evaluate the viability of a given solution for the problem in hand. In traditional electronics, for example, terms such as functional correctness (which may involve many sub-functions) as well as area and power costs are general metrics applied. Other metrics applied may relate to the issue being resolved e.g. reliability.

Although power issues are important in traditional electronics, such metrics are only starting to appear in EH work [41]. However, the application of area metrics, especially gate counts, is often applied, but may be questioned. There are two key issues here: The majority of EH solutions are intended for reconfigurable technology and this focus on evolving small circuits is in contrast to the relatively large resources available. Further, although area count may show the efficiency of EH compared to a traditional technique, for a given generalised solution, one can easily say that a good traditional designer can often match the evolved solution given such an optimisation task. It should be borne in mind, therefore, that the power of evolution as an optimisation technique is when the search space is very large, when the number of parameters is large, and when the objective landscape is non-uniform. Thus for such small evolved circuits, gate area—although possibly interesting—is not a metric that should be used to illustrate the viability of the solution.

A further challenge with metrics, highlighted in the above description, is the ability to compare solutions between traditional and evolved designs. A traditional designer needs to know whether a given evolved solution is comparable to the traditional designs in terms of the particular evaluation criteria the designer is interested in. A criticism of EH may certainly thus be directed to the use of "number of generations" as a metric, which has no relevance to any performance metric in traditional design.

A common design feature addressed in EH, and also part of the grand challenges described by the ITRS roadmap 2009 [42], is fault tolerance, i.e. reliability. Traditionally, the reliability metric measures the probability that a circuit is 100% functional, and thus says nothing about how dis-functional the circuit is when it is not 100% functional. To apply such a metric as a fitness criterion to EH would not provide sufficiently fine-grained circuit feedback needed to separate potential solutions and, therefore, limits evolution's search to a rougher search. More commonly in EH, reliability is expressed in terms of how correct a solution is, on average, across a spectrum of faults, where the faults may be explicitly defined [43] or randomly generated [39]. Such a metric provides evolution with more information about practical solutions and thus supports evolution of solutions. However, as shown in [44], since the traditional metric is designed with traditional design techniques in mind and the EH metric is designed with evolution in mind, the metrics favour the design technique for which the metric was designed. As such, a comparison is difficult. It is thus important to find metrics that cross the divide between traditional and evolved designs.

5.4 Realistic Environments or Models

One area of interest for EH is device production challenges. Defects arising during production lead to low yield i.e. fewer "correct devices" and thus more faulty devices—devices that pass the testing process, but where undetected defects lead to faults during the device's lifetime. There are many ways that evolution may be applied in an attempt to correct or tolerate such defects. However, the challenge remains: a need for real defect data or realistic models. Fault models are difficult to define due to the strong dependence between the design itself and the actual production house due to the complex physical effects that lead to defects.

It should be noted, however, that this challenge is being faced by the traditional design community too. Closer collaboration with industry, illustrating the power of EH is needed to secure realistic data and thus provide the basis for further research interest in this area and more realistic results.

In addition, application areas may be found where EH can solve issues with limited data where traditional techniques struggle. In [45] a future pore-network CA machine is discussed. The machine would apply an EA to the sparse and noisy data from rock samples, evolving a CA to produce specified flow dynamic responses.

Rather than relying on models, EH can be applied to the creation of robust designs where realistic faults may be applied to the actual hardware and evolution enabled to create robust designs. Zebulum et al. [46] address the issue of environmental models. Although such models have become more refined, design and environmental causes are significant sources of failure in today's electronics. The work addresses the space environment and focuses on improvements in design and recovery from faults, without the need for detection, through the application of EH. In this case, the underlying hardware is the FPTA, described in Sect. 3, and faults are injected through opening switches to simulate the stuck-at fault model.

A further example may be seen in Thomson's work [12] on evolving robust designs in the presence of faulty FPGAs where the extreme conditions are physically imposed on the actual chips. The goal is to create robust designs that tolerate elements in the "operational envelope"—variations in temperature, power-supply voltages, fabrication variances and load conditions—by exposing the circuits to elements of such an operational window during fitness evaluations. Similarly, the work of [47] involves placing an FPGA and a Reconfigurable Analogue Array (RAA) chip in an extreme temperature chamber to investigate fault recovery in a realistic extreme temperature scenario.

6 Newer Approaches

For the last 10 years a number of researchers have been addressing the scalability problem, focusing on finding ways to simplify the task for evolution in terms of finding more effective ways to explore the search space for more complex applications.

Divide and Conquer

The first set of approaches build on the traditional notion of divide and conquer, providing evolution with simpler sub-problems to solve at a time, reducing the search space for evolution. There are two driving forces behind such approaches in terms of resource usage: (a) reduction in the number of evaluations needed to find a solution, and (b) reduction of evaluation time to evaluate a single individual. The latter point has strong consequences for intrinsic evolvable hardware, as the evaluation time to fully test a design grows exponentially with the number of inputs tested.

Various alternatives may be found in the literature, including Function-level Evolution [30, 48], Increased Complexity Evolution [49, 50] and Bidirectional Evolution [51].

Function-level evolution [30, 48, 52] does not require a specific division of the problem domain. Instead, the focus is on the selection of larger basic units, i.e. adders or multipliers, enabling a scaling up of potential evolved solutions. A study of fitness landscapes by Vassilev and Miller [53] illustrates that by using sub-circuits as building blocks, rather than using simple gates, speed-up of the evolutionary process can be achieved, but at the cost of an increased number of gates. Further, the selection of appropriate components is domain specific and limiting evolution to larger components does limit the search space available to evolution.

In Increased Complexity Evolution (ICE) [49, 50] (also termed Divide and Conquer and Incremental Evolution), each basic unit may be a simple gate or higher-level function. These blocks from the units for further evolution, incrementally increasing the complexity of the task at each evolutionary run, i.e. the block evolved in one run can be applied as a basic block in a subsequent run. As such, it is a bottom-up approach to increasing complexity, and can be said to build on the work of Gomex and Miikulainen [54] on incremental evolution and the biological fact that humans perform more and more difficult tasks on route to the adult state [55]. However, it is quite a challenge to find ways in which an EH design may be split into simple tasks that can be incrementally increased, and also to find suitable fitness measures for each stage. A possible solution is to divide the application based on the number of outputs [56]. However, such a solution is only advantageous in the case of multiple outputs.

Bidirectional Evolution (BE) [51] is similar to increased complexity evolution except that incremental evolution is applied both to decompose the problem into the sub tasks and then to incrementally make the tasks more challenging. That is, an automatic identification of sub-tasks is provided for through the application of output and Shannon decomposition.

In both ICE and BE, dividing the application by the number of outputs does not reduce the number of test vectors required to fully test the circuits since the number of inputs is unchanged. As such, although a reduction in the number of evaluations may be obtained, the challenge of fitness evaluation remains. Generalised Disjunction Decomposition [57, 58] is a method that can be applied to improve BE by decomposing the problem based on inputs rather than outputs.

Development and Modularisation

A second set of approaches are inspired by nature's way of handling complexity, i.e. that of artificial development. Here a genome is more similar to DNA, far from a blueprint of the organism to be developed, rather a somewhat dynamic recipe for the organism that can be influenced by external factors in the growth and specialisation of the cells of the organism. Evolution operates on the DNA-like circuit description of the genome, and in the evaluation phase, each individual DNA is developed into a fully-grown phenotype (a circuit in electronics). The introduction of development to an EA, provides a bias to modular iterative structures that already exist in real-world circuit designs whilst allowing evolution to search innovative areas of the search space [59]. Further, introducing development to an evolutionary process may be said to enhance exploration, rather than optimisation of solutions [60].

Although the idea of artificial development already existed in the literature [61–66] when the field of evolvable hardware picked up this attractive notion for improving scalability, algorithms for development were few and far between, and far from thoroughly researched. A number of EH research groups are involved in developmental approaches to EH [20, 59, 67–73], including theoretical and practical investigations into the feasibility of development as an EH technique [74–77]. However, there is much research to be conducted before such a technique can be said to really be moving along the path towards an efficient technique for EH.

The concept of iterative structures need not just be obtained through developmental approaches but, similar to Koza and Rice's Automatically Defined Programs (ADFs) [23], EAs for EH may be designed in a way so as to protect partial solutions that may contribute to the final solution. In this way, instead of defining the functions, as in Function-level evolution, such functions are generated by evolution and protected and encapsulated as a function for re-use. The importance of such reuse in EH is discussed in [70]. Embedded Cartesian Genetic Programming (ECGP) [78, 79] is an EA that may be applied to EH enabling modularisation.

Fitness Evaluation Improvements

The third set of approaches address the evaluation of individuals, a key concern in intrinsic evaluation. From the earliest days of EH, filtering of signals has been shown to be a successful application area. Here, only a single fitness test is required for each individual where the frequency characteristics of the evolved system are compared to the test signals. However, in many applications involving multiple input circuits, the number of evaluations per individual is exponentially related to the number of inputs.

One approach to this issue is to focus on ways to reduce the number of test vectors required whilst being able to provide a reliable evaluation method that allows for prioritising of individuals. Such an approach is not new in that it is an approach commonly applied in neural networks and classification systems. That is, the set of input vectors are divided into a set of training vectors and a set of test vectors. Creating EH through the application of Neural Networks may use a similar process, verifying the evolved circuits by applying the test vectors.

In [80] a random sample of input vector tests for digital truth tables is applied to assess the evolvability of functionality with such limited evaluations. Unfortunately, the results show that with the setup involved, little merit is found in such an approach, and it is clear that the EA needed the complete truth table to provide sufficiently detailed information to evolve a 100% fit individual.

Instead of considering an incomplete search, another possibility is to identify applications that require a single evaluation similar to that of signal filtering. In [81] it is shown that if the circuit sought could be interpreted as a linear transformation, i.e. that the basic elements are linear operators (add, shift, subtract etc.), then only a single test vector is needed. The work investigates a multiple constant multiplier. Although not a complex application in its own right, it is a useful component in more complex circuits. Further investigation is needed to identify more complex applications for such a technique.

Newer Materials

As discussed in Sect. 2, various digital and electronic platforms have provided the basis for EH research. Specialised devices have on the one hand provided devices more tuned to the needs of BIAs, whilst on the other hand provided an experimental platform and results with very limited access and applicability to the broader community, except in the case of Cell Matrix where both the device and supporting software are available.

In general, non-specialized devices have provided a more common platform, enabling more exchange of knowledge between different research groups and different experiments. However, today's technology is far from suitable for BIAs. The early work of Thompson [82] has already shown that evolution is capable of not only exploiting the underlying physics of FPGAs, and has further shown that evolution creates a design solution for a task that the chip was not designed to handle—that of a tone discriminating circuit [83]. That is, evolution explores areas of the design space beyond that which the device is designed for. However, as may be seen from the success stories highlighted in Sect. 4, where all the experiments are based on non-standard devices, there are many challenges to be faced in achieving EH on standard components.

The seminal work of [82] does, however, provide a proof of concept that BIAs may be applied to search areas of the design space beyond that which today's electronic technology is designed to meet. Similarly, the work of Harding and Miller applied evolution to liquid crystal, exploiting the underlying physics [84] to achieve various functions (a tone discriminator [85] and a robot controller [86]). Here, evolution is applied to an unconventional computation medium treated as a black box, causing functionality to emerge from within the liquid crystal. That is, evolution exploits the structure and inherent dynamics of the liquid crystal to achieve functionality.

Under the realms of the MOBIUS project, Mahdavi and Bentley [87] have investigated Smart materials for robotic hardware. Nitonal Wires (an alloy made up of Nickel and Titanium) are used as the muscle hardware of a robotic snake, applying an EA to evolve the activations for the wires controlling the snake-like movement. Further, inherent adaptation to failure is seen after one of the wires broke during experimentation. Nitonal is a shape memory alloy that has two forms depending on temperature, such that changing from one temperature to the other causes a phase transformation and the alloys are super elastic and have shape memory. Such an adaptive solution (EA + shape memory alloy) may be useful not only for snake-like robots but for devices in varying environments that may need to change their morphology accordingly.

A further medium suggested by Olteam [88] is Switchable Glass (smart windows), controlling the transmission of light through windows by the application of variable voltages. Included under such a classification is liquid crystal. Other possibilities include Electrochromic devices and suspended particle devices—rods (particles) suspended in a fluid. The former relies on chemical reactions where the translucency (visibility) of the glass may be adjusted through the application of various voltages. The latter relies on field effects where the rods align to a given electric field and varying the voltage can control the amount of light transmitted.

7   The Future

Although the newer approaches described in Sect. 6 are still facing many challenges, these illustrate that benefits can be gained by looking again at biology and even combining both traditional and biologically-inspired techniques. Nature does not have all the answers for us since we are in fact designing an artificial organism, i.e. a hardware device/circuit.

In Sect. 4, the scalability challenge is discussed. It is important that researchers continue to reach out to both natural and artificial inspirations for solutions to our challenges. However, for real progress it is important that researchers address the fundamental issues with any new technique and keep the scalability issue in mind. A new technique may be interesting to work with, but the real issues to be solved and how these should be approached to scale up the technique are critical for the future of the field.

Applications

Looking to potential applications has always been a fundamental issue for the EH community. One possible approach is to turn the attention away from complex applications to applications that may not necessarily be inherently complex but in some way provide challenges to more traditional techniques. There are many opportunities within design challenges that should be given further attention, including fault tolerance [89] and fault detection and repair [90–92]. Although already a focus in EH, such challenges provide many opportunities for further discoveries. Moving from faults to testing—a significant challenge facing the electronic industry—provides many opportunities such as in test generation [93] and testing methodology benchmark circuits [94, 95]. Surprisingly, although a major focus within the electronic industry, power usage [96] is only recently gaining attention in the EH field. A further increasing challenge in industry is yield due to the challenge of production defects [78, 97].

Within such design challenges, certain application areas provide further challenges. For example, in space, manned and unmanned vehicles will require not only to tolerate single faults—a traditional fault tolerance scenario—but also multiple faults [98]. Space provides hard application areas for any technique whether traditional or non-traditional, especially with respect to robustness and adaptivity requirements. Here the EH approach is not being applied to "reinvent the wheel" but rather applied in the search for novel or more optimal solutions where traditional techniques have either failed or are still challenged. Of course, the inherent real-time constraints are inherently hard for such stochastic search algorithms.

One area that is still relatively untouched, but where EH has a real future, is in adaptive systems. Such adaptivity may be seen in the light of failures, i.e. adaption to faults or adaptive functionality. The goal is to adapt in real time to environmental changes requiring changes in functionality and/or stability of functionality. Two key application areas already in focus are both digital and analogue challenges in space applications, where access to reconfiguration is limited, and also in autonomous robots reacting in real time to environmental influences. There are many challenges facing the field within true adaptive design, i.e. evolving a new design and replacing the old design in real time.

Instead of focusing on the challenge of achieving adaption "on-the fly", another option is to adapt to pre-defined solutions. A newer approach to adaptive solutions that may be said to capture some of the features of EH together with more traditional approaches to adaptive design is Metamorphic solutions proposed by Greenwood and Tyrrell [99]. Such an approach borrows the key elements from evolvable design with the exception that many of the adaptive solutions are pre-designed. Similar to more traditional adaptive design, real-time adaptivity is sought through the process of selection from a pool of pre-designed solutions. However, Metamorphic Systems extends the concept of pre-defined solutions by allowing for components implemented in various technologies.

Platforms

Evolvable Hardware has mainly focused on digital and analogue electronic platforms that may be far from the most suitable medium for BIAs. Other mediums are of course viable candidates for EH. Miller and Downing [100] propose that the effectiveness of EAs can be enhanced when the underlying medium has a rich and complicated physics. Further, such new technologies and even new ways to compute may be devised, possibly even at the molecular level, providing many advantages over conventional electronic technology.

It is important to note from the work of Harding and Miller [85, 86] that (a) a black box approach can achieve computation, (b) the choice of material was biased by the commercial availability of the material and that electrical contacts were present, (c) the liquid crystal in fact stopped working, possibly indicating limited reconfigurability and (d) it is still a black box: how it worked no one knows.

Such black box approaches will always be limited due to natural scepticism to the application of BIAs, exploiting materials of any kind, where verification of the functionality achieved is all but impossible. Conventional design starts with an abstract computation model or framework for the design and a technology is created to meet the requirements of the model, enabling a simple transformation from design to implementation. On the other hand, an evolved design is focused on a different mapping: that is a mapping between inputs and outputs of the design itself and pays little attention to either the computation model or the technology in hand (except where some form of optimization of the design is required). Thus, the difficulty of identifying the computational model behind an evolved design is perhaps not so surprising. One approach to address such an understanding is to study the computational properties of the design through theoretical approaches as in [101]. A further approach suggested by Stepney is that evolutionary techniques may be applied, not only to achieving functionality from a black box but also as a technique to identify such computational characteristics of the underlying material [102].

Looking to the ITRS Roadmap 2009 [42] we see that possible future materials for mainstream electronics include materials such as carbon-based nano-electronics, spin-based devices and ferromagnetic logic. Investigation of the suitability of such materials for the application of BIAs would be well worth investigation.

A further medium, not currently being investigated within the field of Evolvable Hardware, but highly relevant, is the within the field of synthetic biology where synthetic DNA is being generated to form logical functions. Further, libraries of such functions are under construction, with the goal of creating biological circuits. BIAs may be able to play a key role in solving some of the challenges ahead. Rather than following the synthetic biology approach to achieving computation in a fashion similar to today's electronics, a different approach would be to apply BIAs directly to the DNA substrate to enable computation to emerge.

Other possible platforms include molecular computing, optical systems and mixed mechanical-electrical systems. However, to really succeed in many such areas we need collaborators: (a) to help identify the problems, and (b) help in the subject specific issues (for example what wet-ware is and is not possible).

8 Summary

While evolvable hardware does have its place, we now know that, at least in terms of silicon design, it is much more limited than was previously suggested. However, we still have a research area that is full of possibilities and potential. Turning the lessons learnt and technological advances made into real progress is still a challenge.

One thing is certain: if we want to keep the field of evolvable hardware progressing for another 15 years, we need to take a fresh look at the way we think about evolvable hardware: about which applications we should be applying it to; about where it will actually be of real use; about where it might finally be accepted as a valid way of creating systems; and about what medium it should be based in.

We wish us all luck in this, the second generation of evolvable hardware.

References

1. Yao, X., Higuchi, T.: Promises and challenges of evolvable hardware. IEEE Trans. Syst. Man Cybern. Part C 29(1), 87–97 (1999)
2. Higuchi, T., Iwata, M., Kajitani, I., Iba, H., Furuya, T., Manderick, B.: Evolvable hardware and its applications to pattern recognition and fault tolerant systems. Towards Evol. Hardware Evol. Eng. App. LNCS 1052, 118–135 (1996)

3. Scott, S.D., Samal, A., Seth, S.: HGA: a hardware based genetic algorithm. In: Proceedings of ACM/SIGDA 3rd International symposium on FPGA's, pp. 53–59 (1995)
4. Salami, M., Cain, G.: Implementation of genetic algorithms on reprogrammable architectures. In: Proceedings of the Eighth Australian Joint Conference on Artificial Intelligence (AI'95), pp. 121–128, (1995)
5. Yao, X.: Evolutionary artificial neural networks. Int. J. Neural Syst. **4**(3), 203–222 (1993)
6. Yao, X., Liu, Y.: Evolving artificial neural networks for medical applications. In: Proceedings of 1995 Australia-Korea Joint Workshop on Evolutionary Computation, pp. 1–16, (1995)
7. Yao, X., Liu, Y.: Towards designing artificial neural networks by evolution. In: Proceedings of International Symposium. on Artificial Life and Robotics (AROB), pp. 265–268, 18–20 Feb (1996)
8. Yao, X., Liu, Y.: Evolving artificial neural networks through evolutionary programming. In: The Fifth Annual Conference on Evolutionary Programming, pp. 257–266. MIT Press (1996)
9. Rosenman, M.A.: An evolutionary model for non-routine design.In: Proceedings of the Eighth Australian Joint Conference on Artificial Intelligence (AI'95), pp. 363–370. World Scientific Publ. Co., Singapore (1995)
10. Davis, L.: Handbook of Genetic Algorithms. Van Nostrand Reinhold, New York, NY 10003 (1991)
11. Higuchi, T. et al.: Real-world applications of analog and digital evolvable hardware. IEEE Trans. Evol. Comput. **3**(3), 220–235 (1999)
12. Thompson, A.: On the automatic design of robust electronics through artificial evolution. In: Proceedings of the International Conference on Evolvable Systems: From Biology to Hardware, pp. 13–24 (1998)
13. Walker, J.A., Hilder, J.A., Tyrrell, A.M.: Evolving variability-tolerant CMOS designs. In: International Conference on Evolvable Systems: From Biology to Hardware, pp. 308–319 (2008)
14. Stepney, S., Smith, R.E., Timmis, J., Tyrrell, A.M.: Towards a conceptual framework for artificial immune systems. Artif. Immune Syst. LNCS **3239**(2004), 53–64 (2004)
15. ispPAC30 Data Sheet, Lattice Semiconductor Corporation. (2001). http://www.latticesemi.com/lit/docs/datasheets/pac/pacover.pdf
16. Stoica, A., Keymeulen, D., Thakoor, A., Daud, T., Klimech, G., Jin, Y., Tawel, R., Duong, V.: Evolution of analog circuits on field programmable transistor arrays. In: Proceedings of NASA/DoD Workshop on Evolvable Hardware (EH2000), pp. 99–108 (2000)
17. Langeheine, J., Becker, J., Folling, F., Meier, K., Schemmel, J.: Initial studies of a new VLSI field programmable transistor array. In: Proceedings 4th Int'l. Conference on Evolvable Systems: From Biology to Hardware, pp. 62–73 (2001)
18. Virtex Field Programmable Gate Arrays Data Book Version 2.5, Xilinx Inc. (2001)
19. User manual and Tutorials for the CELL MATRIX MOD 88
20. Sanchez, E., Mange, D., Sipper, M., Tomassini, M., Perez-Uribe, A., Stauffer, A.: Phylogeny, ontogeny, and epigenesis: three sources of biological inspiration for softening hardware. Evol. Syst. Biol. Hardw. ICES **96**, 35–54 (1996)
21. Tyrrell, A.M., Sanchez, E., Floreano, D., Tempesti, G., Mange, D., Moreno, J.M., Rosenberg, J., Villa, A.E.P.: POEtic tissue: an integrated architecture for bio-inspired hardware. In: Proceedings of 5th International Conference on Evolvable Systems, pp. 129–140. Trondheim (2003)
22. Tyrrell, A.M., Greensted, A.J.: Evolving dependability. ACM J. Em. Technol. Comput. **3**(2), Article 7, 1–20 (2007)
23. Koza, J.: Genetic Programming II: Automatic Discovery of Reusable Programs. MIT Press, Cambridge, MA (1994)
24. Koza, J., Keane, M., Streeter, M.: What's AI done for me lately? genetic programming's human-competitive results. IEEE Intell. Syst. **18**(3), 25–31 (2003)

25. Koza, J., Yu, J., Keane, M.A., Mydlowec, W.: Use of conditional developmental operators and free variables in automatically synthesizing generalized circuits using genetic programming. In: Proceedings of the Second NASA/DoD Workshop on Evolvable Hardware, pp. 5–15 (2000)
26. Keane, M., Koza, J., Streeter, M.: Automatic synthesis using genetic programming of an improved general-purpose controller for industrially representative plants. In: Stoica,A. (ed.) Proceedings of the 2002 NASA/DOD Conference on Evolvable Hardware, pp. 113–122 (2002)
27. Streeter, M., Keane, M., Koza, J.: Routine duplication of post-2000 patented inventions by means of genetic programming. In: Foster, J. et al. (eds.) Genetic Programming: 5th European Conference, EuroGP 2002, pp. 26–36 (2002)
28. Koza, J., Jones, L.W., Keane, M.A., Streeter, M.J., Al-Sakran, S.H.: Toward automated design of industrial-strength analog circuits by means of genetic programming. In: Genetic Programming Theory and Practice II, Chap. 8, pp. 121–142 (2004)
29. Takahashi, E., Kasai, Y., Murakawa, M., Higuchi, T.: Post fabrication clock-timing adjustment using genetic algorithms. In: Evolvable Hardware, pp. 65–84. Springer (2006)
30. Murakawa, M., Yoshizawa, S., Kajitani, I., Furuya, T., Iwata, M., Higuchi, T.: Hardware evolution at function level. Int. Conf. Parallel Problem Solv. Nature PPSN **1996**, 62–71 (1996)
31. Kajitani, I., Hoshino, T., Kajihara, N., Iwata, M., Higuchi, T.: An evolvable hardware chip and its application as a multi-function prosthetic hand controller. In: Proceedings of 16th National Conference on Artificial Intelligence (AAAI-99), pp. 182–187 (1999)
32. Stoica, A., Arslan, T., Keymeulen, D., Duong, V., Gou, X., Zebulum, R., Ferguson, I., Daud, T.: Evolutionary recovery of electronic circuits from radiation induced faults. CEC **2004**, 1786–1793 (2004)
33. Linden, D.: Optimizing signal strength in-situ using an evolvable antenna system. In: Proceedings of the 2002 NASA/DOD Conference on Evolvable Hardware, pp. 147–151 (2002)
34. Lohn, J.D., Hornby, G., Rodriguez-Arroyo, A., Linden, D., Kraus, W., Seufert, S.: Evolutionary design of an X-Band antenna for NASA's space technology 5 mission. In: 3rd NASA/DoD Conference on Evolvable Hardware, pp. 1–9 (2003)
35. Minsky, M.L., Papert, S.A.: Perceptrons. MIT Press, Cambridge, MA (1969)
36. Grossberg, S.: Contour enhancement, short-term memory, and constancies in reverberating neural networks. Stud. Appl. Math. **52** 213–257 (1973)
37. Bryson, E., Ho, Y.C.: Applied optimal control: optimization, estimation, and control. Blaisdell Publishing Company (1969)
38. Rumelhart, D.E., Hinton, G.E., Williams, R.J.: Learning representations by back-propagating errors. Lett. Nature Nature **323**, 533–536 (1986)
39. Hartmann, M., Lehre, P.K., Haddow, P.C.: Evolved digital circuits and genome complexity. NASA Int. Conf. Evol. Hardw. **2005**, 79–86 (2005)
40. Ziv, J., Lempel, A.: A universal algorithm for sequential data compression. IEEE Trans. Inf. Theory **IT-23**(3), 337–343 (1977)
41. Kobayashi, K., Moreno, J.M., Madrenas, J.: Implementation of a power-aware dynamic fault tolerant mechanism on the ubichip platform. In: International Conference on Evolvable Systems: From Biology to Hardware (ICES10), pp. 299–399 (2010)
42. International Technology RoadMap for Semiconductors (2009)
43. Thompson, A.: Evolutionary techniques for fault Tolerance. In: International Conference on Control, pp. 693–698 (1996)
44. Haddow, P.C., Hartmann, M., Djupdal, A.: Addressing the metric challenge: evolved versus traditional fault tolerant circuits. In: The 2nd NASA/ESA Conference on Adaptive Hardware and Systems, pp. 431–438 (2007)
45. Yu, T., Lee, S.: Evolving cellular automata to model fluid flow in porous media. In: 2002 NASA/DoD Conference on Evolvable hardware, pp. 210–217 (2002)

46. Zebulum, R.S., et al.: Experimental results in evolutionary fault recovery for field programmable analogue devices. In: Proceedings of the NASA/DOD International Conference on Evolvable Hardware, pp. 182–186 (2003)
47. Stoica, A., et al.: Temperature-adaptive circuits on reconfigurable analog arrays. IEEE Aerospace Conf. **2007**, 1–6 (2007)
48. Kalganova, T.: An extrinsic function-level evolvable hardware approach. Genetic Program. Lect. Notes Comput. Sci. **1802**, 60–75 (2004)
49. Torresen, J.: A scalable approach to evolvable hardware. In: The International Conference on Evolvable Systems: From Biology to Hardware, (ICES98), pp. 57–65 (1998)
50. Torresen, J.: Scalable evolvable hardware applied to road image recognition. In: The second NASA International Conference on Evolvable Hardware, pp. 245–252 (2000)
51. Kalganova, T.: Bidirectional incremental evolution in extrinsic evolvable hardware. In: The second NASA/DoD Workshop on Evolvable Hardware, pp. 65–74 (2000)
52. Liu, W., Murakawa, M., Higuchi, T.: ATM cell scheduling by functional level evolvable hardware. In: Proceedings of the First International Conference on Evolvable Systems, pp. 180–192 (1996)
53. Vassilev, V.K.: Scalability problems of digital circuit evolution: evolvability and efficient design. In: Proceedings of the 2nd NASA/DoD Workshop on Evolvable Hardware, pp. 55–64 (2000)
54. Gomaz, F., Miikulainen, R.: Incremental evolution of complex general behaviour. In: Special Issue on Environment Structure and Behaviour, Adaptive Behaviour, vol. 5, Issue 3, 4, pp. 317–342. MIT Press (1997)
55. Brooks, R.A., et al.: Alternative essences of intelligence. In: Proceedings of the 15th National Conference on Artificial Intelligence (AAAI-98), pp. 961–967. AAAI Press (1998)
56. Hong, J.H., Cho, S.B.: MEH: modular Evolvable Hardware for designing complex circuits. In: IEEE Congress on Evolutionary Computation, pp. 92–99 (2003)
57. Stomeo, E., Kalganova, T., Lambert, C.: Generalized decomposition for evolvable hardware. IEEE Trans. Syst. Man Cybern. Part B **36**(5), 1024–1043 (2006)
58. Stomeo, E., Kalganova, T.: Improving EHW performance introducing a new decomposition strategy. In: 2004 IEEE Conference on Cybernetics and Intelligent Systems, pp. 439–444 (2004)
59. Gordon, T., Bentley, P.J.: Towards development in evolvable hardware. In: Proceedings of the NASA/DoD Conference on Evolvable Hardware, pp. 241–250 (2002)
60. Bentley, P.J.: Exploring component-based representations? The secret of creativity by evolution. In: Fourth International Conference on Adaptive Computing in Design and Manufacture, pp. 161–172 (2000)
61. Gruau, F.: Neural network synthesis using cellular encoding and the genetic algorithm. PhD Thesis, France (1994)
62. Kitano, H.: Designing neural networks using genetic algorithm with graph generation system. Complex Syst. **4**, 461–476 (1990)
63. Bentley, P.J., Kumar, S.: Three ways to grow designs: a comparison of embryogenies for an evolutionary design problem. In: Genetic and Evolutionary Computation Conference (GECCO 99), pp. 35–43 (1999)
64. Kitano, H.: Building complex systems using development process: an engineering approach. In: Evolvable Systems: From Biology to Hardware, ICES. Lecture Notes in Computer Science, pp. 218–229. Springer (1998)
65. Siddiqi, A.A., Lucas, S.M.: A comparison of matrix rewriting versus direct encoding for evolving neural networks. In: Proceedings of the 1998 IEEE International Conference on Evolutionary Computation, pp. 392–397 (1998)
66. Eggenberger, P.: Creation of neural networks based on development and evolutionary principles. In: Proceedings of the International Conference on ANNs, pp. 337–342 (1997)
67. Hemmi, H., Mizoguchi, J., Shimohara, K.: Development and evolution of hardware behaviours. Towards Evol. Hardw. LNCS **1062–1996**, 250–265 (1996)

68. Ortega, C., Tyrrell, A.M.: A hardware implementation of an embyonic architecture using virtex FPGAs. In: Evolvable Systems: From Biology to Hardware, ICES. Lecture Notes in Computer Science, pp. 155–164 (2000)

69. Haddow, P.C., Tufte, G., ven Remortel, P.: Shrinking the genotype: L-systems for EHW? In: International Conference on Evolvable Systems: From Biology to Hardware, pp. 128–139 (2001)

70. Koza, J., Keane, M.A., Streeter, M.J.: The importance of reuse and development in evolvable hardware. In: Proceedings of the 2003 NASA/DoD Conference on Evolvable Hardware, pp. 33–42 (2003)

71. Miller, J.F., Thomson, P.: A developmental method for growing graphs and circuits. In: Proceedings of the 5th International Conference on Evolvable Systems (ICES03), pp. 93–104 (2003)

72. Tufte, G., Haddow, P.C.: Towards development on a silicon-based cellular computing machine. J. Natural Comput. **4**(4), 387–416 (2005)

73. Liu, H., Miller, J.F., Tyrrell, A.M.: Intrinsic evolvable hardware implementation of a robust biological development model for digital systems. In: Proceedings of the 2005 NASA/DoD Conference on Evolvable Hardware, pp. 87–92 (2005)

74. van Remortel, P., Ceuppens, J., Defaweux, A., Lenaerts, T., Manderick, B.: Developmental effects on tunable fitness landscapes. In: Proceedings of the 5th International Conference on Evolvable Systems, ICES2003, pp. 117–128 (2003)

75. Roggen, D., Federici, D.: Multi-cellular development: is there scalability and robustness to gain? In: Proceedings of Parallel Problem Solving from Nature 8, PPSN2004, pp. 391–400 (2004)

76. Lehre, P.K., Haddow, P.C.: Developmental mappings and phenotypic complexity. In: Proceedings of the Congress on Evolutionary Computation (CEC2003), pp. 62–68 (2003)

77. Tufte, G.: Phenotypic developmental and computation resources: scaling in artificial development. Genetic Evol. Comput. Conf. **2008**, 859–866 (2008)

78. Walker, J.A., Miller, J.F.: The automatic acquisition, evolution and re-use of modules in Cartesian genetic programming. IEEE Trans. Evol. Comput. **12**(4), 1–21 (2008)

79. Walker, J.A., Miller, J.F.: Evolution and acquisition of modules in Cartesian genetic programming. In: Proceedings of 7th European Conference on Genetic Programming (EuroGP 2004). Lecture Notes in Computer Science, vol. 3003, pp. 187–197 (2004)

80. Miller, J.F., Thomson, P.: Aspects of digital evolution: geometry and learning. In: Proceedings of the International Conference on Evolvable Systems: From Biology to Hardware, pp. 25– 35 (1998)

81. Vazilicek, Z., et al.: On Evolutionary synthesis of linear transforms in FPGA. In: International Conference on Evolvable Systems: From Biology to Hardware 2008. LNCS, vol. 5216, pp. 141–152 (2008)

82. Thompson, A.: An evolved circuit, intrinsic in silicon, entwined with physics. In: 1st International Conference on Evolvable Systems 1996, Springer, pp. 390–405 (1996)

83. Lohn, J., Hornby, G.: Evolvable hardware using evolutionary computation to design and optimize hardware systems. IEEE Comput. Intel. Mag. 19–27 (2006)

84. Harding, S.L., Miller, J.F., Rietman, E.A.: Evolution in materio: exploiting the physics of materials for computation. Int. J. Unconv. Comput. **4**(2), 155–194 (2008)

85. Harding, S.L., Miller, J.F.: Evolution in materio: a tone discriminator in liquid crystal. Congress Evol. Comput. **2004**, 1800–1807 (2004)

86. Harding, S.L., Miller, J.F.: Evolution in Materio: investigating the stability of robot controllers evolved in liquid crystal. In: The International Conference on Evolvable Systems: From Biology to Hardware, pp. 155–164 (2005)

87. Mahdavi, S.H., Bentley, P.: Evolving motion of robots with muscles. In: Applications of Evolutionary Computing. LNCS 2003, vol. 2611, pp. 149–155 (2003)

88. Oteam, M.: Switchable glass: a possible medium for evolvable hardware. In: First NASA/ESA Conference on Adaptive Hardware and Systems, pp. 81–87 (2006)

89. Thompson, A.: Hardware evolution: automatic design of electronic circuits in reconfigurable hardware by artificial evolution. Distinguished Dissertation Series. Springer (1998)
90. Garvie, M., Thompson, A.: Evolution of combinational and sequential on-line self diagnosing hardware. In: Proceedings of the 5th NASA/DoD Workshop on Evolvable Hardware, pp. 177–183 (2003)
91. Lohn, J.D., Larchev,G.V., Demara, R.F.: A genetic representation for evolutionary fault recovery in Virtex FPGAs. In: Proceedings of the 5th International Conference on Evolvable Systems: From Biology to Hardware (ICES), pp. 47–56 (2003)
92. Zhang, K., Demara, R.F., Sharma, C.A.: Consensus-based evaluation for fault isolation and on-line evolutionary regeneration. In: Proceedings of the 6th International Conference on Evolvable Systems: From Biology to Hardware (ICES05), pp. 12–24 (2005)
93. Corno, F., Cumani, G., Reorda, M.S., Squillero, G.: Efficient machine-code test-program induction. In: Proceedings of the Congress on Evolutionary Computation (CEC), IEEE, pp. 1486–1491 (2002)
94. Pecenka, T., Kotasek, Z., Sekanina, L., Strnadel, J.: Automatic discovery of RTL benchmark circuits with predefined testability properties. In: Proceedings of the NASA/DoD Conference on Evolvable Hardware, pp. 51–58 (2005)
95. Pecanka, T., Sekanina, L., Kotasek, Z.: Evolution on synthetic RTL benchmark circuits with predefined testability. ACM Trans. Design Auto. Electron. Syst. 13(3), 1–21 (2008)
96. Kobayashi, K., Moreno, J.M., Madreas, J.: Implementation of a power-aware dynamic fault tolerant mechanism on the Ubichip platform. In: International Conference on Evolvable Systems: From Biology to Hardware, pp. 299–309 (2010)
97. Djupdal, A., Haddow, P.C.: Evolving efficient redundancy by exploiting the analogue nature of CMOS transistors. In: Fourth International Conference on Computational Intelligence, Robotics and Autonomous Systems (CIRAS), pp. 81–86 (2007)
98. Keymeulen, D., Zebulum, R.S., Jin, Y., Stoica, A.: Fault-tolerant evolvable hardware using field-programmable transistor arrays. IEEE Trans. Reliab. 49(3), 305–316 (2000)
99. Greenwood, G., Tyrrell, A.M.: Metamorphic systems: a new model for adaptive systems design. In: Proceedings of the Congress on Evolutionary Computation, pp. 3261–3268 (2010)
100. Miller, J.F., Downing, K.: Evolution in materio: looking beyond the silicon box. In: NASA/DoD Conference on Evolvable Hardware (EH'02), pp. 167–178 (2002)
101. Sekanina, L.: Evolvable hardware: from applications to implications for the theory of computation. Unconv. Comput. LNCS 5715, 24–36 (2009)
102. Stepney, S.: The neglected pillar of material computation. Physica D 237(9), 1157–1164 (2008)

Bridging the Gap Between Evolvable Hardware and Industry Using Cartesian Genetic Programming

Zdenek Vasicek

Abstract Advancements in technology developed in the early nineties have enabled researchers to successfully apply techniques of evolutionary computation in various problem domains. As a consequence, a new research direction referred to as evolvable hardware (EHW) focusing on the use of evolutionary algorithms to create specialized electronics has emerged. One of the goals of the early pioneers of EHW was to evolve complex circuits and overcome the limits of traditional design. Unfortunately, evolvable hardware found itself in a critical stage around 2010 and a very pessimistic future for EHW-based digital circuit synthesis was predicted. The problems solved by the community were of the size and complexity of that achievable in fifteens years ago and seldom compete with traditional designs. The scalability problem has been identified as one of the most difficult problems that researchers are faced with and it was not clear whether there existed a path forward that would allow the field to progress. Despite that, researchers have continued to investigate how to overcome the scalability issues and significant progress has been made in the area of evolutionary synthesis of digital circuits in recent years. The goal of this chapter is to summarize the progress in the evolutionary synthesis of gate-level digital circuits, and to identify the challenges that need to be addressed to enable evolutionary methods to penetrate into industrial practice.

1 Introduction

Advancements in technology developed in the early 1990s have enabled researchers to successfully apply techniques of evolutionary computation in various problem domains. Higuchi et al. [11] and Thompson [30] demonstrated that evolutionary algorithms (EAs) are able to solve non-trivial hardware-related problems. The achievements presented in the seminal paper of Higuchi et al. motivated other scientists to intensively explore a new and promising research topic. As a consequence, a

Z. Vasicek (✉)
Faculty of Information Technology, Centre of Excellence IT4Innovations,
Brno University of Technology, Brno, Czech Republic
e-mail: vasicek@fit.vutbr.cz

© Springer International Publishing AG 2018
S. Stepney and A. Adamatzky (eds.), *Inspired by Nature*, Emergence,
Complexity and Computation 28, https://doi.org/10.1007/978-3-319-67997-6_2

new research direction referred to as *Evolvable hardware* (EHW) has emerged [6]. Evolvable hardware, a field of evolutionary computation, focuses on the use of evolutionary algorithms to create specialized electronics without manual engineering. The vision of EHW is to replace expensive and sometimes unreliable designers and develop robust, flexible and survivable autonomous systems. EHW draws inspiration from three fields: biology, computer science and electronic engineering.

Several schemes have been developed for classifying EHW [6]. Usually, two research areas are distinguished: *evolutionary circuit design* and *evolvable circuits*. In the first case, evolutionary algorithms are used as a tool employed to design a system that meets a predefined specification. For example, genetic programming can be used to discover an area-efficient implementation of a circuit whose function is specified by a truth table. In the second case, an evolutionary algorithm represents an inherent part of an evolvable circuit. The resulting adaptive system is autonomously reconfigured to provide a degree of self-adaptive and self-repair behaviour.

In the context of the circuit design, EHW is an attractive approach as it provides another option to the traditional design methodology: to use evolution to design circuits for us. Moreover, a key strength of the EHW approach is that it can be applied to designing the circuits that cannot be fully specified *a priori*, but where the desired behaviour is known. Another often emphasized advantage of this approach is that circuits can be customized and adapted for a particular environment. For example, if we know that some input combinations in our target application occur with relatively low probability, we can take this information into account and evolve a circuit that shows better parameters such as reduced size, delay or power consumption.

There are several works devoted mainly to the evolvable circuits and adaptive systems [7, 12, 25, 31]. However, there is no similar work systematically mapping the history and progress of evolutionary circuit design. Such a situation may suggest that evolutionary circuit design represents an area in which only marginal results have been achieved. Let us restrict ourselves to the evolution of digital circuits represented at the gate-level, i.e. circuits implemented using common (typically two-input) gates. Considering the complexity of circuits routinely used in industry, we have to admit that only little was done before 2010 even if many novel and promising approaches were published, as surveyed in Sect. 2.1. However, this area has been undergoing a substantial development in the last five years, and many competitive and real-world results have been obtained (Sect. 2.3).

Before we proceed further, it is necessary to clarify what we mean by a complex circuit, as researchers in EC usually have a somewhat illusory notion about circuit complexity. In theoretical computer science, the complexity of Boolean functions is expressed as the *size* or *depth* of corresponding combinational circuits that compute them. The size of a circuit is the number of gates it contains; its depth is the maximal length of a path from an input gate to the output gate. In evolutionary computation, complexity is typically assessed according to the number of input variables, because the number of variables significantly impacts the scalability. In practice, the number of gates is usually considered as a complexity measure. Since the goals of the early pioneers of EHW were to evolve complex circuits and overcome the limits of traditional design, it is necessary to take the parameters of real circuits into account. The electronic design automation (EDA) community relies heavily on public benchmarks to evaluate the performance of academic and commercial design tools, because benchmarking is essential to develop effective methodologies. Hence we can analyse characteristics of benchmark circuits to provide a real picture regarding the circuit complexity. The first widely accepted benchmark suite ISCAS-85 consisting of ten industrial designs was introduced by Brglez and Fujiwara [3]. The circuits included in this benchmark set contain from 32 to 207 inputs and were originally implemented using 160–3512 logic gates. With advancements in technology, several updates have been introduced. The most recent updates are LGSynth93 [17] and IWLS 2005 [1]. LGSynth93 consists of 202 circuits with up to 24,658 gates and 1,465 registers; IWLS2005 contains 84 designs with up to 185,000 registers and 900,000 gates collected from different websites with open-source IP cores. A combinational benchmark suite was introduced by Amaru et al. [2]. This suite is designed to challenge modern logic optimization tools, and consists of 23 natively combinational circuits that are divided into three classes: arithmetic circuits, random/control circuits, and difficult benchmarks. The arithmetic benchmarks have up to 512 inputs and more than a thousand gates; the difficult benchmark set consists of circuits with more than million gates and more than 100 inputs. As is evident, a *complex circuit* is a circuit having at least a hundred inputs and consisting of thousands of gates.

The goal of this chapter is to summarize the progress in the evolutionary synthesis of gate-level digital circuits, and to identify the challenges that need to be addressed to enable evolutionary methods to penetrate into industrial practice. The rest of this chapter is organized as follows. Section 2 briefly maps the history of evolutionary circuit design. Among others, it summarizes the key results that have been achieved in the area of evolutionary design of digital circuits for the whole existence of EHW. Then, challenges identified in recent years and that should be addressed in near future are mentioned in Sect. 3. Concluding remarks are given in Sect. 4.

2 Evolutionary Design of Digital Circuits

2.1 First Generation EHW

Gate-level evolution was addressed only rarely before 2000. The first results in the area of digital circuit synthesis were reported by Koza [15], who investigated the evolutionary design of the even-parity problem in his extensive discussions of the standard genetic programming (GP) paradigm. Although the construction of an optimal parity circuit is a straightforward process, parity circuits are considered to be an appropriate benchmark problem within the evolutionary computation community when a small set of gates (AND, OR, NOT) is used.

Thompson [30] used a form of direct encoding loosely based on the structure of an FPGA in his experiment with evolution of a square wave oscillator. A genetic algorithm was employed by Coello et al. [4], to evolve various 2-bit adders and multipliers. Miller demonstrated that evolutionary design systems are not only able to rediscover standard designs, but they can, in some cases, improve them [18, 22]; he was interested in the evolutionary design of arithmetic circuits and digital filters.

A new evolutionary algorithm, Cartesian genetic programming (CGP), was introduced by Julian F. Miller in 2000.[1] Miller designed this approach to address two issues related to the efficiency of common genetic programming: poor ability to represent digital circuits, and the presence of bloat. This variant of GP is called 'Cartesian' because it represents a program using a two-dimensional grid of nodes. The genotypes are encoded as lists of integers of fixed length that are mapped to directed acyclic graphs (DAGs) rather than the usual trees. For more details about CGP and its applications, we refer the reader to the collection edited by Julian F. Miller [20].

CGP has been used to demonstrate that evolutionary computing can improve results of conventional circuit synthesis and optimization algorithms. As a proof of concept, small arithmetic circuits were considered originally. A 4-bit multiplier is the most complex circuit evolved in this category [38]. However, the problems addressed by the EHW community have remained nearly of the same complexity since then. The most complex combinational circuit directly evolved during the first two decades of EHW consists of tens of gates and has around 20 inputs [29]. It is clear that those results could barely compete with conventional circuit design tools producing circuits counting thousands of gates and hundreds of inputs.

One of the goals of the early pioneers of EHW was to evolve complex circuits, overcome the limits of traditional design, and find ways how to exploit the vast computational resources available in today's computation platforms. Unfortunately, nobody has been able to approach this goal, except Koza [16], who reports tens of human-competitive analogue circuits obtained automatically using genetic programming. Since 2000, many researchers have invested enormous effort in proposing new ways to simplify the problem for evolution in terms of finding more effective approaches to explore the search space for more complex applications. Many novel techniques, including decomposition, development, modularisation and even new representations, have been proposed [20, 27, 29, 40]. Despite this, only a little progress has been achieved and the gap between the complexity of problems addressed in industry and EHW continues to widen as the advancements in technology develop. This further supports the belief that evolutionary search works better for analogue circuits than for digital circuits, possibly due to the fact that analogue behaviours provide relatively smoother search spaces [28].

In order to address the increasing complexity of real-world designs, some authors escape from the gate-level representation and use function-level evolution. Instead of simple gates, larger building blocks such as adders and multipliers are employed. Many and even patentable digital circuits have been discovered, especially in the area of digital signal processing [20, 25, 37]. One of the most complex circuits evolved by means of the function-level approach is a random shot noise image filter with

---

[1] Cartesian genetic programming grew from a method of evolving digital circuits developed by [22]. However the term *Cartesian genetic programming* first appeared in [19], and was proposed as a general form of genetic programming in [21].

25 inputs consisting of more than 1,500 two-input gates when synthesised as 8-bit circuit [37]. Despite this work, EHW found itself in a critical stage around the year 2010, and it was not then clear whether there existed a path forward that would allow the field to progress [8]. The scalability problem has been identified as one of the most difficult problems that researchers are faced with in the EHW field and one that should be, among others, addressed by the second generation of EHW.

2.2 Scalability Issues

Poor scalability typically means that evolutionary algorithms are able to provide solutions to small problem instances only, and a partially working solution is usually returned in other cases. The scalability problem can primarily be seen from two perspectives: *scalability of representation* and *scalability of fitness evaluation*. From the viewpoint of the scalability of representation, the problem is that long chromosomes are usually required to represent complex solutions. Long chromosomes, however, imply large search spaces that are typically difficult to search. The scalability of fitness evaluation represents another big challenge. The problem is that complex candidate solutions might require a lot of time to be evaluated. As a consequence, only a small fraction of the search space may be explored in a reasonable time. This problem is especially noticeable in the case of the evolutionary design of digital circuits where the time required to evaluate a candidate circuit grows exponentially with the increasing number of inputs. One possibility how to overcome this issue is to reduce the number of test cases utilised to determine the fitness value. However, this approach is not applicable in the case of gate-level evolution [14]. There is a high chance that a completely different circuit violating the specification is evolved even if only a single test case is omitted.

Both mentioned scalability issues have independent effects that are hard to predict in practice. Consider, for example, the evolutionary design of gate-level multipliers from scratch. It is relatively easy to evolve a 4-bit (i.e. 8-input) multiplier, yet a huge computational effort is required to discover a 5-bit multiplier, even though the time required to evaluate a candidate solution increases only 4-fold and the number of gates is approximately doubled. Experiments with various accelerators have revealed that the number of evaluations has to be increased more than 170-fold to discover a valid 5-bit multiplier [13].

It is believed that the scalability of representation is the root cause that prevents evolutionary algorithms from handling complex instances. This hypothesis is supported mainly by the fact that the number of evaluations needed to discover a digital circuit is significantly higher compared to Koza's experiments with analogue circuits. In addition, the community has been unable to substantially improve the scalability of gate-level evolutionary design, even when various FPGA-based, GPU-based and CPU-based accelerators have been employed [13, 20, 36]. There is no doubt that the size of the search space grows enormously as the complexity of problem instances increases. On the other hand, only little is known about the phenotypic search space of digital circuits. It may be possible, for example, that the number of valid solutions proportionally increases with the increasing size of search space.

In addition to the previously discussed scalability issues, *scalability of specification* is sometimes mentioned [35]. Let us assume that the previously mentioned problems do not exist. Even then, we will not be able to design complex circuits in reasonable time. The problem is that the frequently used specification in the form of truth table does not itself scale. The amount of memory required to store the whole truth table grows exponentially with the increasing number of variables.

2.3   Second Generation EHW

Despite pessimism of the EHW community, its researchers have continued to investigate how to overcome the scalability issues. Vasicek and Sekanina [34] demonstrate that it is feasible to optimize complex digital circuits when the common truth-table-based fitness evaluation procedure is replaced with formal verification. The method exploits the fact that efficient algorithms have been developed in the field of formal verification that enable us to relatively quickly decide whether two circuits are functionally equivalent. Compared to the state-of-the-art synthesis tools, the authors reported a 37.8% reduction in the number of gates (on average) across various benchmark circuits having 67–1408 gates and 22–128 inputs.

An interesting feature of the approach is that it focuses solely on the improvement of fitness evaluation efficiency. In order to reduce the time needed to determine the fitness value, a method routinely used in the area of logic synthesis known as combinational equivalence checking is employed. But it is not the equivalence checking alone that enables speedups of several orders of magnitude. The approach benefits from a tight connection between the evolutionary algorithm and combinational equivalence checking. Since every fitness evaluation is preceded by a mutation, a list of nodes that are different for the parent and its offspring can be calculated. This list can be used to determine a cone of influence (COI)—the set of outputs and gates that have to be compared with the reference circuit—and only these outputs and gates are checked. In addition to that, a parental circuit serves simultaneously as a reference.

The efficiency of the proposed method has been further improved by Vasicek [33], who combines a circuit simulator with formal verification in order to detect the functional non-equivalence of the parent and its offspring. In contrast with previously published work, an extensive set of 100 real-world benchmarks circuits is used to evaluate the performance of the method. The least complex circuit has 106 gates, 15 primary inputs, and 38 outputs. The most complex circuit, an audio codec controller, has 16,158 gates, 2,176 inputs, and 2,136 outputs. Half of the benchmark circuits have more than 50 primary inputs and more than a thousand gates. On average, the method enables a 34% reduction in gate count, and the evolution is executed for only 15 min.

A very pessimistic future for EHW-based digital circuit synthesis was predicted by Greenwood and Tyrrell [7]. Despite that, five years later and after little more than

15 years of EHW, Vasicek and Sekanina [34] proposed an approach easily overcoming the previous empirical limitation of evolutionary design represented by a digital circuit having about twenty inputs and up to hundred gates. At the same time, several doubts and questions emerged. Since only the scalability of fitness evaluation was addressed, it is questionable whether the scalability of representation is really the key issue preventing the handling of complex instances. We believe that this is not the case for at least the evolutionary optimization of digital circuits. Hence, addressing the problem of scalability of representation in the context of evolutionary optimization of digital circuits seems not to be urgent. The extensive experimental evaluation presented by Vasicek [33] has revealed another and probably more severe issue of evolutionary optimization of digital circuits: the enormous inefficiency of the evolutionary algorithm itself. The ratio between the number of acceptable candidate solutions and invalid candidate solutions is worse than 1:180 in average. A candidate solution is invalid if it is functionally incorrect, i.e. violates the requirements given by the specification. So approximately 99.5% of the runtime is wasted by generating and evaluating invalid candidate circuits that lie beyond the desired space of potential solutions.

If we want to keep the EHW field progressing, it seems that the researchers should firstly turn their attention away from tweaking the parameters of various evolutionary algorithms, and begin to combine evolutionary approaches with state-of-the-art methods routinely used in the field that is addressed with evolution. This claim is based mainly on the fact that the results reported by Vasicek [33] were obtained by means of the standard variant of genetic programming and standard settings recommended in the literature. Although CGP is considered as one of the most efficient methods for evolutionary design and optimization of digital circuits [20], there is no guarantee that it offers the best possible convergence. It is not even clear whether the poor efficiency discussed in the previous paragraph is caused by CGP, or if it is a common problem of all evolutionary approaches. It is unfortunately nontrivial to point out the algorithm that achieves the best performance. Firstly, many authors are evaluating algorithms using artificial problems such as simple parity circuits instead of real benchmark circuits. Secondly, only evolutionary design of simple circuits has been addressed. Hence, the efficiency of the evolutionary algorithm and its improvement represents an open question for future investigations.

The approach of [34] has sometimes been criticised that it has to be seeded with an already working circuit. and thet it cannot be used to evolve digital circuits from scratch. At first sight, this objection seems to be legitimate. However, it is important to realise and accept how circuits are specified in conventional tools. Conventional circuit synthesis and optimization usually starts from an unoptimized fully functional circuit supplied by a designer. Such a circuit is specified in the form of a netlist, typically using low-level (e.g. BLIF specification) or high-level (e.g. Verilog or VHDL language) hierarchical description. In some cases, truth tables are directly used, but to specify only simple circuits. So the seed required by evolution can easily be obtained from the specification, for example by transforming the Sum-of-Products representation into a logic network or synthesizing Verilog description to a gate-level netlist.

An interesting question, at least from the theoretical point of view, is whether complex digital circuits can be evolved from randomly seeded initial populations, i.e. whether a truly evolutionary circuit design is possible for complex circuits. Firstly, it is necessary to accept the fact that this approach can hardly compete with conventional state-of-the-art synthesis tools if the time of synthesis is considered. Obtaining a fully functional solution from a randomly seeded population consumes a considerable time because the approach exploits the generate-and-test principle, and no additional knowledge about the problem is available. However, there are reasons why it makes sense to develop the aforementioned evolutionary approach. Firstly, the circuit design problem can serve as a useful test problem for the performance evaluation and comparison of genetic programming systems. Secondly, the approach seems to be suitable for adaptive embedded systems, because running standard circuit design packages is usually infeasible on such systems.

In order to address the evolution of complex circuits from scratch, Vasicek and Sekanina [35] propose a method that exploits the fact that the specification can be given in the form of binary decision diagram (BDD). Although there are some pathological cases of circuits for which BDDs do not scale well, BDDs are known to be an efficient tool for representation and manipulation with digital circuits. Vasicek and Sekanina [35] use BDDs to determine Hamming distance between a candidate circuit and specification. This requires only a few milliseconds, even for circuits having more than 40 inputs. Compared to the well-optimized common CGP, the proposed method achieves a speedup of four to seven orders of magnitude when circuits having from 23 to 41 inputs are considered. As a proof of concept, 16 nontrivial circuits outside of the scope of well-optimized standard CGP are considered. Correctly evolved circuits are reported in twelve cases. In addition, evolution is able to improve the results of conventional synthesis tools. For example, evolution discovered a 28-input circuit having 57% fewer gates than the result obtained from the state-of-the-art synthesis tool. An average gate reduction of 48.7% is reported for all evolved circuits.

We can conclude that it has been experimentally confirmed that it is beneficial to employ EAs even in the evolution of digital circuits from scratch. In many cases, more compact solutions may be obtained, because the EA is not biased to the initial circuits that must be provided in the case of conventional synthesis. However, we have to accept and tolerate the overhead of evolution manifesting itself in the higher demand for computational resources. For example, at most three hours are required to evolve the circuits investigated by Vasicek and Sekanina [35]. On the other hand, this represents an unimportant problem when a theoretical point of view is considered. There are more fundamental questions that we should deal with.

Firstly, how is it possible that evolution successfully discovered a 28-input circuit (frg1), yet no solution was provided for 23-input circuit (cordic)? Since the 23-input circuit consists of fewer gates than the 28-input circuit, this issue is evidently not connected with poor scalability of representation. We can exclude even the problem of scalability of fitness evaluation, because the time required to evaluate the candidate solutions is nearly the same in both cases. Hence, there must be some other issue that prevents the discovery of some circuits. Secondly, the notion of circuit complexity remains an open question. Computation effort (derived from the number of input combinations and the number of gates) is used to measure the circuit complexity. Some of the circuits that are not evolved, however, exhibit a lower complexity value compared to the successfully evolved circuit. Thirdly, the paper does not offer an answer to the question related to the most complex circuit that could be directly evolved. The work was intended as a proof-of-concept only. Finally, it is worth noting that only the scalability of fitness evaluation is targeted. Does it mean that the scalability of representation is an unimportant problem, at least to some circuit size? [9] state that it is becoming generally accepted that there needs to be a move away from a direct, or one-to-one, genotype-phenotype mapping to enable evolution of large complex circuits. This conjecture seems to be wrong at least for the evolutionary optimisation of existing gate-level circuits, because large genotypes encoding circuits with thousands of gates can be directly optimised using CGP [33].

3   Open Challenges

Significant progress has been made in the area of evolutionary synthesis of digital circuits in recent years. Despite that, we are still facing many challenges that need to be properly addressed in order to enable evolutionary circuit design to become a competitive and respected synthesis method.

3.1   *Evolutionary Synthesis and Hardware Community*

One of the problems with the EHW community is that the practical point of view is usually in second place. The typical goal is to evolve some circuit and nobody questions whether there is a real need to do so. Although there is no doubt that much work has been done in the last more than two decades of EHW, there is a barrier preventing EHW being fully accepted by hardware community. Let us mention the main aspects that are usually criticised by the "traditional" circuit design community.

The evolutionary design of digital circuits is sometimes criticised due to its inherently non-deterministic nature. To understand this, it is necessary to realize how the conventional synthesis tools are implemented. The tools typically start with a netlist represented using a network, for example in the form of and-inverter graph (AIG). Then, deterministic transformations are performed over the AIG. Since the transformations are local in nature, the network may be refined by their repeated application. The solution quality can be improved in this way at the expense of run-time. Designers have a guarantee that the optimisation process ends. In addition, they have a certain degree of guarantee that a circuit of some quality will be obtained when, say, one hundred iterations are employed.

Another handicap of the evolutionary approaches is that they are time consuming and generally not scalable when compared with methods of logic synthesis. While the problem of scalability of evaluation has been partly eliminated, the method still suffers from long execution times due to the inefficiency of the generate-and-test approach. Even though the recent results indicate that more complex circuits may be evolved, it will probably be necessary to accompany the proposed methods with a suitable partitioning in order to reduce the problem complexity. It makes no sense to optimize complex netlists describing the whole processors or other complex circuits at the gate-level. The conventional tools usually benefit from a hierarchical description of a given digital circuit.

Another issue is the relevance of the chosen optimisation criteria for industrial practice. Originally, the number of gates was the only criterion that was optimized within the evolutionary community. In many cases, not only the number of gates, but also delay and area on a chip, play an important role. In addition, power efficiency has emerged as one of the most critical goals of hardware design in recent years [10]. Logic synthesis is a complex process that has to consider several aspects that are in principle mutually dependent. Two basic scenarios are typically conducted in practice: optimizing the power and/or area under some delay constraints, or optimizing the delay possibly under some power and/or area constraints. In order to efficiently determine the delay, area and power dissipation of a given circuit, so as to avoid running of time-consuming SPICE simulator and yet obtain results exhibiting a reasonable accuracy, several algorithms have been developed to estimate the delay as well as power dissipation of digital circuits [10]. The evolutionary community should adopt them and combine them with evolution.

Although the attitude of hardware community to evolutionary techniques seems to be a rather sceptical, it does not mean that evolutionary synthesis has no chance to be accepted at hardware conferences. Interestingly, there is a new research area—approximate computing—in which evolutionary approaches seem to be accepted [26]. Conventional synthesis tools are not constructed to perform the synthesis of approximate circuits, and no golden design exists for an approximate circuit. In the case of approximate synthesis, it is sufficient to design a circuit that responds correctly for a reasonable subset of input vectors, provided that the worst-case (or average) error is under control. Because of the nature of approximate circuits (in fact, partially working circuits are sought) and principles of evolutionary circuit design (evolutionary-based improving of partially working circuits), evolutionary computing seems to be the approach of the first choice.

3.2   Efficiency of Cartesian Genetic Programming

Some problems, such as adopting relevant optimisation criteria, are a matter of time, but there are fundamental issues that need to be addressed. Since its introduction, CGP has been considered to be one of the most efficient methods for evolutionary design and optimization of digital circuits. However, the experiments with complex Boolean problems revealed that the evolutionary mechanisms of CGP should be revised, because there are cases for which CGP performs poorly.

It seems that the evolutionary *optimization* of digital circuits exhibits a completely different behaviour when compared to the evolutionary *design* of digital circuits. Let us give one example supporting this claim. Redundancy is believed to play an important role in evolutionary computation, as it represents a powerful mechanism for searching the fitness landscape. Two forms of redundancy have been identified in CGP [32]. Firstly, CGP employs a genotype-phenotype mapping that removes the genes coding the nodes that are inactive, i.e. the nodes that do not participate in creating the output values. This form of redundancy is referred to as *explicit genetic redundancy*. Secondly, there is *implicit genetic redundancy*. This refers to the situation when there active genes, ones that are decoded into the phenotype, that do not contribute to the behavior of the phenotype under common fitness functions that measure the distance between a candidate circuit and its specification. Both forms of genetic redundancy, together with the replacement strategy, encourages genetic drift, as individuals can be mutated without affecting their fitness. The replacement strategy in CGP enables these individuals to replace the old parent. It is believed that genetic drift strongly aids the escape from local optima. Recently, the theory that the effectiveness of evolutionary search is correlated with the number of available nodes has been disproved [32]. Despite that, it has been shown that explicit genetic redundancy offers a significant advantage. Vasicek [33] reports extensive experimental evaluation that reveals, however, that neutral mutations and redundancy surprisingly offer no advantage when using CGP to optimise complex digital circuits. This result does not mean that neutral mutations are of no significance generally, but they are less important in the case of circuit optimisation. Hence, the role of neutrality and redundancy in CGP represents still an open problem that it is not well understood, especially in the context of evolutionary synthesis of complex circuits.

CGP was originally designed to evolve digital circuits represented using a two-dimensional array of nodes. An open question is whether there is an evidence that this representation improves the efficiency of the evolutionary algorithm. Nowadays, this limitation seems to be an unnecessary constraint [32] and many researchers prefer to represent the circuits by means of a linear array of nodes. While both arrangements have the ability to represent any DAG, the main advantage of the linear encoding is that it substantially reduces the number of nodes required to encode a circuit, because the number of columns and rows of the two-dimensional array must respect circuit parameters such as the circuit depth and width. On the other hand, the two-dimensional array of nodes has the ability to explicitly limit the circuit depth. DAGs are encoded in a linear genome of integer values in CGP. Each node in the DAG is represented by a tuple of genes. One gene specifies the function that the node applies to its inputs, and the remaining genes encode where the node takes inputs from. Nodes can take input from primary circuit inputs or any node preceding them in the linear genome. This restriction prevents the creation of cycles and offers several advantages compared to the genetic programming (e.g. it allows CGP to reuse sub-circuits), but it introduces a positional bias that has a negative impact on CGP's ability to evolve certain DAGs [5]. Nodes that could be connected without creating a cycle may still be prevented from forming that connection because of the given genome ordering. This effect has been investigated only on the evolutionary design of few small circuits, and it will be necessary to perform an additional study focused on the evolutionary optimization. Vasicek [33] did not register any significant degradation in performance of the evolutionary optimisation process related to the positional bias during the experiments with complex circuits. The manifestation of CGP's positional bias thus remains in this context an open question. Goldman and Punch [5] show that this bias is connected with an inherent pressure in CGP that makes it very difficult to evolve individuals with a higher number of active nodes. Such a situation, called the length bias, is prevalent as the length of an individual increases. We have not explicitly investigated this problem in the context of evolutionary design of complex circuits, however, this conclusion seems to be valid. The most complex circuit that has been evolved using BDDs consists of around 155 gates [35]. Fortunately, the length bias is related to evolutionary design only, and does not affect evolutionary optimization.

It is natural to expect that the previously mentioned issues can degrade the performance of CGP in some way. We are convinced, however, that the most critical part of CGP is the process of generating candidate solutions. In order to generate a new population, CGP utilizes a single genetic operator, point mutation. This operator modifies a certain number of randomly chosen genes in such a way that the value of a selected gene is replaced with a randomly generated, yet valid, new one. The range of valid values depends on the gene position and can be determined in advance. This scheme seems to be extremely inefficient. Let us give two facts supporting our claim. Firstly, a population containing only two individuals (i.e. parent and a candidate solution) provides the best convergence when a fixed number of evaluations is considered [34]. Secondly, Vasicek [33] observes that approximately 180 candidate solutions need to be generated to obtain a single valid candidate solution, i.e. an individual with the same or better fitness. Both observations suggest that the evolutionary-based approach requires the generation of a large number of candidate solutions to compensate the poor performance of the mutation operator. The poor ability to generate an acceptable candidate circuit is emphasised by the fact that the smallest possible population size exhibits the best performance. Assuming that our claims are correct, we could also explain why we have pushed forward the limits of evolution when we focus only on the improvement of the scalability of fitness evaluation. The reason is that formal methods such as BDD and SAT achieve speedup of a few orders of magnitude compared to standard CGP. We believe that this part of CGP offers a great potential to significantly improve not only the convergence but also scalability of evolutionary synthesis. It is clear that a form of informed mutations benefiting from problem domain knowledge should be employed, instead of blind random changes. The question is how we should accomplish that.

3.3 Deceptive Fitness Landscape

The notion of a fitness landscape has become an important concept in evolutionary computation. It is well known that the nature of a fitness landscape has a strong relationship with the effectiveness of the evolutionary search. There are three aspects forming the structure of the fitness landscape: the encoding scheme, the fitness function, and the connectivity determined by the genetic operators. The structure can be specified in terms of two characteristics: ruggedness and neutrality of landscapes. Ideally, small gene changes should lead to small changes in the phenotype and, consequently, a small change in fitness value. A landscape exhibiting such a behaviour is said to be smooth. In contrast, if small movements result in large changes in fitness, then the landscape is considered rugged.

[39] study the structure of the fitness landscape by evolving 2-bit multipliers represented at the gate level. The landscape is modelled as a composition of three sub-landscapes as discussed above. The experiments reveal not only that the landscapes are irregular but also that their ruggedness is very different. The conclusion is that the fitness landscapes of digital circuits are quite challenging for evolutionary search. Unfortunately, only a little is known about the phenotypic search space of digital circuits, despite the fact that every search algorithm provides implicit assumptions about the search space [8].

If we ignore the problematic scalability issue, it is fair to say that CGP performs very well in general. However, our experiments with evolution of circuits having more than 15 inputs reveals that there are instances that are extremely hard to evolve. What is even worse, there are trivial problems for which the evolutionary search completely fails. One of them is the well-known and well-studied evolutionary design of parity circuits. In the case of the 30-input parity circuit, for example, the evolutionary search is stuck, and no correct circuit is evolved, provided that a complete set of gates (including XOR gate) is employed. To understand this phenomenon, the fitness landscape needs to be investigated. Under common conditions, there is nearly 100% probability that a randomly generated individual receives a high fitness score whose value equals $HD(F) = 2^{29}$ (i.e. 50% of the worst-case Hamming distance between the correct solution and individual capturing a Boolean function F). Among many others, constant value (i.e. $F = 0$), identities (i.e. $F = x_i$), or XOR of two input variables (i.e. $F = x_i \oplus x_j$) represent Boolean functions exhibiting $HD(F) = 2^{29}$. Even XOR over 29 input variables (i.e. $F = x_1 \oplus \ldots \oplus x_{29}$) has the same Hamming distance. In fact, every Boolean function F implemented as k-input XOR ($k = 2, \ldots, 29$) exhibits $HD(F) = 2^{29}$. Considering phenotypic space, there is a huge disproportion between the number of Boolean functions with Hamming distance equal to 2^{29} and the rest. Hence there is a high chance that such a candidate solution is produced by mutation. In addition to that, it is necessary to consider also the complexity of the correct and a partially working solution. While the correct parity circuit consists of 29 XOR gates, a circuit with Hamming distance equal to 1, for example, consists of a significantly larger number of gates.[2] It is then no surprise that it is extremely hard to generate a Boolean function consisting of more gates yet having better fitness. It

---

[2]Such a circuit can be implemented using a 2-input multiplexer that provides the correct output value for all input combinations except of a single combination for which the inverse output value is given. The implementation consisting of two-input gates evidently requires more gates than common parity circuit.

can be argued that evolution is an incremental process utilising neutral genetic drift that can help build a circuit consisting of more gates. Of course, but there is no pressure that helps to escape from this trap. If we analyse the number of gates during evolution, it oscillates around a few gates only.

One of the possibilities for how to eliminate this problem is to disable XOR gates. The presence of XOR gates in the function set, however, represents the main benefit for the evolutionary approach. While the conventional circuit synthesis approaches are not fully capable to perform the XOR decomposition, evolution is surprisingly able to do that very well. This is typically the main reason why such a huge reduction in the gate count is achieved in comparison with state-of-the-art synthesis tools [34]. Hence, it seems to be more beneficial to combine the Hamming distance with some additional metric (e.g. the number of used variables) to encourage the selection pressure and smooth the fitness landscape.

Evidently, a fitness function based solely on the Hamming distance produces a really deceptive fitness landscape. The question is whether parity represents a singularity or whether there exist a whole class of problems with similar behaviour. It is known that parity is a typical example of symmetric Boolean functions. The symmetric functions are Boolean functions whose output value does not depend on the permutation of the input variables. It means that the output depends only on the number of ones in the input. A unique feature of this class of Boolean functions is that a more compact representation can be utilized instead of the truth table having 2^n rows. Each symmetric Boolean function with n inputs can be represented using the $(n + 1)$-tuple, whose i-th entry $(i = 0, \dots, n)$ is the value of the function on an input vector with i ones.

4   Final Remarks

When Julian F. Miller, the (co)inventor of Cartesian genetic programming, published his paper devoted to the evolutionary design of various gate-level circuits such as adders and multipliers where he demonstrated the strength of evolutionary approach [23][3], he not only motivated researchers around the world to further develop their investigations of evolutionary design of digital circuits, but also inspired many others to engage in a relative young and promising research area: evolvable hardware. Julian brought the evolvable hardware community an efficient method for modelling and evolution of digital circuits. Despite the fact that CGP has been evaluated on small problem instances only and there is no guarantee that it will work on different problems, CGP has become very popular in EHW. Since its introduction, CGP is still considered to be one of the most efficient methods for evolutionary design and optimization of digital circuits.

---

[3]This paper is currently the most cited paper in the journal of Genetic Programming and Evolvable Machines.

We can say without any exaggeration that Julian gave rise to a problem that seems to be an endless source of research opportunities, challenges and questions. After little more than 15 years of CGP's existence, there exist many open questions that are still waiting to be addressed. While there are questions, such as the role of neutrality and importance of redundancy, that accompany CGP since its introduction, progress in various areas, such as evolutionary synthesis of logic circuits, has gradually revealed further questions that were not properly addressed in the past. Many problems have emerged within the last five years, during the experiments with evolutionary synthesis of complex gate-level circuits. The primary reason is that many researchers dealt only with small circuit instances in the past. Also, circuit optimization was originally out of the main focus of EHW community, because the researchers addressed the problem of circuit design.

Cartesian genetic programming, and its variants, is not the only contribution of Julian F. Miller. He is also a pioneer of an unconventional computing paradigm known as evolution-in-materio [24]. Julian is still providing us with revolutionary and inspiring ideas.

Acknowledgements This work was supported by The Ministry of Education, Youth and Sports of the Czech Republic from the National Programme of Sustainability (NPU II); project IT4Innovations excellence in science - LQ1602.

References

1. Albrecht, C.: IWLS 2005 benchmarks. Technical Report, published at 2005 International Workshop on Logic Synthesis (June 2005)
2. Amaru, L., Gaillardon, P.-E., De Micheli, G.: The EPFL combinational benchmark suite. In: 24th International Workshop on Logic & Synthesis (2015)
3. Brglez, F., Fujiwara, H.: A neutral netlist of 10 combinational benchmark circuits and a target translator in Fortran. In: Proceedings of IEEE International Symposium Circuits and Systems (ISCAS 85), pp. 677–692. IEEE Press, Piscataway, N.J. (1985)
4. Coello, C.A.C., Christiansen, A.D., Aguirre, A.H.: Automated design of combinational logic circuits by genetic algorithms. In: Artificial Neural Nets and Genetic Algorithms: Proceedings of the International Conference in Norwich, U.K., 1997, pp. 333–336. Springer, Vienna (1998). doi:10.1007/978-3-7091-6492-1_73. ISBN 978-3-7091-6492-1
5. Goldman, B.W., Punch, W.F.: Analysis of Cartesian genetic programming's evolutionary mechanisms. IEEE Trans. Evol. Comput. **19**(3):359–373 (2015). doi:10.1109/TEVC.2014.2324539. ISSN 1089-778X
6. Gordon, T.G.W., Bentley, P.J.: On evolvable hardware. In: Soft Computing in Industrial Electronics, pp. 279–323. Physica-Verlag, London, UK (2002)
7. Greenwood, G.W., Tyrrell. A.M.: Introduction to evolvable hardware: a practical guide for designing self-adaptive systems. In: IEEE Press Series on Computational Intelligence. Wiley-IEEE Press (2006). ISBN 0471719773
8. Haddow, P.C., Tyrrell, A.: Challenges of evolvable hardware: past, present and the path to a promising future. Genet. Progr. Evol. Mach. **12**, 183–215 (2011)
9. Haddow, P.C., Tufte, G., van Remortel, P.: Shrinking the genotype: L-systems for EHW? In: Proceedings of the 4th International Conference on Evolvable Systems: From Biology to Hardware. LNCS, vol. 2210, pp. 128–139. Springer (2001)

10. Hassoun, S., Sasao, T. (eds.): Logic Synthesis and Verification. Kluwer Academic Publishers, Norwell, MA, USA (2002)

11. Higuchi, T., Niwa, T., Tanaka, T., Iba, H., de Garis, H., Furuya, T.: Evolving hardware with genetic learning: a first step towards building a Darwin machine. In: Proceedings of the 2nd International Conference on Simulated Adaptive Behaviour, pp. 417–424. MIT Press (1993)

12. Higuchi, T., Liu, Y., Yao, X. (eds.): Evolvable Hardware. Springer Science+Media LLC, New York (2006)

13. Hrbacek, R., Sekanina, L.: Towards highly optimized Cartesian genetic programming: from sequential via SIMD and thread to massive parallel implementation. In: Proceedings of 2014 Annual Conference on Genetic and Evolutionary Computation, pp. 1015–1022. ACM, New York, NY, USA (2014). doi:10.1145/2576768.2598343

14. Imamura, K., Foster, J.A., Krings, A.W.: The test vector problem and limitations to evolving digital circuits. In: Proceedings of the 2nd NASA/DoD Workshop on Evolvable Hardware, pp. 75–79. IEEE Computer Society Press (2000)

15. Koza, J.R.: Genetic Programming: On the Programming of Computers by Means of Natural Selection. MIT Press, Cambridge, MA (1992)

16. Koza, J.R.: Human-competitive results produced by genetic programming. Genet. Progr. Evol. Mach. **11**(3–4), 251–284 (2010)

17. McElvain, K.: LGSynth93 benchmark set version 4.0 (1993)

18. Miller, J.F.: Digital filter design at gate-level using evolutionary algorithms. In: Proceedings of the Genetic and Evolutionary Computation Conference, GECCO 1999, pp. 1127–1134. Morgan Kaufmann (1999a)

19. Miller, J.F.: An empirical study of the efficiency of learning Boolean functions using a Cartesian Genetic Programming approach. In Banzhaf, W., Daida, J., Eiben, A.E., Garzon, M.H., Honavar, V., Jakiela, M., Smith, R.E. (Eds), Proceedings of the Genetic and Evolutionary Computation Conference, vol. 2, pp. 1135–1142. Morgan Kaufmann, Orlando, Florida, USA. 13–17 July (1999b). ISBN 1-55860-611-4

20. Miller, J.F.: Cartesian Genetic Programming, Springer (2011)

21. Miller, J.F., Thomson, P.: Cartesian genetic programming. In: Proceedings of the 3rd European Conference on Genetic Programming EuroGP2000. LNCS, vol. 1802, pp. 121–132. Springer (2000)

22. Miller, J.F., Thomson, P., Fogarty, T.: Designing electronic circuits using evolutionary algorithms. arithmetic circuits: a case study. In: Genetic Algorithms and Evolution Strategies in Engineering and Computer Science, pp. 105–131. Wiley (1997)

23. Miller, J.F., Job, D., Vassilev, V.K.: Principles in the evolutionary design of digital circuits— part I. Genet. Progr. Evol. Mach. **1**(1), 8–35 (2000)

24. Miller, J.F., Harding, S.L., Tufte, G.: Evolution-in-materio: evolving computation in materials. Evol. Intell. **7**(1):49–67 (2014). doi:10.1007/s12065-014-0106-6. ISSN 1864-5917

25. Sekanina, L.: Evolvable components: from theory to hardware implementations. In: Natural Computing Series. Springer (2004)

26. Sekanina, L., Vasicek, Z.: Evolutionary computing in approximate circuit design and optimization. In: 1st Workshop on Approximate Computing (WAPCO 2015), pp. 1–6 (2015)

27. Shanthi, A.P., Parthasarathi, R.: Practical and scalable evolution of digital circuits. Appl. Soft Comput. **9**(2), 618–624 (2009)

28. Stoica, A., Keymeulen, D., Tawel, R., Salazar-Lazaro, C., Li, W.-T.: Evolutionary experiments with a fine-grained reconfigurable architecture for analog and digital CMOS circuits. In: Proceedings of the 1st NASA/DOD workshop on Evolvable Hardware, EH 1999, pp. 76–84. IEEE Computer Society, Washington, DC, USA (1999)

29. Stomeo, E., Kalganova, T., Lambert, C.: Generalized disjunction decomposition for evolvable hardware. IEEE Trans. Syst. Man Cybern. Part B **36**(5), 1024–1043 (2006)

30. Thompson, A.: Silicon evolution. In: Proceedings of the First Annual Conference on Genetic Programming, GECCO '96, pp. 444–452. MIT Press, Cambridge, MA, USA (1996)

31. Trefzer, M.A., Tyrrell, A.M.: Evolvable Hardware: From Practice to Application. Springer, Berlin, Heidelberg (2015)

32. Turner, A.J., Miller, J.F.: Neutral genetic drift: an investigation using Cartesian genetic programming. Genet. Progr. Evol. Mach. **16**(4):531–558 (2015). doi:10.1007/s10710-015-9244-6. ISSN 1573-7632
33. Vasicek, Z.: Cartesian GP in optimization of combinational circuits with hundreds of inputs and thousands of gates. In: Proceedings of the 18th European Conference on Genetic Programming—EuroGP. LCNS 9025, pp. 139–150. Springer International Publishing (2015)
34. Vasicek, Z., Sekanina, L.: Formal verification of candidate solutions for post-synthesis evolutionary optimization in evolvable hardware. Genet. Progr. Evol. Mach. **12**(3), 305–327 (2011)
35. Vasicek, Z., Sekanina, L.: How to evolve complex combinational circuits from scratch? In: Proceedings of the 2014 IEEE International Conference on Evolvable Systems, pp. 133–140. IEEE (2014)
36. Vasicek, Z., Slany, K.: Efficient phenotype evaluation in Cartesian genetic programming. In: Proceedings of the 15th European Conference on Genetic Programming. LNCS 7244, pp. 266–278. Springer (2012)
37. Vasicek, Z., Bidlo, M., Sekanina, L.: Evolution of efficient real-time non-linear image filters for fpgas. Soft Comput. **17**(11):2163–2180 (2013). doi:10.1007/s00500-013-1040-8. ISSN 1433-7479
38. Vassilev, V., Job, D., Miller, J.F.: Towards the automatic design of more efficient digital circuits. In: Lohn, J., Stoica, A., Keymeulen, D., Colombano, S. (eds.) Proceedings of the 2nd NASA/DoD Workshop on Evolvable Hardware, pp. 151–160. IEEE Computer Society, Los Alamitos, CA, USA (2000)
39. Vassilev, V.K., Miller, J.F., Fogarty, T.C.: Digital circuit evolution and fitness landscapes. In: Proceedings of the Congress on Evolutionary Computation, vol. 2. IEEE Press, 6-9 July (1999)
40. Zhao, S., Jiao, L.: Multi-objective evolutionary design and knowledge discovery of logic circuits based on an adaptive genetic algorithm. Genet. Progr. Evol. Mach. **7**(3), 195–210 (2006)

Designing Digital Systems Using Cartesian Genetic Programming and VHDL

Benjamin Henson, James Alfred Walker, Martin A. Trefzer
and Andy M. Tyrrell

Abstract This chapter describes the use of biologically inspired Evolutionary Algorithms (EAs) to create designs for implementation on a reconfigurable logic device. Previous work on Evolvable Hardware (EHW) is discussed with a focus on timing problems for digital circuits. An EA is developed that describes the circuit using a Hardware Description Language (HDL) in a Cartesian Genetic Programming (CGP) framework. The use of an HDL enabled a commercial hardware simulator to be used to evaluate the evolved circuits. Timing models are included in the simulation allowing sequential circuits to be created and assessed. The aim of the work is to develop an EA that is able to create time dependent circuity using the versatility of a HDL and a hardware timing simulator. The variation in the circuit timing from the placement of the logic components, led to an environment with a selection pressure that promoted a more robust design. The results show the creation of both combinatorial and sequential circuits.

B. Henson
Communication Technologies Group, Department of Electronic Engineering,
University of York, York YO10 5DD, UK
e-mail: bth502@york.ac.uk

J.A. Walker (✉)
The Digital Centre, School of Engineering and Computer Science,
University of Hull, Hull HU6 7RX, UK
e-mail: j.a.walker@hull.ac.uk

M.A. Trefzer · A.M. Tyrrell
Intelligent Systems and Nanoscience Group, Department of Electronic Engineering,
University of York, Heslington, York YO10 5DD, UK
e-mail: martin.trefzer@york.ac.uk

A.M. Tyrrell
e-mail: andy.tyrrell@york.ac.uk

© Springer International Publishing AG 2018
S. Stepney and A. Adamatzky (eds.), *Inspired by Nature*, Emergence,
Complexity and Computation 28, https://doi.org/10.1007/978-3-319-67997-6_3

1 Introduction

Evolution has become the focus of attention for engineers looking for a way of optimising a solution given a very large problem space. This is akin to a life form being adapted to an environment with many external factors affecting the ability of the organism to survive.

Genetic Algorithms (GAs): GAs generally take the form of an optimisation algorithm where the input is represented as a "chromosome" or "genotype" which is often an abstraction of the parameters for the system. The representation is critical because it sets out the problem space, with the permissible chromosome alterations dictating how it is searched.

Genetic Programming (GP): GPs differ from Genetic Algorithms as they not only evolve the parameters, but also the method for solving a problem. The idea with a GP is to evolve a program that is able to run and produce more complex behaviour. The parameters to run the program are inherent in the evolutionary process.

Evolvable Hardware (EHW): is the application of evolutionary principles to the development, calibration, and maintenance of electronic hardware.

Developments in EHW may be considered in two main areas. *Extrinsic*, where electronic hardware design parameters are optimised with the use of an Evolutionary Algorithm (EA), these values are then used to either calibrate an existing design or create a design that is later fabricated to the specification. *Intrinsic*, where EAs are used to adapt hardware to create new designs and/or new conditions, this might be a change in the requirements for a device or new operating conditions for a piece of equipment.

With the advent of reconfigurable logic devices there became a new avenue for intrinsic EHW [33]. These devices are made up of a large array of basic logic functions, often described in terms of a configurable look up table (LUT). Circuits are created by not only defining the LUT entries but also the connections between them. A Field Programmable Gate Array (FPGA) is one such reconfigurable device which consist of many blocks with LUTs and latch elements which may be programmed along with the connections between them.

Figure 1, shows a configurable block (slice) from a Xilinx Virtex 5 device, it is a collection of blocks such as these that make the logic expressions. The configuration for the LUTs and the connections are loaded into the FPGA with a binary configuration file. In early devices, the configuration and the corresponding effect in the hardware was documented. This meant that it was possible to perform genetic algorithm operations directly on the configuration bit string; allowing a close relationship between the genotype (the configuration bit string) and the phenotype (the resultant logic). As technology progressed more complex devices became available. Also, there was a requirement for FPGA manufacturers to make the devices more secure for the increasingly complex and expensive intellectual property that was embodied in the firmware. Therefore, the configuration streams became encrypted and the relationship between the HDL and the logic that was created was obscured. As the devices became more complex, the rules for a permissible configuration became

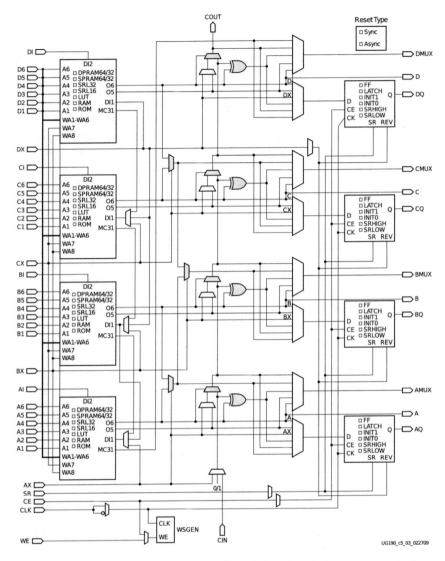

Fig. 1 A slice (Xilinx reconfigurable element) from a Xilinx Virtex 5 device [40]. The slice contains four 5-input LUTs, these can be seen on the left. The flip-flops that can be used are shown on the right

stricter. Again many of these rules were obscured by synthesis tools. Despite these problems, commercial FPGAs are still a viable platform for research, as they offer a relatively cheap and reliable reconfigurable device. Later generation FPGAs also had more resources available not only in terms of reconfigurable logic but also embedded processors. This not only enabled an FPGA to implement a solution found using an EA but also the evolutionary loop, including evaluation and generation of new candidate solutions [22, 44].

Thompson et al. [32] used an FPGA in his work, however, he considered the analogue characteristic of the device and how they changed with different configurations. Using different configurations, Thompson was able to exploit the device characteristics to build a tone discriminator [30]. One of the aims for their work was to remove "design abstractions and the accompanying constraints" [32]. However as Thompson et al. point out "The exploitation of all of the hardware's physical properties must be traded against sensitivity to variations in them" [31]. In other words there is a danger of optimising too much for a particular device, meaning that a design could not be transferred to another device because of the fabrication variations between them. This is difficult to guard against for true intrinsic EHW where each candidate is tested on the hardware to implement it. Variation may be introduced in the stimulus to allow for the selection to create some tolerance to changing conditions. Indeed this is an argument in favour of simulation in that idealised components may be modelled with variation within tolerances. In some applications it is correcting for variation in devices that is the subject for the EA. Takahashi et al. [29] for example, used a GA to adjust for clock skew. They evolved the values for clock delay lines within the integrated circuit, improving the overall delay (and hence throughput) for specific circuits such as a multiplier. This idea of calibration, or selection of components using EAs is being investigated further as a solution to the problems associated with the increased miniaturisation of silicon devices. As devices get smaller there can be significant variation in the characteristic individual transistors. A layered approach is the aim of the PAnDA project [35, 37], using evolutionary techniques the underlying transistors will be optimised to build an FPGA. In support of this project, a timing simulation was developed from a model of the physics of the transistors and incorporated in to a HDL representation [2]. A more generalised form of single device adaptation has been examined for deep space missions, where the circuitry available is fixed at launch. Evolution in this context could be ongoing as the aim would be to adapt to things such as changing characteristics of the components as they aged, or environmental conditions such as extreme temperatures. For greater flexibility, analogue reconfigurable devices have been designed. These devices are Field Programmable Analogue Arrays (FPAAs) [5, 13] and Field Programmable Transistor Arrays (FPTAs) [17, 28]. Each contain analogue components, such as op-amps or transistors with firmware controlled switches connecting them to the circuit. Experiments by NASA use the devices to adapt to changing conditions, in particular the extreme temperature variations found in space environments [28].

2 Evolving Circuits

Broadly speaking digital circuits may be classified into two groups; *combinatorial*, where the output is derived from the inputs only, and once the transitions have propagated through the circuit then the output is stable. The second is *sequential*, this is where a circuit transitions from one stable state to another and it is this state that determines the output. There have been attempts to evolve circuits for each of these two areas. Using genetic programming, Koza [15] developed boolean functions to build behavioural circuit designs. He showed that it was possible to evolve binary circuit with a GP representation, with a logic function at each of the nodes and binary inputs at the leaves of the tree. However, this approach did not transfer well to the intrinsic evolution of circuits due to the varying size of the tree that could be created. This lead to structures that were better suited to a hardware representation with a fixed size of phenotype [19, 23]. Representations such as Cartesian Genetic Programming proposed by Miller [19] where there is a grid of functions that is of a fixed size with configurable connections made between them. In this way the problem of an expansion in the size of a GP tree representation was avoided and there could be a predictable mapping onto hardware [20].

These approaches were able to evolve circuits, and by examination, features could be extracted such as a carry chain in an adder. One of the major challenges is in the scaling of problems [8]. As a problem becomes more complex, the logic resources required to solve it becomes greater. With more resources comes a large number of potential configurations in a larger search space. For complex problems, the evaluation of the result and therefore the fitness may also become more complex and therefore use more resources. For digital logic circuits, the evaluation time grows exponentially with the number of inputs to the circuit, and has recently seen the adoption of formal equivalence checking methods using SAT solvers to manage the circuit complexity and minimise runtime [25].

2.1 Combinatorial

To try to alleviate the scaling problem one avenue of research is to divide the problem up into smaller circuit to evolve. A strategy by Kalganova [12], uses a system where a limited attempt is made to evolve the desired circuit. From this an analysis is made and a new problem created based on parts of the truth table that is not yet solved. This process is then repeated, decomposing the problem into smaller simpler parts. Once a functioning circuit exists then it is optimised further. The method allowed more complex circuits to be evolved by having a reduced search space for each partition. However, to some extent the method for dividing the problem will guide any search, constraining it and reducing the ability of the algorithm to find a novel solution for the circuit as a whole. An alternative method is to try to select reusable parts of the genotype and use them as modules. Koza [15] proposed this as a way of reusing GP tree sections terming them Automatically Defined Functions (ADFs). This idea has also been developed for other forms of Genetic Programming, such as Embedded Cartesian Genetic Programming [36]. The portion of the genotype selected can be made at random, this maintains the simplicity of an evolutionary algorithm in that no special knowledge of the problem is required.Use of this method has shown an improvement in performance for more complex problems [36].

Instead of evolving a circuit directly another type of EA evolves rules for building circuits. This can be seen in natural systems where things are built up of arrays of smaller components such as cells. DNA contains information on the type of cells but also rules as to how to arrange them to make more complex structures. Developmental systems is where there is a rule based mapping between genotype and phenotype. It is this expression of evolved rules that Gordon et al. sought to emulate mapping rules into an FPGA cell representation [6]. The rules associated with Lindenmayer systems can also be the basis for EAs. L-systems have been used to build a network of logic functions and interconnects, as demonstrated in the work of Haddow et al. [7].

2.2   Sequential

Sequential circuits need to have memory element incorporated into the circuit to record the state. Work by Soliman and Abbas [27], constructed a grid layout for functions allowing a single level of connection. The last layer of the grid was allocated to have flip-flop elements, the output from the grid was then fed back as the inputs. The fitness was with reference to the correct output of the circuit at each of the stable states. Creating the circuit in this manner means the circuit style is prescribed in the layout with a register element before each output. This would, however, be more in keeping with the registered style of a more conventional FPGA design. Soliman et al. then compared to evolving the elements of each combinatorial section of the circuit independently. This was a viable strategy, however, some knowledge of the circuit is needed to translate a design into combinatorial elements, more specifically, what would be expected at each stage of a pipeline. This could be easy to determine if all of the registered signals formed part of the output, however for more complex circuits this could prove more of a challenge.

Soleimani et al. [26] used a similar idea of evolving the combinatorial sections of the circuit, with a D-type flip-flop storing the result as either an output or a value that would be feed back into the circuit. However, unlike Soliman et al. the fitness function did not just assess the outputs from the combinatorial circuits or the final output, but looked at the state transitions for the circuit as a whole.

2.3   HDLs in the Evolutionary Loop

One difficulty with intrinsic evolution of hardware is framing the problem in a way that may be physically realised while providing enough flexibility so that a large enough search space can be explored, thus finding a solution.

For the evolution of digital circuits there is the possibility of using commercially available parts. Reconfigurable devices such as FPGAs are very versatile, however, they are designed for a more conventional circuit design techniques. This means that any EHW system needs to fit within these available capabilities. One way in which hardware has been placed in the evolutionary loop is to translate any candidate solution into a Hardware Description Language (HDL) and use the tools supplied by the FPGA manufacturers to build the design. A HDL is often used as the intermediate step between a more abstract representation of the problem which can be used in a Evolutionary Algorithm (EA). Figure 2 illustrates the different evolutionary loops incorporating HDLs.

One example of loop 1 (Fig. 2) implementation is work by Damavandi and Mohammadi [4] who used an EA to create a serial counter. VHDL code was created that was then tested using a ModelSim simulator [18]. The VHDL implementation was a way to simulate a circuit and not targeted at an eventual implementation in hardware. This means that the EA was constrained to use an idealised discrete time.

Tree representations for GP have also been used with HDLs [9, 24], however, the implementations suffer with problems of creating a valid tree structure that can be implemented in the fixed structure of a reconfigurable device. Cullen [3] used Verilog for Grammatical Evolution. This is where each decision based on the grammar of the language is represented as a tree of choices. For example, a two input AND might be selected out of a set of logical operators as the basis for a statement. Given that operator, three additional selections need to be made to complete the statement, these being the signals for the two inputs and the connection for the output. These statements then form the genotype with the genes directly represented in the language with only valid statements being created. Another representation for an EA using VHDL statements was presented by Kojima and Watanuki [14] in their work creating system controllers. There are many similarities to Cullen's grammatical approach in that the VHDL language was used directly as the genotype. The language was further encoded with reference to statement constructs but the elements such as signal declaration or process statement each formed part of the genotype. The Kojima et al.

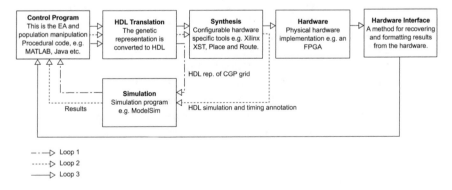

Fig. 2 The different evolutionary loops encompassing HDLs

representation also had additional rules about connections that ensured that feedback loops were not created, this way information only moved in one direction though the circuit.

Loop 3 (Fig. 2) is implemented in the work of [24], Popp et al. where a GP was used to create the candidate solution which was then synthesised for evaluation. It was found that the time taken to synthesise the design made it impractical as a design tool. Although computing power has increased since publication, the complexity and capabilities of the target devices has also increased making for a more complex synthesis process. Later work by Chu et al. [11], used a full Altera FPGA tool chain and FPGA [1] to create a DC motor controller. The feedback in the evolutionary loop was provided by a microcontroller monitoring the motor's movements. Two of the authors of this paper developed the system further using a grid representation of functions for the genotype. They use this evolutionary loop to investigate fault recovery and tolerance for evolved designs [39].

2.4   Development of GA and CGP Representations

The original form of Genetic Programming (GP) used a tree based graph to represent the program. A function such as add or divide would be placed at a node and the operands would be the terminals of the tree. Tree based GP has been successfully used to evolve electronic circuits [15], however this representation has difficulties that need to be considered. Firstly, in order to evaluate an individual the evolved program must be executed, therefore care must be taken to ensure that an individual is viable in terms of creating a valid program. This problem is addressed with Grammatical Evolution (GE) where the rules for the grammar of a language determine the things that are permissible in crossover and mutation [3]. Another problem is that the size of the GP tree tends to grow as the evolution continues, this is commonly termed bloat. This can lead to a very large structure to be evaluated. This can be problematic in the fixed physical resources of a hardware implementation. Strategies can be adopted to reduce bloat, either pruning the tree at evaluation or introducing negative evolutionary pressure, however, the size of the finished solution can still be difficult to estimate. An additional consideration of a pruned tree structure is that all of the sections of the tree contribute to the result, that is there are no redundant sections. It is suggested by Miller [20] that having redundant or inactive sections of the genotype allow the algorithm to move around the search space more effectively and prevent stagnation in the search. This is termed neutral drift, when there are different genotypes (although not necessarily phenotypes) that can produce the same fitness score. Indeed Miller goes on to say that bloat in the GP tree representation is a manifestation of neutral drift in the search [20].

Cartesian Genetic Programming (CGP) is based on a grid structure of functions with input and output that can be connected together to form circuits. Figure 3 shows the basic form of a CGP representation. Each node has a function which can be one of a set (or alleles) that will be used for the evolution. For example, an evolutionary

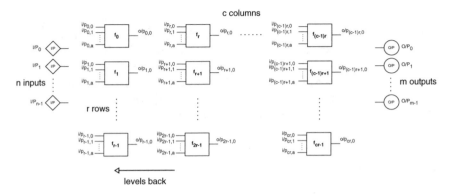

Fig. 3 The basic structure of a CGP grid. Diagram redrawn from [34]

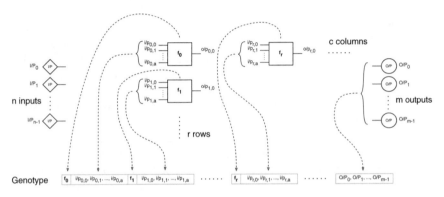

Fig. 4 The basic structure of a CGP genotype. Diagram redrawn from [34]

experiment to derive a logic circuit could use the logical operators AND, OR, NOT as the function set. Each node has a set of inputs and outputs and these would be connected in accordance with rules set out for the experiment. There are also a set of inputs and outputs to the grid as a whole, these would be the interface to the environment. The nodes are numbered from 0 starting in the top left corner and then down the rows and along the columns. Levels are the number of columns away from the current column and it is used to define the range the connections can make. The outputs of the grid may connect to the output of any function or directly to an input. The CGP grid can have any shape in terms of the numbers of rows and columns, however it is common for there to be a single row with the number of columns being the estimate of the number of functions required to fulfil the task. This is an important consideration because the larger the grid the larger the search space with the corresponding increase in computational effort. Figure 4 shows a grid encoded as a genotype in a general form.

An important aspect to recognise in CGP representation is that although all of the inputs to a function need to be connected the output does not, that is, there are inactive functions that do not contribute to the output [34]. This means that there can be parts of the genotype that can hold information about how to process the input that are not used, however, they can become active with the simple reconnection of the functions output. These non-active regions are still subject the evolutionary process of crossover and mutation although they may not affect the output, and hence the fitness. This is an example of neutral drift [20, 21]. The fixed size of the CGP genotype makes it ideal for hardware implementation.

3 Proposed Method

The system described here uses an evolutionary cycle to derive digital circuits that may be realised in electronic hardware. The intention is to use VHDL as a practical method for using commercial FPGAs and the tool chain that accompany them. Also, to include a timing model as part of the environment that the design evolves in. The aim initially is to evolve combinatorial circuit and then to evolve timing dependent sequential circuits. The intention being to investigate the effect of timing variation on the type of circuit evolved.

The evolutionary loop, illustrated in Fig. 2 consists of a controlling program that builds an initial population, where each individual is a candidate circuit and has a Cartesian Genetic Programming (CGP) representation. This population is then translated into a Hardware Description Language (HDL). In this form it can be submitted to hardware specific tools that allow the circuit to be simulated to a given level of detail (loop 1, Fig. 2). Alternatively, the representation may be synthesised into firmware for a reconfigurable device (loop 2, Fig. 2). The candidate solution is evaluated and a fitness value obtained. The results from the evaluation are then passed back to the controlling program where they are used to create the next generation, at which point the cycle begins again. The process is stopped when a solution is reached or some other stopping criterion is met; for example a generation limit or, an unchanging fitness for a number of generations.

3.1 *Physical Representation*

Although a Hardware Description Language (HDL) is able to have a degree of abstraction, there is a point, if a design is to be realised, that it must be specific about the details of the implementation. Field Programmable Gate Arrays (FPGAs) are reconfigurable devices that are able to implement logic expressions directly in hardware. FPGAs have their origins in Programmable Logic Devices which, in their simplest form had a switch matrix that connected a Sum Of Products (SOP) boolean equation, consisting of AND and OR gates. This interconnection developed further to allow more complexity, but also RAM was used to allow the logic functions to be programmed via a look up table (LUT) element.

The difficulty for modern commercial FPGA devices when used for evolutionary algorithms is how to represent a circuit, it is here that using a HDL provides an intermediate step. The level of abstraction means that components (in the HDL sense) could be instantiations of manufacturer specific logic such as a LUT, or it could be an adder that would be built by the synthesiser later in the process. Alternatively, pre-evolved sections could be placed as components. This does mean that the synthesis tools are included in the environment that the GA is evolved in; this being to a greater or lesser degree depending on the abstraction. Depending on the settings, the synthesis tools have rules about placement, routing etc., that would also effect an individuals fitness, for example if the timing was a major consideration.

For EHW a simulation could be as simple as the calculation of logic functions, giving a behavioural description. Alternatively, it could be a much more sophisticated simulation of physical parameters such as propagation delay given a specific temperature of the silicon. Within the field of EHW, simulations are a useful tool for extrinsic evolution. They may be used where the physical realisation and testing of hardware is not practical for each individual. In addition to this there is an ability within a simulator to control some of the parameters that are of interest and may be difficult to reproduce in physical experiments. A simulator could also be used to control the environment that an EA evolves within. For example a simpler environment model could be used at first, with the corresponding improvement in evaluation time, moving to more complex world models as the fitness improved. Simulators are the basis for extrinsic EHW in that they are the means of evaluating a candidate solution without using the physical hardware. This could be potentially useful where the system may be damaged by the testing of poor solutions, one example of this is the evolution of motor controllers, where a poor solution could easily damage not only the motor but any components it drives. Of course one of the features of intrinsic EHW, particularly concerning fault tolerance, is that it can potentially cope with unforeseen faults, or faults that may be difficult to identify or measure. If a simulation was easy to create under these circumstances then it could be easier to correct a fault using more conventional means.

3.2 Hardware Description Language and VHDL

Hardware Description Languages are used to convey in a concise text based format the function of a digital, and more recently analogue and mixed signal circuit. Very high speed integrated circuit Hardware Description Language (VHDL) is one such language where the intention is to describe circuits that may be implemented in many different ways. VHDL is designed to be a way of comprehensively describing a circuit. Not only the behaviour of the circuit is described but also certain aspects of the implementation, such as the time delay associated with physical components. It is this generality that makes VHDL a good candidate as an intermediate language for evolutionary designs.

Listing 3.1 VHDL listing showing the instantiation of a LUT primitive [42]. The connections show two input and an output. The INIT value implements an AND function.

```
1      -- LUT5: 5-input Look-Up Table with general output
2      -- Virtex-5/6, Spartan-6
3      -- Xilinx HDL Libraries Guide, version 11.2
4      LUT5_inst : LUT5
5      generic map (
6         INIT => X"00000008") -- Specify LUT Contents
7      port map (
8         O  => output0,
9         I0 => input0,
10        I1 => input1,
11        I2 => '0',
12        I3 => '0',
13        I4 => '0' );
14     -- End of LUT5_inst instantiation
```

Listing 3.2 VHDL listing showing the implementation of the AND function but leaving the synthesis tool to infer the correct implementation

```
1      output0 <= input0 AND input1;
```

3.2.1 VHDL and FPGAs

VHDL has been adopted as a method of describing the circuit that would be implemented in a FPGA [1, 43]. The circuit can be represented on different levels of abstraction with a greater or less influence of the synthesis tools on the implementation. These levels can be very specific, controlling the individual components that are manufactured into the fabric of the integrated circuit. For instance specifying the contents of a particular Look Up Table (LUT), or instantiating a larger block such as a multiplier. On another level a circuit can be specified on the behavioural level, with state machines and boolean expressions.

Listing 1 shows a fragment of VHDL code that could be used to instantiate a LUT within a slice. It can be seen that control can be at a very low level. Listing 2 illustrates a fragment of code that has a boolean expression that, depending on the exact synthesiser implementation, could be used to infer the same logic as Listing 1. The LUT is a common way to built logic expressions in FPGAs, however there are other components, such as multipliers and large RAM elements, that are more manufacturer specific. Again, these can be instantiated and configured directly or they can be inferred from a different abstraction level.

3.2.2 VITAL and Specific Hardware Description

A fundamental aspect of FPGA design is the ability to simulate the behaviour and physical implementation of a circuit. For simulation, FPGA manufacturers provide a set of libraries that conform to the IEEE VHDL Initiative Towards ASIC Libraries (VITAL) standard. These libraries are designed to promote the development of

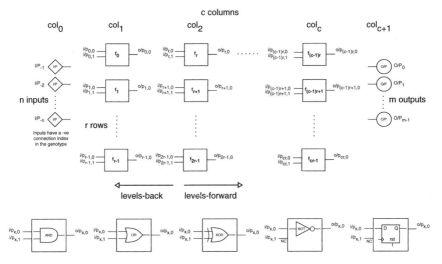

Fig. 5 The structure of a CGP grid for the implementation. There is the addition of forward levels that allowed the feedback required for sequential circuits. Also, the labelling of the columns and inputs are shown as they are represented in the genotype. The diagram also shows the available functions and the connections within the nodes. Each node has two inputs, however for the NOT gate and D-type flip-flop the second input is Not Connected (NC). Diagram modified from [34]

accurate simulation models for VHDL [10] and provide both behavioural and timing models for specific silicon circuits within the device.

For combinatorial circuits, which do not have the potential for feedback loops, there is no need to simulate the timing. This makes the simulation loop much faster. For sequential circuits, where there is potential for feedback loops that do not contain any flip-flop elements, the propagation delay of the signals must be modelled.

3.3 CGP Representation and Evolutionary Strategies

For experimentation a CGP grid is constructed with a set of nodes that map onto a specific function. For the combinatorial circuit experiments the functions are the boolean operators; AND, OR, NOT, XOR. The connections can only be made backwards through the levels. For the sequential circuit experiments the extra function of a D-type flip-flop (DFF) is added. Also, connections can be made forward through the levels, thus it is possible to create feedback loops. For the input and output connections, the inputs are considered to be in column 0, and the outputs are considered to be in column $c + 1$, where c is the total number of CGP columns. They then followed the rules for the levels of connection forwards and backwards for those columns. Figure 5 shows the structure of the CGP implementation.

3.3.1 Fitness Calculation

The fitness calculations for the candidate solutions are based on bitwise summation and the Hierarchical If and only If (HIFF) [16, 38]. The HIFF fitness measure uses a set weighted bitwise comparisons to give a single fitness value. The comparisons are made over blocks of bits the size which is a 2^n where $n \in (0, \log_2 K)$ is the level in the hierarchy, and K total number of output bits.

Algorithm 1 Hierarchical If and only If (HIFF)

Input: $A = \{a_1, \ldots, a_K\}$, K length bit string as output by candidate individual, $B = \{b_1, \ldots, b_K\}$,
    K length target bit string
Initialise: $F_{-1} = 0$
 1: **for** $n = 0$ to K **do**
 2:    $l = 2^n$
 3:    $M_{-1} = 0$
 4:    **for** $p = 0$ to K/l **do**
 5:        $M_p = M_{p-1} + g\{a_{l(p-1)+1}, \ldots a_{lp}; b_{l(p-1)+1}, \ldots b_{lp}\}$
 6:    **end for**
 7:    $F_n = F_{n-1} + lM_p$
 8: **end for**

$g\{a, b\}$ is a comparison operator returning 1 if $a = b$, else 0. F_n is the final fitness value.

One issue with the HIFF function however, is that the solution to the problem needs to have a number of bits that is a power of 2. Dsepending on the problem, adding dummy bits to the target bit pattern leads to a much larger search space. An alternative is to pack the output from the EWH before the fitness is calculated. This is the strategy used for the experiments with uneven outputs.

3.3.2 Level of Detail

One of the features of VHDL with FPGAs, is the ability to use components that can describe the desired hardware at different levels of detail. For this system, illustrated in Fig. 6, the structure is as follows: A top level component holds the whole population. This is convenient primarily for the synthesis of the components; most designs require a top level to specify the connections to the pins. This module consisted of a top interface to the outside world and a list of the individuals, with their connections bringing the signals to the outside. The next level is the individual in the population. Within this each of the genes are represented, that is, the connections and the functions at each node. The final level is the implementation of the functions themselves. Within VHDL this may be specified as boolean expression leaving the synthesiser to create the appropriate logic. Alternatively, a device specific component may be instantiated and configured to give the appropriate output, in a Xilinx FPGA this would use one or more LUTs and a flip-flip element if required. The implementation could be defined further by specifying the precise placement of the components

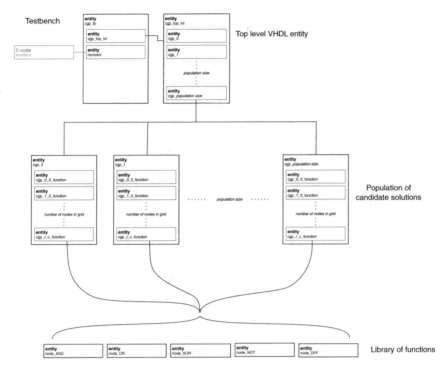

Fig. 6 The structure of the VHDL implementation

within the device. It is the level of specification for how the design is created, that, to a greater or lesser extent, means that the FPGA synthesis tools become part of the environment that a designed individual needed to be successful in. A example of what this could mean is that for a design that very loosely defines component placement, the place and route (P&R) tools would have a large effect on the environment. Each compile would place components in a different position and change the timing. This would mean that individuals with marginal timing would not maintain a consistent fitness, giving them a poorer average fitness for this varying environment. The experiments detailed in this chapter use two input logic gates AND, OR, XOR, an inverter and a D-type flip-flop. These are relatively simple functions, however, if an interface can be defined then there is no reason that these could not be more complex expressions. The flexibility of the CGP layout means that the functions can have any number of connections, therefore other functions can incorporated, such things as multipliers that are provided as set blocks from the manufacturer. It could also mean that smaller evolved blocks could be incorporated as functions within the same VHDL structure.

3.3.3   Intrinsic, Extrinsic, Hardware and Simulators

To create an evolutionary loop a simulator was used as the means of evaluating the circuits. Figure 2, show three workflows. The first shows the combinatorial evolution, with the simulator simply evaluating the logic expressions. Provided a circuit can accommodate any propagation delay and any glitches can be tolerated, then it may be expected that this design will work if implemented in hardware.

For the second, workflow timing is a critical part of the correct function of the circuit, therefore the simulation needs to be more detailed. The FPGA synthesis tools need to be incorporated into the evolutionary loop and an estimate of operating temperature and voltage need to be put into the model. The FPGA tools report worst case timing models [41] whereas absolute timing would be needed to describe the correct operation of a combinatorial loop.

Finally, loop three is the workflow that incorporates the physical FPGA into the loop. This involves the full FPGA tool chain, the FPGA to implement the design and support hardware to generate inputs and evaluate the output from the candidate solution. This would create a design optimised for that particular device, under the particular conditions. If the solution needed to be tolerant to changing conditions then these would need to be present during the evolution.

4   Experimental Setup

4.1   Environment

The experiments were performed using a PC; Intel Core i5 CPU 650, 3.20 GHz with 3 GB of RAM. The software version for the main tools are; Linux version 2.6.34.1 (gcc version 4.4.4 (GCC)), Xilinx ISE Linux 13.2 (Xilinx ISE Linux 11.3, used for the 4 bit counter), Mentor Graphics ModelSim SE 6.5c. The 4 bit counter experiment used Xilinx ISE Linux 11.3 and targeted Xilinx Virtex 5 LX 50—speed grade 1 (xc5vlx50t-ff1136-1). All of the other evolutions used Xilinx ISE Linux 13.2 and targeted Xilinx Virtex 6 LX 75—speed grade 1 (xc6vlx75t-1-ff784). The target device is important for the type of primitives used and the models used to build the simulation. The speed grade effects the timing models for the device. The time mentioned for each run is simply the wall clock time, and is included to give an indication of the time needed for simulation or synthesis and simulation.

4.2   CGP Parameters

The general parameters used for all experiments are shown in Table 1. The problem specific parameters where extra resources were required either in the CGP genotype

Table 1 General parameters for CGP

Parameter	Value
No. of rows in CGP grid	1
No. of individuals in a population	96
No. of elite individuals	1
Tournament size for selection	5
No. of mutations (% genotype length)	5
Maximum no. of generations	10 000
Simulation clock period (ns)	10

or in the evaluation process, are described in Table 2. The function set used in the evolution of the Combinatorial circuits is AND, OR, XOR and NOT. The function set used for evolving the sequential circuits uses a D-type Flip-Flop in addition to the four functions from the Combinatorial function set. The sampling frequency for the output is 100 MHz. The output samples are taken at the falling edge of the clock and the input vectors are changed at the rising edge. This gave 5 ns of propagation time.

Table 2 Problem specific parameters for CGP

Parameter	Combinatorial			Sequential	
	4-bit parity	2-bit adder	3-bit adder	3-bit counter	4-bit counter
No. of columns in CGP grid	20	50	80	50	50
No. of levels forward (columns)	–	–	–	50	50
No. of levels back (columns)	20	50	80	50	50
No. of inputs to the CGP design	4	4	6	1	1
No. outputs from the CGP design	1	3	4	3	4
No. of test vectors to be generated	16	16	64	16	32

For each circuit a single run of the algorithm is shown with the fitness-generation plots being of the best individual.

5 Combinatorial Circuit Results

5.1 4-Bit Even Parity Generator

For the initial experiment a relatively simple problem is chosen, this was to find the even parity of four inputs with the functions AND, OR, XOR and NOT. The inclusion of XOR makes a solution easier to find, as even parity may be generated by the XORing of all of the inputs.

The maximum fitness possible using the HIFF measure is 64. A solution was found after only 8 generations and the time taken was 10 min, 28 s. Figure 7 shows the expressed phenotype of the evolved circuit, which is identical to the conventional design.

5.2 2-Bit Adder

A 2 bit adder was evolved as the second combinatorial circuit. The HIFF function was used as the fitness measure, as the circuit output only three bits a zero is prefixed to each output to make the total number of output bits a power of 2.

The maximum fitness with the HIFF measure was 384. The number of mutation cycles for the circuit were 7. The final generation was 1526 and the time taken was 9 h 17 min 54 s.

Figure 8 shows the final evolved circuit. In group A the OR and the two NOT gates can be removed leaving just the XOR adding the LSB. The carry can be seen as the AND gate labelled C. Group B shows the logic for the second bit. Finally, the gates labelled C and D make up the one path of logic for the third bit. There was some extra logic generated, 5 extra gates, however there wasn't any evolutionary pressure

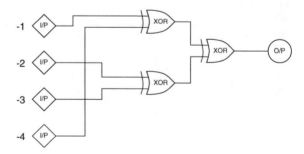

Fig. 7 The circuit diagram for and evolved 4-bit even parity, using the logic function set AND, OR, XOR, NOT

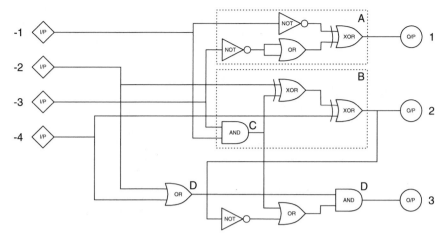

Fig. 8 The circuit diagram for an evolved 2 bit adder, using the logic function set AND, OR, XOR, NOT

to evolve smaller circuits simply to create a complete solution, which is achieved. Figure 9 shows the number of generations against the best fitness for the evolution.

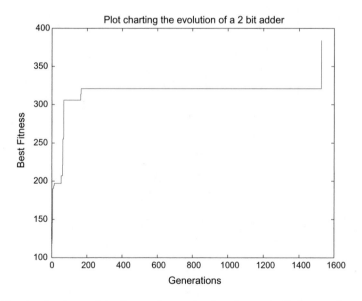

Fig. 9 Plot showing the number of generations against best fitness; the final generation was 1526

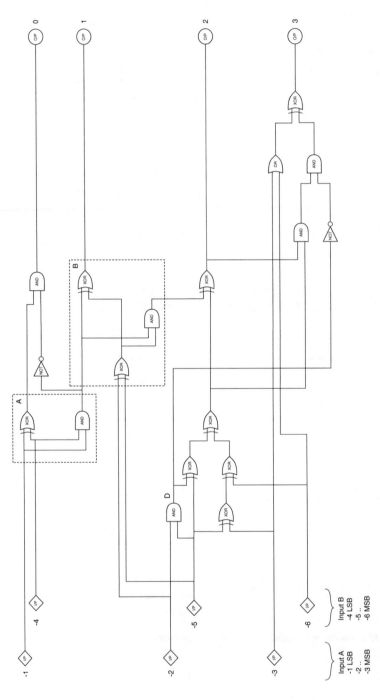

Fig. 10 The circuit diagram for an evolved 3 bit adder, using the logic function set AND, OR, XOR, NOT

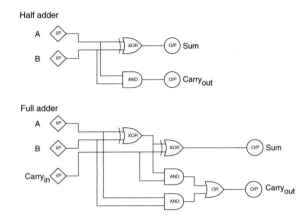

Fig. 11 The circuit diagram showing the conventional design for a half and full adder with ripple carry

5.3 3-Bit Adder

The maximum fitness with the HIFF measure was 2048. The number of mutation cycles for the circuit were 71. The final generation was 4562 and the time taken was 1 days 7 h 3 min 25 s.

The final evolved circuit for the 3 bit adder is shown in Fig. 10. Similarities can be seen between the evolved circuit and a conventional design for a ripple adder shown in Fig. 11. In group A a half adder can be seen summing the two LSBs, the following

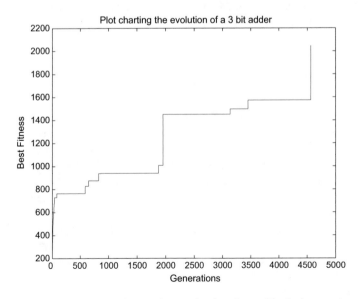

Fig. 12 Plot showing the number of generations against best fitness. The final generation was 4562

NOT and AND gates before Output 0 are not required. Group B shows elements of the full adder with the carry in from group A being accommodated. In addition half of the carry out signal is generated, with the other carry being generated with Gate D. For output bit 2 there is again a XOR tree combining the carry in signals. The final output bit being generated from the MSBs and a calculated value from Output 3. An observation from this circuit is that although there were 80 nodes that could have been used, the final circuit only had 18 gates, this is because previous result from less significant bits are reused. Also, there are modules such as group B that could conceivably be used again. Figure 12 shows the number of generations against the best fitness for the evolution.

6 Sequential Circuits Results

In order to evolve sequential circuits a D-type flip-flop (DFF) was added to the function set. All of the DFF are clocked at the same time at a clock frequency of 100 MHz, in this case. Although the DFF should start in a state specified (given that FPGAs have a defined startup state), a asynchronous reset signal was also added. In the simulation a short pulse is applied before the clock started, this guaranteed that all of the DFF outputs were cleared. Combinatorial loops are allowed (and preserved through synthesis) and no restrictions are made as to where the DFFs are placed. Each of the candidate solutions, including the elite were re-synthesised every generation. This means that there was some variation in the placement and therefore the timing between runs. This variation in the implementation can be considered as part of the environment that the design evolved in.

6.1 3-Bit Counter

A counter was chosen as a sequential circuit to evolve. Although an input is added to the circuit it is held high for all of the test vectors, this means that effectively a pattern generator was evolved. The HIFF function was used as the fitness measure, again the output was prefixed to make the total number of output bits a power of 2.

The maximum fitness with the HIFF measure (packed) was 384. The number of mutation cycles for the circuit were 52. The final generation was 85 and the time taken was 5 h 36 min 12 s.

The final evolved circuit is shown in Fig. 13. The input is always held high so point A would be high allowing the circuit to run. The final circuit is a pattern generator, with delay loops and the initial conditions generating the sequence. Group B toggles the output with a single clock delay and group C toggles the output with a clock delay of 2. A more interesting interaction can be seen between the 2nd and 3rd bits with the circuit in group D. This uses only three more DFF to toggle the 3rd output with a period of 4 clocks. The evolution was performed comparing the circuit output with

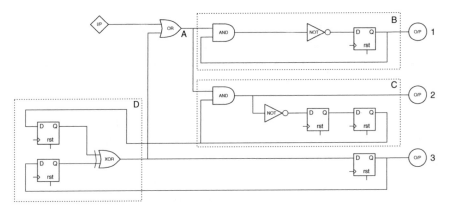

Fig. 13 The circuit diagram for an evolved 3 bit counter, using the logic function set AND, OR, XOR, NOT, DFF. NOTE: the clock and reset connections have been removed for clarity

two cycles of counting (0–15). It can be seen the final circuit will continue to cycle. Another point to notice is that although combinatorial loops were allowed, none exist in the final circuit. Having DFFs in the circuit led to a more stable solution, and given the few levels of logic between flip-flops the circuit is likely to meet the desired timing regardless of the layout. This is important given that there was variation in the timing for the circuit due to changes in the placement of components within the device. This variation in timing can be seen in the evolution. Figure 14 shows

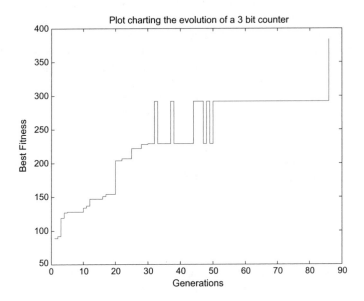

Fig. 14 Plot showing the number of generations against best fitness for the evolved 3 bit counter

Table 3 Table showing the variation in output from the 3 bit counter

Target	Generation		
	30	31	32
000	000	000	000
001	001	001	001
010	010	010	110
011	111	011	011
100	100	100	100
101	101	101	101
110	110	110	110
111	111	111	111
000	100	100	100
001	101	101	101
010	110	110	110
011	111	111	111
100	100	100	100
101	101	101	101
110	110	110	110
111	111	111	111
Fitness	229	292	229

the best fitness for the population through the evolution. Even though the algorithm implemented elitism there is variation in the best fitness. Table 3 shows the output from the circuit for generations 30, 31 and 32, the variation is in the 3rd bit (lines 3–4). The results are created from the same individual and the genotype is unchanged. Figure 15 shows the circuit for this individual. Similarities with the final circuit can be seen; point A is constant and groups B and C are the final implementation. The 3rd bit does contain DFF in some loops driving it, however there are combinatorial loops also contributing to it making it susceptible to timing changes. Indeed in a real implementation the timing would be susceptible to environmental aspects such as temperature and voltage. This feature is not very useful in this instance, which is perhaps why it disappeared, however if the aim was to monitor the environment then this sort of circuit timing would show a change.

6.2 4-Bit Counter

The maximum fitness with the HIFF measure was 896. The number of mutation cycles for the circuit were 297. The final generation was 1264 and the time taken was 2 days 20 h 13 min 28 s.

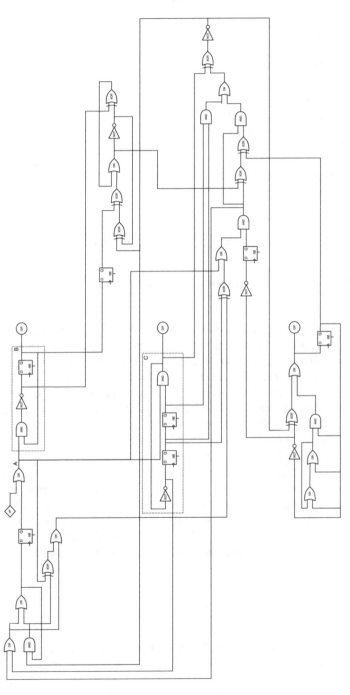

Fig. 15 The circuit diagram for an elite individual for the evolved 3 bit counter at generation 30, 31 and 32. The logic function set AND, OR, XOR, NOT, DFF is used. NOTE: the clock and reset connections have been removed for the clarity of the diagram

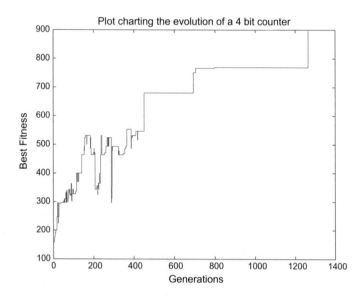

Fig. 16 Plot showing the number of generations against best fitness for the evolved 4 bit counter

The change in the best fitness through the evolution of the 4 bit counter is shown Fig. 16. Again, the instability of the best fitness can be seen in the earlier generations and then later becoming more stable. Examining the circuit shown in Fig. 17 again there aren't any combinatorial loops that contribute to the output. With reference to the circuit in Fig. 17. In this design the input is left out completely (point A), therefore the counter simply runs from the initial conditions, this would seem reasonable since the input did not change. Group B show the toggling circuit for the LSB, however unlike the three bit counter, the 2nd bit is dependent on the 1st. This can be seen in group C. The 3rd bit is again independent with a 4 bit shift register running correctly from the initial conditions, this can be seen in group D. The final 4th bit has a more complex relationship with the rest of the circuit, however, it is also a fully synchronous circuit. For the first three bits it can be seen that the pattern would repeat. For the final bit it is not immediately clear that the pattern would repeat over more than two count cycles. Group E show an example of a combinatorial loop that only drove itself and is therefore superfluous to the circuit for the output bits.

7 Conclusion

The developed system produced viable combinatorial and sequential circuits, with all the experiments able to produce the defined truth table. However, for sequential circuits there is a need for a target output that is better able to encapsulate the intention of the circuits. For instance, the counters required two count cycles for the

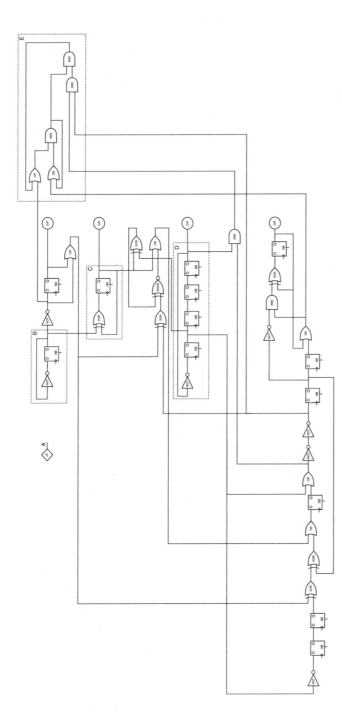

Fig. 17 The circuit diagram for an evolved 4 bit counter, using the logic function set AND, OR, XOR, NOT, DFF. NOTE: the clock and reset connections have been removed for the clarity of the diagram

target output, but with so many flip-flop memory elements, only sequence generators for those two specific outputs were produced with no guarantee (or evolutionary pressure) that this sequence would continue to repeat. Further work may be the development of a fitness that promotes dependency of one output on another where it is appropriate; for the example of a counter, making a design have a higher fitness if the more significant bits are dependent on the lower ones.

The evolution of sequential circuits in an environment with simulated timing gave interesting results as to the stability of the final circuits. The use of flip-flops in sequential circuits appeared to remove less stable combinatorial loops even though there was no explicit evolutionary pressure to do so; purely the variation in the timing meant that more stable circuits were maintained in the gene pool.

Using a hardware description language was a useful tool to gain access to the full range of simulation software available for FPGA design, in particular detailed timing models. In addition, the use of VHDL gave a practical means of implementation that fitted well with the CGP architecture.

Acknowledgements This work is part funded by the EPSRC PAnDA project (EP/I005838/1) and the EPSRC Platform Grant: Bio-inspired Adaptive Architectures and Systems (EP/K040820/1).

References

1. Altera corp. website. http://www.altera.com (2011). Accessed 5 March 2011
2. Campos, P.B., Trefzer, M.A.,Walker, J.A., Bale, S.J., Tyrrell, A.M.: Optimising ring oscillator frequency on a novel FPGA device via partial reconfiguration. In: 2014 IEEE International Conference on Evolvable Systems (ICES), pp. 93–100. IEEE (2014)
3. Cullen, J.: Evolving digital circuits in an industry standard hardware description language. In: Simulated Evolution and Learning, pp. 514–523 (2008)
4. Damavandi, Y., Mohammadi, K.: Co-evolution for communication: an EHW approach. J. Univ. Comput. Sci. **13**(9), 1300–1308 (2007)
5. Dmello, D.R., Gulak, P.G.: Design approaches to field-programmable analog integrated circuits. In: Field-Programmable Analog Arrays, pp. 7–34. Springer (1998)
6. Gordon, T., Bentley, P.: Towards development in evolvable hardware. In: Proceedings of the NASA/DoD Conference on Evolvable Hardware 2002, pp. 241–250. IEEE (2002)
7. Haddow, P., Tufte, G., Van Remortel, P.: Shrinking the genotype: L-systems for EHW? In: Evolvable Systems: From Biology to Hardware, pp. 128–139 (2001)
8. Haddow, P.C., Tyrrell, A.M.: Challenges of evolvable hardware: past, present and the path to a promising future. Genet. Program. Evolvable Mach. **12**(3), 183–215 (2011)
9. Hemmi, H., Mizoguchi, J., Shimohara, K.: Development and evolution of hardware behaviors. In: Towards Evolvable Hardware, pp. 250–265 (1996)
10. IEEE: IEEE standard for VITAL ASIC (application specific integrated circuit) modeling specification. IEEE Std 1076.4-2000, pp. 0_1–420 (2001). doi:10.1109/IEEESTD.2001.93351
11. Jie, C., Qiang, Z., Guo-liang, D., Liang, Y.: The implementation of evolvable hardware closed loop. In: 2008 International Conference on Intelligent Computation Technology and Automation (ICICTA), vol. 2, pp. 48–51. IEEE (2008)
12. Kalganova, T.: Bidirectional incremental evolution in extrinsic evolvable hardware. In: Evolvable Hardware 2000, pp. 65–74. IEEE (2000)
13. Kemerling, J.C., Greenwell, R., Bharath, B.: Analog-and mixed-signal fabrics. Proc. IEEE **103**(7), 1087–1101 (2015)

14. Kojima, K., Watanuki, K.: Supporting VHDL design for air-conditioning controller using evolutionary computation. In: Proceedings of the 17th IFAC World Congress (IFAC'08), pp. 12, 318–312, 323 (2008)

15. Koza, J.: On the programming of computers by means of natural selection, vol. 1. MIT Press (1996)

16. Kuyucu, T., Trefzer, M., Greensted, A., Miller, J., Tyrrell, A.: Fitness functions for the unconstrained evolution of digital circuits. In: 2008 IEEE Congress on Evolutionary Computation (IEEE World Congress on Computational Intelligence), pp. 2584–2591. IEEE (2008)

17. Langeheine, J., Becker, J., Fölling, S., Meier, K., Schemmel, J.: Initial studies of a new VLSI field programmable transistor array. In: International Conference on Evolvable Systems, pp. 62–73. Springer (2001)

18. Mentor graphics corporation ModelSim website. http://www.mentor.com (2011). Accessed 26 Aug 2011

19. Miller, J.: An empirical study of the efficiency of learning Boolean functions using a Cartesian genetic programming approach. In: Proceedings of the Genetic and Evolutionary Computation Conference, vol. 2, pp. 1135–1142. Citeseer (1999)

20. Miller, J.: What bloat? Cartesian genetic programming on Boolean problems. In: 2001 Genetic and Evolutionary Computation Conference Late Breaking Papers, pp. 295–302. Citeseer (2001)

21. Miller, J., Smith, S.: Redundancy and computational efficiency in Cartesian genetic programming. IEEE Trans. Evol. Comput. **10**(2), 167–174 (2006). doi:10.1109/TEVC.2006.871253

22. Mrazek, V., Vasicek, Z.: Acceleration of transistor-level evolution using Xilinx Zynq platform. In: 2014 IEEE International Conference on Evolvable Systems (ICES), pp. 9–16. IEEE (2014)

23. Poli, R.: Evolution of graph-like programs with parallel distributed genetic programming. In: Proceedings of the Seventh International Conference on Genetic Algorithms, pp. 346–353. Citeseer (1997)

24. Popp, R., Montana, D., Gassner, R., Vidaver, G., Iyer, S.: Automated hardware design using genetic programming, VHDL, and FPGAs. In: 1998 IEEE International Conference on Systems, Man, and Cybernetics, 1998, vol. 3, pp. 2184–2189. IEEE (1998)

25. Sekanina, L., Vasicek, Z.: Functional equivalence checking for evolution of complex digital circuits. In: Trefzer, M., Tyrrell, A. (eds.) Evovable Hardware: From Practice to Application, Chap. 6, pp. 175–190. Springer (2015)

26. Soleimani, P., Sabbaghi-Nadooshan, R., Mirzakuchaki, S., Bagheri, M.: Using evolutionary strategies algorithm in the evolutionary design of sequential logic circuits. Int. J. Comput. Sci. Issues (2011)

27. Soliman, A., Abbas, H.: Synchronous sequential circuits design using evolutionary algorithms. In: Canadian Conference on Electrical and Computer Engineering 2004, vol. 4, pp. 2013–2016. IEEE (2004)

28. Stoica, A., Keymeulen, D., Zebulum, R., Thakoor, A., Daud, T., Klimeck, Y., Tawel, R., Duong, V.: Evolution of analog circuits on field programmable transistor arrays. In: Proceedings of the Second NASA/DoD Workshop on Evolvable Hardware 2000, pp. 99–108. IEEE (2000)

29. Takahashi, E., Kasai, Y., Murakawa, M., Higuchi, T.: Post-fabrication clock-timing adjustmentusing genetic algorithms. IEEE J. Solid State Circuits **39**(4), 643–650 (2004)

30. Thompson, A.: Exploring beyond the scope of human design: automatic generation of FPGA configurations through artificial evolution. In: 8th Annual Advanced PLD and FPGA Conference, pp. 5–8 (1998)

31. Thompson, A., Harvey, I., Husbands, P.: Unconstrained evolution and hard consequences. In: Towards Evolvable Hardware, pp. 136–165 (1996)

32. Thompson, A., Harvey, I., Husbands, P.: The natural way to evolve hardware. In: 1996 IEEE International Symposium on Circuits and Systems ISCAS'96, 'Connecting the World', vol. 4, pp. 37–40. IEEE (2002)

33. Trefzer, M.A., Tyrrell, A.M.: Evolvable Hardware: From Practice to Application. Springer (2015)

34. Tyrrell, A., Walker, J., Trefzer, M., Lones, M., Tempesti, G.: Evolvable Hardware (EHW) Lecture Course and Assessment (2011)
35. Walker, J.: Overcoming variability through transistor reconfiguration: evolvable hardware on the PAnDA architecture. In: Trefzer, M., Tyrrell, A. (eds.) Evolvable Hardware: From Practice to Application, Chap. 5, pp. 153–174. Springer (2015)
36. Walker, J., Miller, J.: The automatic acquisition, evolution and reuse of modules in Cartesian genetic programming. IEEE Trans. Evol. Comput. 12(4), 397–417 (2008)
37. Walker, J., Trefzer, M., Bale, S., Tyrrell, A.: PAnDA: a reconfigurable architecture that adapts to physical substrate variations. IEEE Trans. Comput. 62, 1584–1596 (2013)
38. Watson, R.A., Hornby, G.S., Pollack, J.B.: Modeling building-block interdependency. In: International Conference on Parallel Problem Solving from Nature, pp. 97–106. Springer (1998)
39. Wu, H., Chu, J., Yuan, L., Zhao, Q., Liu, S.: Fault-tolerance simulation of brushless motor control circuits. In: Applications of Evolutionary Computation, pp. 184–193 (2011)
40. Xilinx: Virtex-5 FPGA User Guide—UG190 (2010)
41. Xilinx: Synthesis and Simulation Design Guide—UG626 (2011)
42. Xilinx: Virtex-5 Libraries Guide for HDL Designs—UG621 (2011)
43. Xilinx inc. website. http://www.xilinx.com (2011). Accessed 5 March 2011
44. Zhang, Y., Smith, S.L., Tyrrell, A.M.: Digital circuit design using intrinsic evolvable hardware. In: 2004 NASA/DoD Conference on Evolvable Hardware, pp. 55–62. IEEE (2004)

Evolution in Nanomaterio: The NASCENCE Project

Hajo Broersma

Abstract This chapter describes some of the work carried out by members of the NASCENCE project, an FP7 project sponsored by the European Community. After some historical notes and background material, the chapter explains how nanoscale material systems have been configured to perform computational tasks by finding appropriate configuration signals using artificial evolution. Most of this exposition is centred around the work that has been carried out at the MESA+ Institute for Nanotechnology at the University of Twente using disordered networks of nanoparticles. The interested reader will also find many pointers to references that contain more details on work that has been carried out by other members of the NASCENCE consortium on composite materials based on single-walled carbon nanotubes.

1 A Bit of History

This chapter describes parts of the research that was carried out within the framework of the FP7 project NASCENCE (Nanoscale Engineering for Novel Computation Using Evolution) [4], funded by the European Community and running from 1 November 2012 to 31 October 2015.

This research was motivated by earlier work of Julian F. Miller and his coworkers (See, e.g., [17–19, 34, 35]), and Julian was also involved as one of the very active partners within the project. He was the major inspirer of the ideas that led to the birth of the NASCENCE project; he also invented the acronym. In my role as the coordinator of the project it is with great pleasure that I contribute to this volume on the occasion of Julian's 60th birthday.

The title of this chapter is a slight (nano)variation on the term *evolution-in-materio* (EIM) that was coined by Miller and Downing in [34]. The birth of the NASCENCE project originates from an inspiring seminar talk entitled "Evolution

H. Broersma (✉)
Faculty of Electrical Engineering, Mathematics and Computer Science,
CTIT Institute for ICT Research, MESA+ Institute for Nanotechnology,
University of Twente, Enschede, The Netherlands
e-mail: h.j.broersma@utwente.nl

© Springer International Publishing AG 2018
S. Stepney and A. Adamatzky (eds.), *Inspired by Nature*, Emergence,
Complexity and Computation 28, https://doi.org/10.1007/978-3-319-67997-6_4

in materio: Using evolution to get matter to compute" that Julian F. Miller gave at Durham University, UK, on 2 March 2010. Almost at the same time, a "Big Ideas" project proposal entitled "Computational Carbon" on this topic was composed by members of two Durham research groups, one in Electronics led by Mike Petty, and the other one in Algorithms and Complexity led by myself. The purpose of these "Big Ideas" was to stimulate collaboration within the School of Engineering and Computing Sciences, a merger between the School of Engineering and the Department of Computer Science. Although our "Big Ideas" project did not get funded, it formed the basis for the NASCENCE proposal. The latter was submitted to the European Community at the time I had already left Durham to return to my former employer, the University of Twente in The Netherlands. Apart from the group at Durham led by Mike Petty and the group led by me in Twente, the NASCENCE project also involved groups led by Julian F. Miller in York, by Gunnar Tufte in Trondheim, and by Jürgen Schmidhuber in Lugano.

2   A Bit of Background

The key idea behind the NASCENCE project was to use nanoscale materials as a black-box for computation, and to apply artificial evolution in the form of genetic algorithms to control the external configuration stimuli on the material in order to find configurations that would enable the material to perform certain computational tasks. More details on the essential concepts follow in the next section, but they are basically the same as in the earlier works of Julian F. Miller and his coworkers mentioned above, except for the down-scaling to nanoscale materials.

The advancement of nanotechnology offers promising opportunities for alternatives to digital computation. These alternatives might be one of the solutions to cope with the possible problems that have been identified with respect to the further miniaturisation and increasing energy consumption of digital circuitry and transistors.

Current digital computers are based on Turing's abstract model of computation that has been around since the 1930s [52], together with a computer architecture described as an outline of a machine to perform this type of computation proposed by Von Neumann in [44]. The latter formed the foundation of modern stored program computers. This type of computation is possible at the scales we see today through the invention of the transistor, and the miniaturisation, integration and cheap production techniques enabled by the advancement of nanotechnology. Since the computations are based on the abstraction to strings of zeros and ones that are realised by low and high valued electrical signals, the common term for this type of computation is digital computation. Although there existed forms of analogue computation before the introduction of digital computation, the latter is usually also referred to as classical or conventional computation. It reflects a top down approach: first designing an architecture and the functionality of a computational device on a drawing board, and then realising it by building it from purpose-built components.

The approach of EIM is more bottom up, as we explain shortly, and this EIM is one of many different alternative approaches that are gathered under the umbrella term "unconventional computing". Since one of the key ingredients of EIM is the use of artificial, computer-controlled evolution, it is also part of a wide research area known as bio-inspired computation. It is inspired by the many complex processes that take place in living systems, and that can be seen as computations. These processes have been enabled through a long adaptive, selective and uncontrolled procedure of Darwinian evolution. The key ingredients of this evolution, like cross-over, mutation, natural selection and survival of the fittest have been mimicked in computer science and operations research, and led to an optimisation technique that we refer to as computer-controlled evolution. This concept of computer-controlled evolution is known under a variety of other terms, e.g., evolutionary algorithms [13], genetic algorithms [20], and genetic programming [24, 47]. Before addressing more details about the use of computer-controlled evolution in this particular setting, it is convenient to first give a brief overview of the NASCENCE project.

3 NASCENCE in a Nutshell

The conceptual ideas and ingredients of the NASCENCE project are illustrated in Fig. 1. The key idea is to use a sample of nanomaterial on a micro-electrode array (the Material Disc in Fig. 1) as a black-box for computational tasks. The way to manipulate the black-box is to change the state of the material, by using the micro-electrode array to interact with the matter through programmable external stimuli, usually consisting of analogue signals, like voltages or currents. These signals are supplied by and read out by a purpose-built interface (as shown in the middle of Fig. 1). This interface is equipped with analogue-digital converters and connected to a digital computer (a laptop or PC) to enable computer-controlled evolution. The computer program that is executed on this digital device acts as a discovery engine (see Fig. 1) for finding configurations of the external stimuli that cause the material to perform the target computational task. Once suitable configurations have been determined for certain specified tasks, the given sample of nanomaterial on the micro-electrode array should be capable to function as a stand-alone reconfigurable device for these tasks.

Apart from the experimental work, another aim of the NASCENCE project was to model, simulate and explore this way of unconventional computation, and analyse and explain the observed phenomena.

This set-up led to a natural breakdown of the project work within NASCENCE into the following nine Work Packages and their key performance indicators (KPIs):

1: Materials processing and evaluation;
   KPI: useful materials and micro-electrode arrays;
2: Interface (hardware);
   KPI: interaction PC/material; controlled stimuli;

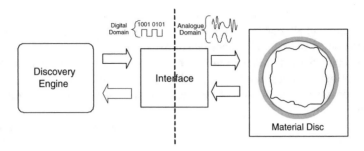

Fig. 1 Schematic conceptual overview of NASCENCE

3: Interface (software);
   KPI: API; genetic algorithms;
4: Computational tasks (experimental);
   KPI: suitable tasks/materials;
5: Computational tasks (simulations);
   KPI: models for simulation/prediction;
6: Evaluation and data mining;
   KPI: functional software for open ended experimentation;
7: Mathematical foundations;
   KPI: theoretical foundation for evolvable systems;
8: Dissemination, collaboration and exploitation;
9: Project management;

In the next sections, the aim is to give a concise overview of the accomplishments of the NASCENCE project. More details can be found in [5] and the papers that are listed there. But we start with some additional remarks about the motivation behind the suggested approach of EIM, and by presenting more details on the use of evolutionary algorithms, with a specific example that was used to evolve a nanoparticle substrate.

4 Why and How to Evolve Dead Matter?

The work within NASCENCE has been inspired on one hand by the great achievements of Nature itself, and on the other hand by the opportunities offered by nanotechnology as a possible alternative to digital computing.

Evolution has done a great job for many living organisms, and it has done so through a conceptually rather easy but smart, albeit very slow bottom up process of natural Darwinian evolution. There seems to be no design involved in this process: the computational instructions that underlie living creatures have not been designed but rather have (been) evolved. Two questions come to mind: Can we use something similar as evolution on 'dead' matter in order to get something useful? Can we do it quickly, via a rapidly converging evolutionary process?

The digital industry has done a great job for computation. It has developed and advanced digital circuitry extremely fast, to very dense packings of millions of transistors of very tiny size, currently in the range of 15 nm. It is not clear whether this will continue, and even the big players in this industry predict that we shall lose control over the production processes within the next decade. Another concern is that we are wasting a lot of energy on the production and use of digital equipment. Moreover, there are fundamental issues that have been raised with respect to the concepts behind digital computation. Due to the abstraction from electrical signal levels to logical zeros and ones, it seems plausible that digital computation is not making use of the full potential opportunities offered by physical systems. According to Conrad this leads us to pay "The Price of Programmability" [10]. Another question then emerges: Is there an alternative way to use (nanoscale) material that exploits the full potential properties of physical systems to solve computational problems rather than restricting them by abstraction to zeros and ones?

Motivated by the above observations, the NASCENCE project has been focusing on using nanoscale material in another, less controlled but hopefully more efficient way. The results that have been obtained so far by the members of the NASCENCE project look promising, as can be concluded from the later sections of this chapter. They show how computer-controlled evolution was used successfully to 'program' several functionalities in nanoscale materials by directly manipulating these physical systems. Although the results should be considered as a proof of principle, it is hoped that the approach taken in NASCENCE will inspire the creation of novel and useful devices in physical systems whose operational principles are not necessarily understood or are hitherto unknown.

The central idea of EIM is that the application of some physical 'configuration' signals to a material can cause it to alter how it affects 'input' signals and produces 'output' signals. The output signals are picked up and a fitness score is assigned, depending on how close the output signals are to the desired response. This fitness is assigned to the member of the population that supplied the configuration signals. For the purpose of using an evolutionary algorithm, the patterns of configuration signals that are applied as external stimuli have to be encoded, mimicking genomes in natural evolution. These artificial genomes enable us to mimic evolution on the codes, in order to find configurations that cause the matter to perform useful computational tasks. This approach is illustrated by the following example that is taken from [3].

4.1　An Illustrative Example: Nanoparticle Networks

Consider Fig. 2, in which the small spheres represent coated gold nanoparticles (NPs) that are trapped on an electrode array to form a fixed NP-network; more details can be found in [3].

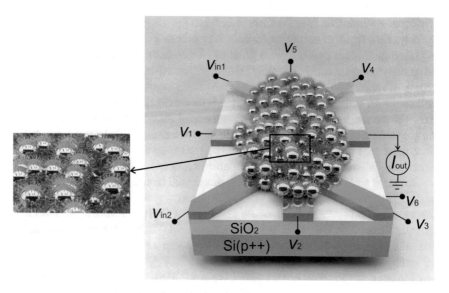

Fig. 2 Artist's impression of a disordered NP-network

Two electrodes are used for applying time-dependent voltage signals (as indicated by V_{in1}, V_{in2} in Fig. 2). All but one of the remaining electrodes are used for applying static control voltages (as indicated by $V_1, \ldots, V_5$ in Fig. 2), and one is used for measuring the resulting time-dependent current (as indicated by I_{out} in Fig. 2). In addition, a static voltage can be applied to the back gate (as indicated by V_6 in Fig. 2). The strongly non-linear switching behaviour of the NPs, that act as single-electron transistors (SETs) under the right circumstances, and their mutual interactions give rise to functionality greatly depending on the control voltages that manipulate the internal state of the NP-network.

At low temperatures, an NP with capacitance C has a charging energy $E = e^2/C$ (where e is the charge on the electron), which is larger than the thermal energy. In this case NPs exhibit a phenomenon that is known as Coulomb blockade [22] and act as SETs. One electron at a time can tunnel when sufficient energy is available (ON state), either by applying a voltage across the SET or by electro-statically shifting its potential. Otherwise, the transport is blocked because of the Coulomb blockade (OFF state).

Measurements in [3] indeed indicated that the low-temperature electron transport in the NP-networks that were used for the experiments is dominated by this Coulomb blockade effect, and that their detailed behaviour strongly depends on the used input and output electrodes, as well as on the static voltages applied to the remaining electrodes. These characteristics lie at the basis of the ability to use these NP-networks for the realisation of computational tasks. Because the current NP-networks have a rather simple structure with only a few configuration electrodes to play with, the aim in [3] was to first look for logic gates. The idea was thus to enable all possible 2-way logic gates with one and the same NP-network, by only varying the configuration voltages, without the necessity to apply a specific design of the NP-network.

So, in this set-up, fixing two of the electrodes as inputs and one as an output, the remaining electrodes together with the back gate have been used to configure the NP-network for the target functionalities, applying a genetic algorithm (GA). More details on the working of a GA, and how it was applied for our purposes, appear in the next section, where the same GA was used for the experiments as well as the simulations. Some additional information on the simulations appears in later subsections, but next we present a short summary of the outcomes of the experimental work in [3].

4.1.1 Logic Gates

In this subsection we give a brief description of the work in [3] where the NP-networks as sketched in Fig. 2 were used to evolve all Boolean logic gates.

In Fig. 3a we see an atomic force micrograph (AFM) image of one of the real NP-networks that were used for the experiments in [3], where the two input electrodes and the output electrode are denoted by V_{IN1}, V_{IN2} and I_{OUT}, respectively. Time-dependent signals in the order of a hundred mV were applied to the input electrodes as illustrated in Fig. 3b, and a time-dependent current in the order of one hundred pA was read from the output electrode. The other five electrodes and the back gate have been used to apply different sets of static configuration voltages. Using a GA, suitable sets of configuration voltages have been found to produce the output functions of Fig. 3c, d. Red symbols are experimental data, solid black curves are expected output signals (matched to the amplitudes of the experimental data). We observe two clear negators (inverters) for the input functions P and Q in Fig. 3c, and we observe a variety of Boolean logic gates in Fig. 3d, including the universal NAND and NOR gate. Supplementary work in [3] reveals that all these gates show a great stability and reproducibility. For the exclusive gates (XOR, XNOR) spike-like features are observed at the rising and falling edges of the (1,1) input, as might have been expected for a finite slope in the input signals. More details can be found in [3].

The remarkable thing here is not that we can produce logic gates using the electrical and physical properties of charge transport in neighbouring NPs. What is remarkable, is that we can do this with one and the same sample of a disordered NP-network in a circular region of about 200 nm in diameter, and by using only six configuration voltages. This shows the great potential for our approach. Note that a similar designed reconfigurable device based on today's transistor technology would require about the same space, and it would also require rewiring of the input signals to multiple inputs.

4.1.2 Behaviour of Gold NPs

The electrical properties and physical effect of Coulomb blockade behind the charge transport in the used gold NP-networks are pretty well understood. As described

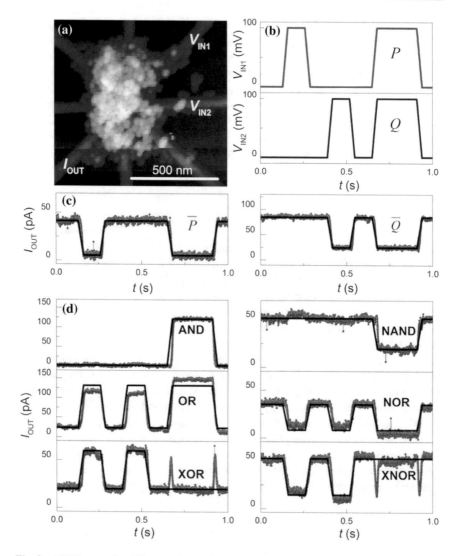

Fig. 3 AFM image of an NP-network (**a**), the input voltages in mV applied to V_{IN1} and V_{IN2} (**b**) and the different logic outputs in pA read from I_{OUT} (**c** and **d**) [3]

above, under suitable energy conditions and restrictions, the charge transport is governed by the Coulomb blockade effect [22, 53]. The particles act as SETs with a high ON/OFF ratio and strong non-linear behaviour. This makes them potentially good candidates for interesting nontrivial functionalities. As an illustration, in Fig. 4 we included some I-V characteristics of one of the NP-networks we used in [3].

The nonlinear behaviour is very clear from the figures. We also observed a special form of nonlinearity usually referred to as negative differential resistance (NDR), as

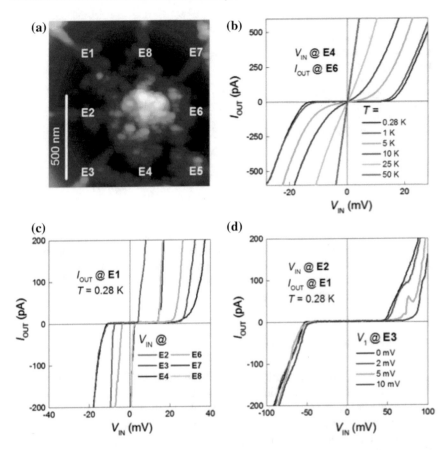

Fig. 4 AFM image of an NP-network (**a**), the input voltages in mV applied to electrode E_4 and the output current in pA read at electrode E_6 at different temperatures (**b**) the outputs in pA read from E_1 for different input electrodes at 0.28 K (**c**), and the effect of a static voltage applied at E_3 on the I–V curves of a fixed pair at 0.28 K (**d**) [3]

shown in Fig. 5: a gate that was evolved to be a negator (inverter) for 0 mV$< V_{IN} <$ 100 mV exhibits NDR within the considerably larger range -50 mV$< V_{IN} <$100 mV.

This behaviour is interesting and we think it plays a key role in the potential evolvability of more complex functions. For instance, if we compare an XOR with an OR, then it can be observed that the OR could in principle be based on simple linear behaviour, where a high input signal gives rise to a high output signal, no matter whether both input signals are high or just one of the input signals. In case of an XOR this is different: we should only have a high output signal if precisely one of the input signals is high and the other is low; two high input signals should yield a low output signal. This is clearly more likely to be a reachable target functionality if the evolvable system exhibits NDR behaviour.

Fig. 5 NDR behaviour of a
gold NP-network: the output
current I_{OUT} increases with
an increasing input voltage
V_{IN} in the interval
$-150\,\text{mV} < V_{IN} < -50\,\text{mV}$,
but the I-V curve bends
down in the region of
$-50\,\text{mV} < V_{IN} < 100\,\text{mV}$ [3]

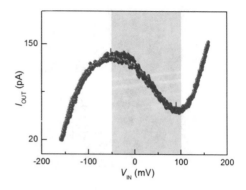

The above experimental results give supporting evidence for our expectation that much more is possible in terms of more complicated functionalities, enabled by such disordered NP-networks with more (smaller) particles and a more sophisticated electrode array and back gate structure. This is part of our current research at the MESA+ Institute for Nanotechnology in Twente.

4.1.3 Simulations of NP-Networks

Apart from the experimental work and results, the NASCENCE consortium has also worked on the theoretical underpinning of the experimental work and on simulations. The latter are based on physical or mathematical models of the material systems. In case of the NP-networks the physics is pretty well understood. In this subsection, a short description of the physical model that underlies the behaviour of jumping electrons in NP-networks are presented, as well as an alternative approach to simulating these NP-networks using neural networks.

As we explained earlier, the charge transport in the NP-networks we have been using in the experiments that have led to [3] is based on a physical phenomenon that is known as the Coulomb blockade effect [22, 53]. The individual gold NPs act as SETs. Electrons can jump between neighbouring particles when the energy conditions are favourable. One electron at a time can tunnel between two particles if sufficient energy is available (ON state), either by applying a voltage across the particle or by electro-statically shifting its potential; otherwise, the transport is blocked due to Coulomb blockade (OFF state). These disordered assemblies of NPs therefore provide an almost random network of interconnected robust, non-linear, periodic switches, as a result of the Coulomb oscillations of the individual NPs. We have observed experimentally that electron transport below 5 K is dominated by Coulomb blockade, and strongly depends on the used input and output electrodes, as well as on the static voltages applied to the remaining electrodes.

Due to the high costs and time consuming experiments involved in the experimental work, it was highly desirable to develop a simulation tool to explore the potential functionalities of such NP-networks without the burden of spending many hours in the lab and wasting expensive resources to look for such functionalities experimentally. In addition, the simulations can also inform us on the minimum requirements that are needed for obtaining the targeted functionality if we were able to produce these NP-networks according to a predetermined design. This could lead to new devices for the digital industry, possibly replacing purpose-built assemblies of transistors. Moreover, simulations can provide us with evidence concerning the scalability of our approach. Simulations can also give us new insights into the dynamics of the charge transport that might lead to a better understanding as to why and how the networks reveal the functionalities we observe. Furthermore, there are many questions on the use of these networks that are difficult to answer experimentally, because there are serious challenges in fabricating examples with smaller central gaps or with more control electrodes using the same area.

The simulation tool we developed in [12] is an extension of existing tools for simulating NP interactions, like SPICE [43] or SIMON [55]. Since the dynamics of our NP-networks is governed by stochastic processes (electrons on particles can tunnel through junctions with a certain probability), there are basically two simulation methods to our disposal: Monte-Carlo Methods and the Master Equation Method [53, 54]. Since the number of particles is large, this rules out the second approach, hence the Monte-Carlo Method is the only suitable candidate. This method simulates the tunneling times of electrons stochastically. To get meaningful results, one needs to run the algorithm in the order of a million times. Doing so, the stochastic process gives averaged values of the charges, currents, voltages, etc. More details on this physical-model based simulation tool can be found in [12].

We have validated our tool for designed systems with small numbers of particles that are experimentally known from the literature, and that have also been simulated before [54]. We have also used our tool to examine other structures of NP-networks. Interestingly, we have shown through simulations that all Boolean logic gates that we evolved experimentally in [3] can be evolved in a regular 4×4 grid consisting of only 16 NPs. We refer to [12] for more details. Currently, we are not aware of any production techniques for constructing these regular grids of NPs.

Although our simulation tool can in principle handle arbitrary systems of any size, scalability is a serious issue if we consider the computation time. Even a parallelised CUDA code we have developed for a GPU does not really solve the problem if we want to simulate networks consisting of hundreds of particles. Moreover, as the networks in [3] cannot be produced according to a predefined specific design, it is not possible to use an accurate physical model for such systems.

With these drawbacks in mind, we have taken an alternative approach. This novel approach is based on training artificial neural networks in order to model and investigate the NP-networks.

Neural networks have proven to be powerful function approximators and have been successfully applied in a wide variety of domains [6, 25, 49, 51]. Being essentially black-boxes themselves, neural networks do not facilitate a better understanding of the underlying quantum-mechanical processes. For that purpose the physical model we described before is more appropriate. But in contrast to physical models, neural networks provide differentiable models and thus offer interesting possibilities to explore the computational capabilities of the nano-material.

Before this exploration can take place, a neural network must first be trained, using data collected from the material. In our case, since we already have a physical model and an associated validated simulation tool for the NP-networks, to show that this approach is useful we can restrict ourselves in the first instance to training data obtained from the simulated material. This gives us the opportunity to predict functionalities in small NP-networks, also networks that have not been fabricated yet, like the 4×4 grid structure we mentioned above. This in turn can inform electrical engineers on the minimum requirements necessary for obtaining such functionalities without the burden of costly and time-consuming fabrication and experimentation.

One of the advantages of the neural network approach is that we do not need to have any detailed information on the structure or physical properties of the material. We only need as many input-output data combinations as we can get from the simulation tool or from measurements on a particular material sample, in order to train a neural network that models this specific sample. The more independent data we use, the more accurate the trained neural network is expected to model the sample.

Another advantage of the neural network approach is that one can optimise the input configuration through gradient descent instead of performing a black-box optimisation. In other words, as soon as we have trained the neural network with sufficiently many input-output combinations, searching for arbitrary functions is very fast and can happen independently of the material or the physical model.

To show that this approach is worthwhile, in [15] we used data obtained from the physical-model based simulations that we mentioned above to train a neural network. We show in [15] that the neural network can model the simulated nano-material quite accurately. The differentiable neural network model of the evolvable NP-network is then used to find logic gates, as a proof of principle.

This shows that the new approach has great potential for partly replacing costly and time-consuming experiments. We are currently also using the neural network approach on real data collected from samples of the NP-networks.

5   Back to the General Method

No matter what matter one uses, the approach of EIM requires a method to optimise or at least find good configurations for a specific target functionality of the evolvable system. But to be able to do this, first of all there is a need for a translation or mapping between the physical domain and the computer domain, and vice versa, as indicated in Fig. 6 that was taken from [5].

In the physical domain, there is a material (in our earlier example an NP-network) to which physical stimuli can be applied (in our example voltages) or measured (in our example currents). In the NP-network example we have just given, the stimuli signals were static voltages and time-dependent voltages and currents, but the type of signals could be different, depending on the type of material and its characteristic

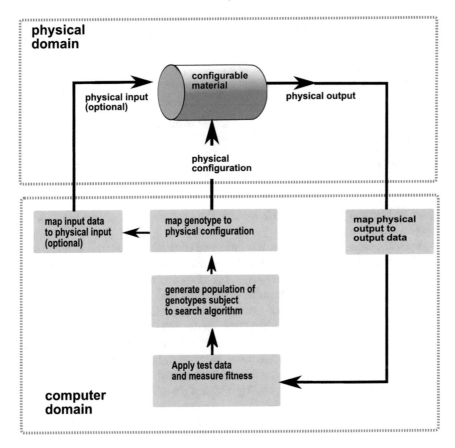

Fig. 6 Concept of EIM: physical and computer domain [5]

behaviour under influence of external stimuli. The signals are either input signals, output signals or configuration signals. With the help of a suitable interface made up of conventional electronics, in the computer domain a digital computer controls the application of physical inputs applied to the material (the time-dependent voltages in our example), the reading of physical signals from the material (the time-dependent currents in our example), and the application of other physical inputs to the material known as configurations (the static voltages in our example). Since we are using a digital computer to control the signals, we need to encode and decode between physical signals and representations of these signals by strings of numerical values (and at a lower level bits), with one entry for each of the applied (or measured) signal. Such strings can be interpreted as genotypes of the state of the physical material when supplied with the according levels of the applied signals. These genotypes of numerical data are stored and manipulated on the computer and, if necessary, again transformed back into configuration signals that physically affect the material, when applied. In order to find suitable signals that induce a target functionality between

---

Algorithm 1 (Genetic Algorithm)

1: Generate an initial generation of size p. Set the number of generations as $g = 0$
2: **repeat**
3:     Calculate the fitness of each member of the generation
4:     Select a number of parents according to quality of fitness
5:     Recombine some, if not all, parents to create offspring genomes
6:     Mutate some parents and offspring
7:     Form a new generation from mutated parents and offspring
8:     Optional: promote a number of unaltered parents from step 4 to the new generation
9:     Increment the number of generations $g \leftarrow g + 1$
10: **until** (g equals the number of generations required) **or** (the fitness is acceptable)

---

inputs and outputs of the evolvable system, the genotypes are subject to a GA that we are going to explain shortly.

Physical output signals, under the influence of certain input and configuration signals, are read from the material and converted to output data in the computer. Comparing this measured output to the ideal output of a target functionality under the given input signals, a fitness value is obtained from the measured output data. This fitness value should be a good measure for how close the evolvable system is to performing the target functionality. This fitness of the measured performance and the corresponding genotype of the current state of the evolvable system is then used in the GA, with the goal to optimise (or at least determine sufficiently good settings of) the configuration signals for the target functionality.

To illustrate the steps in the GA, Algorithm 1 gives the ingredients of a generic GA in pseudocode, adapted from [5].

In Line 1 of the pseudocode, the initial generation of size p corresponds to p (possibly randomly chosen, but usually based on earlier experiments to characterise the typical behaviour of the system) initial genomes that are translated into configuration and input signals, and tried on the evolvable system.

Based on the measured outputs, for all the p choices, a fitness value is obtained in Line 3, and a selection of a number of genomes that correspond to the best fitness values is taken in Line 4.

To obtain new choices for genomes, the next generation of genomes to try on the material system is based on combining and manipulating the entries (genes) of the selected genomes, using principles that mimic natural evolution, like recombination, crossover, mutation, as in Lines 5–7. The "survival-of-the-fittest" principle of Darwinian evolution is implemented by using a form of fitness-based selection that is more likely to choose solutions for the next generation that are fitter rather than poorer. Mutation is an operation that changes a genome by making random alterations to some genes, with a certain probability. This is usually a suitable tool to avoid running into a bad local optimum. Recombination is a process of generating one or more new genomes by recombining genes from two or more genomes. Sometimes, genomes from one generation are promoted directly to the next generation; this is referred to as elitism (see the optional step in Line 8 of the algorithm).

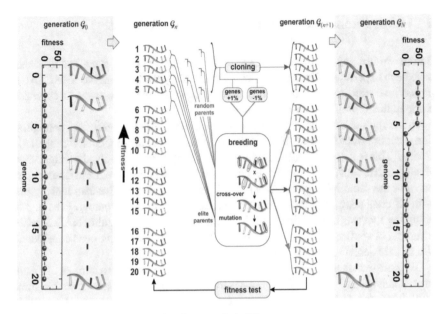

Fig. 7   The GA that was used for the NP-networks in [3]

It is clear that there are many choices involved in specifying a GA for a particular optimisation problem or for a particular evolvable system and target functionality, as in our research. These choices are usually based on trial and error, until a satisfactory convergence to acceptable solutions is observed.

Figure 7 illustrates the GA that was used for the NP-networks in [3]. Like every GA, this GA searches for optimum genomes by iterating over sufficiently many generations. In this specific GA, every generation G_n has a fixed population of 20 genomes (represented by the strands in Fig. 7) that are each composed of six genes (configuration voltages V_1 up to V_6, represented by coloured bars in the figure). The GA starts with a random initial generation G_0, whose fitness is generally low, as indicated in the leftmost column (for a GA search for an AND gate). With a composite cloning-breeding procedure (as illustrated in the middle of Fig. 7; see [3] for details) for evolving generation G_n into G_{n+1}, we eventually reach a final generation G_N whose top genomes have a (much) higher fitness than the top genomes of G_0 and the intermediate generations (as illustrated in the rightmost column of Fig. 7). For more details on the settings of the GA that were chosen for the evolution of Boolean logic in these NP-networks we refer the reader to the supplementary material of [3].

6 Other Examples from the NASCENCE Project

Before we are going to present some of the results for other functionalities that have been experimentally tried out using the EIM approach within the NASCENCE project, we need to point out that there are basically two main ways to approach computational problems with EIM.

In the approach that we took in our earlier example, the GA determines configurations that in principle allow the material to act as a stand-alone reconfigurable logic device. This is a device which, provided with the appropriate evolved configuration signals, carries out the desired target computational mapping. In our earlier example, one and the same NP-network was able to perform all 2-way Boolean logic gates, with different settings of the static configuration voltages, after removing the digital computer that was used to find suitable configuration settings. Within NASCENCE, a number of applications for which we took the same approach have been considered. We shortly describe some of the other applications.

We note here that the term configuration of a material can have a number of meanings. It can merely be the application of physical signals to the material so that some underlying physical properties change, e.g. conductance or resistance. As a result, the material is put into a state that allows the desired computation to take place. This was the case in our earlier NP-network example. However, alternatively it may be that when the physical signals are applied to the material, the material physically changes in some way. For instance, the underlying (electrical) network might be rearranged, or the nanoscale elements could self-organise to a desired state so that the target computational functionality is obtained. An example of the latter is provided by earlier work with liquid crystal of Harding and Miller [16]. In this case, applied configuration signals caused liquid crystal molecules to twist, thus there was a physical change in the material when configuration signals were applied. As a matter of fact, both of the above effects may happen at the same time.

There is another approach in which EIM could be applied with the aid of evolvable systems. In this alternative approach, a material is merely used for the mapping of genomes to fitness values. The material is seen as an assistant in an evolutionary search process. It provides a "black-box" mapping from genome to output data (from which fitness is assessed). The thinking behind this is that the material may provide a more evolvable genotype-to-phenotype mapping, since physical variables can be exploited that could not be exploited if a purely algorithmic mapping was used (as is standard in evolutionary computation). In this type of hybrid system, much of the data required for solving a particular problem would remain on a digital computer. The role of the material would be to improve the search process itself. Thus in this case the material does not necessarily require any input data. Examples of computational problems that can be tackled using this approach are: Travelling Salesman Problem (TSP), Function Optimisation and Bin-packing. The TSP is the well-known problem of determining the shortest tour through a number of cities. Function optimisation is the problem of determining a vector of numbers which minimises a complex function. Bin-packing is the problem of packing a number of items into as few bins as possible, assuming that each bin has a fixed weight capacity. To obtain solutions to such problems using EIM requires that a set of configuration signals are determined that cause the material to output a suitable vector of measured values. The solutions are not necessarily optimal solutions, but usually approximate solutions. Generally, this method is difficult to scale and has at the time only resulted in solutions for rather small instances of the problems.

6.1 Materials and Interfaces Used Within NASCENCE

As we noted before, computational materials may be configured by different kinds of stimuli, e.g., electrical signals, magnetic fields, temperature variations, light, etc. However, it was decided at an early stage to only manipulate electrical signals within the NASCENCE project. Two types of evolvable material systems were constructed for this purpose. Both are based on electrode arrays. A material is deposited in the vicinity of the electrodes. Some of the electrodes are chosen as inputs (if the computational problem demands inputs), some are chosen as outputs, and a number of electrodes are chosen as configuration electrodes. In one system that we showed before, the material consisted of functionalised 20 nm AU NPs interconnected by insulating molecules (1-octanethiols) that were trapped in a circular region (200 nm in diameter) between radial metal (Ti/Au) electrodes on top of a highly doped Si/SiO$_2$ substrate, which functions as a back gate. This device operates at temperatures below 1°K [3]. In the second system, the material deposited was a mixture of single-walled carbon nanotubes (SWCNT) randomly mixed in an insulating material. An example of this other system is shown in Fig. 8. The insulating material was either PMMA/PBMA (Polymethy/butyl methacralate) [42]. The material in the centre is a mixture of SWCNT and PMMA. The concentration of SWCNT is 0.05% by weight. SWCNTs are mixed with PMMA or PBMA and dissolved in anisole (methoxyben-zene). 20 μL of material is drop dispensed onto the electrode array. This is dried at 100°C for 30 min to leave a film over the electrodes. Carbon nanotubes are conducting or semi-conducting and the role of the PMMA/PBMA is to introduce insulating regions within the nanotube network, to create non-linear current versus voltage characteristics.

In order to be able to apply a GA to determine a set of signals that should be applied to the electrode arrays, one requires a hardware interface system between a

Fig. 8 Circular twelve electrode array with a mixture of SWCNT and PMMA in the centre [42]

computer and the material. The hardware system needs to allow a variety of signals to be applied to the electrodes. In the NASCENCE project the signals used were one of the following:

- Digital voltages
- Analogue voltages
- Square-wave signals

One also needs to be able to sample and record voltages and currents detected on electrodes, since from these measurements a fitness value is determined. Thus one needs equipment that allows the user to choose a sampling frequency and store the values (in a buffer). Since it is not known in advance to which electrodes input signals should be applied, generally one needs a way of allowing the GA to choose which electrodes receive inputs (if the computational problem requires inputs) and which electrodes are designated as outputs, and finally which electrodes are the configuration inputs. A variety of different hardware systems have been explored for doing this.

- Digital acquisition cards together with programmable switch arrays [8]
- Mbed microcontrollers with digital to analogue converters [33]
- Purpose built platforms [3, 31].

6.2 Computational Problems

The NASCENCE consortium investigated a diverse range of computational problems. The list of problems with references to the associated papers is given below. We refrain from giving details here, but refer the interested reader to the listed papers instead. Except for the results in [3] that were obtained using the NP-networks we showed before, all other results were obtained by using different versions of the mentioned SWCNT-composites.

1. Logic gates

    a. Two-input single output Boolean functions (e.g. (N)AND, (N)OR, XOR) [3, 23, 31]
    b. Three/Four input single output Boolean functions (e.g. even-3 and 4 parity) [37]
    c. Two-input two-output Boolean functions (e.g. half adder) [3, 23, 33]
    d. Three-input, two-output Boolean functions (e.g. full-adder)

2 Travelling Salesman

    This has no inputs and as many outputs as there are cities [8]

3. Classification

a. Standard machine learning benchmarks (Iris, Lens, banknote): number of inputs equals the number of attributes, number of outputs is equal to the number of classes [7, 38]

b. Frequency classification: this requires one input for carrying the source signal whose frequency is to be classified and two outputs which are used to decide the class of the frequency (high or low) [39, 42]

c. Tone discriminator: this has the same number of inputs and outputs as the frequency classifier [42]

4. Function Optimisation

This has no inputs and as many outputs as there are dimensions in the function to be optimised [41, 42]

5. Bin-Packing

This has no inputs and as many outputs as there are items to be placed into bins [40]

6. Robot control

This has as many inputs as robot sensors and as many outputs as robot actuators (e.g. motors) [36]

7. Graph colouring

This has been looked at with a single input (graph select) and as many outputs as there are nodes to be coloured. Each output selects the colour of the node [30]

The seven classes of problems cover many types of problems involving markedly different numbers of inputs, outputs and number of instances. Some problems like TSP, Function Optimisation and Bin-packing have no inputs. The material acts like a form of genetic programming and via evolved configurations generates an approximate solution from its outputs. This is standard practice in genetic programming. In this type of approach a material is used in the genotype-phenotype mapping. However, one must be careful that in problems that have no inputs, evolution is not merely evolving configuration signals to produce outputs that are desired. In other words, that it is not directly wiring configuration signals to outputs, thus effectively ignoring the material.

In the work on evolving logic gates various functions present much greater difficulty due to their inherent non-linearity. It is well-known that parity functions are difficult to evolve non-linear functions. Indeed, they have been used as benchmark problems in genetic programming for some time.

6.3 Electrical Behaviour of SWCNT-Composites

Whereas the physics behind the non-linear behaviour of the NP-networks that were used within NASCENCE is pretty much understood, the case is quite different for the material samples that are based on SWCNTs. Within NASCENCE several models at different levels of abstraction have been used for describing and analysing the electrical behaviour, and for predicting and simulating the computational capabilities of SWCNT-composites. As mentioned earlier, this SWCNT-based material may be configured by different kinds of stimuli, e.g., electrical signals, magnetic fields, temperature variations, light, etc., but within NASCENCE we have restricted these signals to three types:

- Static voltages;
- Square waves;
- A mixture of the two.

In order to be able to perform any kind of computation with the help of SWCNT-based composites, the input data and the configuration data must allow the exploitation and manipulation of underlying physical properties in the SWCNT-composite material. Moreover, such manipulated properties must be observable and give a measurable response, i.e., output response. As such, the choice of the input signal and configuration types play an important role and define which physical properties are available and utilised. In [30], a comparison between different signal types and their effect has been made for the evolution of solutions for small instances of the graph colouring problem.

In the case of static voltages, the parameters under evolutionary control are typically the electrode to which the signal is applied to, the starting time, the ending time, and the voltage amplitude. In the experiments in [30], the range of amplitude was limited from 0 to 3.3 V. In the case of square wave signals, the evolved parameters are the electrode to which the signal is applied to, the starting time, the ending time, the frequency, and the duty cycle. In the experiments in [30], the square wave amplitude was fixed at 0 and 3.3 V. The results there show that it was possible to evolve satisfactory solutions using all three types of signals (static voltages only, square waves only, and a mixture of static voltages and square waves). However, the choice of signal types influenced the evolutionary results. Square wave signals showed more promising potential than the other signal types, and the ability to produce richer dynamics, as one can also conclude from [45].

In [29], the model used to describe the SWCNT-networks suggests these networks behave as networks of resistors if only static DC voltages are applied. Moreover, it was shown there that TSP problems as introduced and solved in [8] could be solved using a SPICE model of the SWCNT-material as a 'cloud' of resistors [43]. In case of applied square wave signals, the material could be seen as an RC circuit, i.e., the SWCNT-material also holds capacitance. Inspection of evolved solutions in [46] reveals that the exploited physical properties are often unanticipated. In particular, it was observed that evolution was able to create and exploit signal delays, signal inversions and signal canceling. All the mentioned properties may provide a source of non-linearity and rich dynamics that may be potentially exploited for physical implementations of reservoir computing [21, 32] in SWCNT-networks.

We conclude this subsection with a short description of the four models that have been considered within NASCENCE to model and analyse possible behaviour of the SWCNT-composites.

6.3.1 Models of SWCNT-Composites

Modelling of any material system consisting of nanoscale elements may be performed at several abstraction levels, ranging from the low level local interactions between neighbouring elements to the high level black-box behavioural models. Other intermediate levels may also be important, particularly for describing emergence of properties at different intermediate scales. Four different models have been considered and analysed within NASCENCE.

1. *Models based on collective electrodynamics within the SWCNT-composites.*
   Such models are described in more detail in [26], and are based on Ashby's systems theory [1], as introduced in classical cybernetics. Electrical properties exploited for computation arise as an emergent property of the stimulated material. The transport of charge in the SWCNT-composite can be considered as a manifestation of a collective emergent property of electrons in the computing substrate. Within this framework, a future research direction is proposed in which one may consider the manipulation of quantum properties of electrons in the material so that the emerging electromagnetic properties can be used for computation. Another aspect of the proposed framework is to allow the manipulation of other parameters, e.g., temperature, in the description of the system state to create a bigger choice of variables. This approach is related to polymorphic electronics [50], where there may be different functionalities for different operating temperatures.

2. *Models based on DC or AC circuits.*
   Observing the behaviour of SWCNT materials under varying inputs, e.g., static voltages or square wave signals, allows macroscopic modelling of pin-to-pin characteristics with simple RC circuits. In [30], two SPICE models [43] are presented, one for describing the electrical behaviour when stimulated with static voltages applied to input pins, and one for capturing the behaviour when square waves are used as manipulation signals.

3. *Models based on dynamical hierarchies and cellular structure.*
   The approach in [27] aims at modelling conductivity dependence on the concentration of SWCNTs and varying electric potential in the material. The approach is based on two main paradigms: dynamical hierarchies [48] and cellular computation [9]. Each material sample is divided into a grid of cells, in which each cell represents a subarea of the sample with a specific content, i.e., polymer molecules, SWCNT bundles, or electrodes. Each cell behaves according to the physics of the elements it contains and the interactions with neighbouring cells.

Results show that higher concentrations of SWCNTs lead to more percolation paths and consequently more current flow. Different cell shapes may be considered for future research.

4. *Models based on cellular automata.*

The wide variety of problems solved with SWCNT-composites does not give any direct indication of the computational properties and computational power of the materials used. However, it is clear that the materials can be exploited at the computational level required to solve the given task. Cellular automata (CA) offer a broader knowledge of different complexity levels and computational classes, e.g., Wolfram classes [56]. As such, CA models of the material may enable the development of a framework that relates measurable physical properties to abstract CA behaviour. In [14], cellular automata transition tables of different complexities have been evolved in-materio. An interesting future direction is the possibility to evolve universal cellular automata [2, 11] with the SWCNT-composites. In addition, ongoing work attempts to relate the evolved in-materio cellular automata with CA parameters [28], and relate the behaviour of the material to the notion of "edge of chaos".

7   Conclusions

This chapter tried to give the reader an overview of the accomplishments of the EC-funded NASCENCE project, with the relevant historical notes, background and the general concepts and principles. The author centred all material and examples around the work that was done within the framework of NASCENCE at his home university in Twente, notably the CTIT Institute for ICT Research and the MESA+ Institute for Nanotechnology. Details of this work and of the work done at the other partner's institutions can be found by following the pointers to the references that have been provided. The goal of the NASCENCE project was to demonstrate that computer-controlled evolution could exploit the physical properties of material systems based on carbon nanotubes and nanoparticles for solving computational problems. Experimental results have shown that this is indeed a plausible, competitive and efficient method for executing computational functions. Proof of concept has been given on several instances of problems within various complexities, and different number of inputs and outputs. The results are very promising and lay the foundation for further work. Future work includes the investigation of novel materials and larger and more sophisticated systems. One of our future goals is to apply these material systems for real world applications, and to develop stand-alone devices based on these principles. The long term goal of this research is to build information processing devices by exploiting bottom-up architectures without a predefined design and without reproducing individual components.

Acknowledgements The research leading to these results has received funding from the European Community's Seventh Framework Programme (FP7/2007–2013) under grant agreement number 317662. It is with great pleasure that the author of this chapter thanks all the members of the NASCENCE project for the wonderful collaboration in this adventurous endeavour, and in particular Julian F. Miller for his inspiration and positive attitude. Happy birthday, Julian!

References

1. Ashby, W.R.: Design for a Brain, the Origin of Adaptive Behaviour. Chapman & Hall Ltd. (1960)
2. Berlekamp, E.R., Conway, J.H., Guy, R.K.: Winning ways for your mathematical plays, volume 4. AMC **10**, 12 (2003)
3. Bose, S.K., Lawrence, C.P., Liu, Z., Makarenko, K.S., van Damme, R.M.J., Broersma, H.J., van der Wiel, W.G.: Evolution of a designless nanoparticle network into reconfigurable boolean logic. Nat. Nanotechnol. **207**, 1048–1052 (2015). doi:10.1038/NNANO.2015.207
4. Broersma, H., Gomez, F., Miller, J.F., Petty, M., Tufte, G.: Nascence project: nanoscale engineering for novel computation using evolution. Int. J. Unconvent. Comput. **8**(4), 313–317 (2012)
5. Broersma, H.J., Miller, J.F., Nichele, S.: Computational matter: Evolving computational functions in nanoscale materials. In: A. Adamatzky (ed.) Advances in Unconventional Computing Volume 2: Prototypes, Models and Algorithms, pp. 397–428 (2016)
6. Ciresan, D.C., Meier, U., Masci, J., Schmidhuber, J.: A committee of neural networks for traffic sign classification. In: International Joint Conference on Neural Networks (IJCNN), pp. 1918–1921 (2011)
7. Clegg, K., Miller, J., Massey, M., Petty, M.: Practical issues for configuring carbon nanotube composite materials for computation. In: Evolvable Systems (ICES), 2014 IEEE International Conference on, pp. 61–68 (2014)
8. Clegg, K.D., Miller, J.F., Massey, M.K., Petty, M.C.: Travelling salesman problem solved 'in materio' by evolved carbon nanotube device. In: Parallel Problem Solving from Nature - PPSN XIII - 13th International Conference, Proceedings, *LNCS*, vol. 8672, pp. 692–701. Springer (2014)
9. Codd, E.F.: Cellular Automata. Academic Press (1968)
10. Conrad, M.: The price of programmability. In: R. Herken (ed.) The Universal Turing Machine A Half-Century Survey, pp. 285–307. Oxford University Press (1988)
11. Cook, M.: Universality in elementary cellular automata. Complex Systems **15**(1), 1–40 (2004)
12. van Damme, R., Broersma, H., Mikhal, J., Lawrence, C., van der Wiel, W.: A simulation tool for evolving functionalities in disordered nanoparticle networks. IEEE Congress on Evolutionary Computation (CEC 2016), 24–29 July 2016, Vancouver, Canada pp. 5238–5245 (2015)
13. Eiben, A.E., Smith, J.E.: Introduction to Evolutionary Computing. Springer (2003)
14. Farstad, S.: Evolving cellular automata in-materio. In: Master Thesis Semester Project, Norwegian University of Science and Technology, Supervisor: Stefano Nichele, Gunnar Tufte. NTNU (2015)
15. Greff, K., van Damme, R., Koutník, J., Broersma, H., Mikhal, J., Lawrence, C., van der Wiel, W., Schmidhuber, J.: Unconventional computing using evolution-in-nanomaterio: Neural networks meet nanoparticle networks. The Eighth International Conference on Future Computational Technologies and Applications, Future Computing (2016)
16. Harding, S., Miller, J.F.: Evolution in materio: A tone discriminator in liquid crystal. In: In Proceedings of the Congress on Evolutionary Computation 2004 (CEC'2004), vol. 2, pp. 1800–1807 (2004)
17. Harding, S., Miller, J.F.: Evolution in materio. In: R.A. Meyers (ed.) Encyclopedia of Complexity and Systems Science, pp. 3220–3233. Springer (2009)

18. Harding, S.L., Miller, J.F.: Evolution in materio: evolving logic gates in liquid crystal. Int. J. Unconvention. Comput. **3**(4), 243–257 (2007)
19. Harding, S.L., Miller, J.F., Rietman, E.A.: Evolution in materio: exploiting the physics of materials for computation. Int. J. Unconvention. Comput. **4**(2), 155–194 (2008)
20. Holland, J.H.: Adaptation in Natural and Artificial Systems: An Introductory Analysis with Applications to Biology. Control and Artificial Intelligence. MIT Press, Cambridge, MA, USA (1992)
21. Jaeger, H.: The echo state approach to analysing and training recurrent neural networks-with an erratum note. In: German National Research Center for Information Technology GMD Technical Report Bonn, Germany 148, 34 (2001)
22. Korotkov, A.: Coulomb Blockade and Digital Single-Electron Devices, pp. 157–189. Blackwell, Oxford (1997)
23. Kotsialos, A., Massey, M.K., Qaiser, F., Zeze, D.A., Pearson, C., Petty, M.C.: Logic gate and circuit training on randomly dispersed carbon nanotubes. Int. J. Unconvention. Comput. **10**, 473–497 (2014)
24. Koza, J.: Genetic Programming: On the Programming of Computers by Natural Selection. MIT Press, Cambridge, Massachusetts, USA (1992)
25. Krizhevsky, A., Sutskever, I., Hinton, G.E.: Imagenet classification with deep convolutional neural networks. In: Advances in Neural Information Processing Systems (NIPS 2012), p. 4 (2012)
26. Laketić, D., Tufte, G., Lykkebø, O.R., Nichele, S.: An explanation of computation–collective electrodynamics in blobs of carbon nanotubes. In: Proceedings of 9th EAI International Conference on Bio-inspired Information and Communications Technologies (BIONETICS), in press. ACM (2015)
27. Laketić, D., Tufte, G., Nichele, S., Lykkebø, O.R.: Bringing Colours to the Black Box–A Novel Approach to Explaining Materials for Evolution-in-Materio. In: Proceedings of 7th International Conference on Future Computational Technology and Applications. XPS Press (2015)
28. Langton, C.G.: Computation at the edge of chaos: phase transitions and emergent computation. Physica D Nonlin Phenomena **42**(1), 12–37 (1990)
29. Lykkebø, O., Nichele, S., Tufte, G.: An investigation of square waves for evolution in carbon nanotubes material. In: Proceedings of the 13th European Conference on Artificial Life (ECAL2015), pp. 503–510. MIT Press (2015)
30. Lykkebø, O., Tufte, G.: Comparison and evaluation of signal representations for a carbon nanotube computational device. In: Evolvable Systems (ICES), 2014 IEEE International Conference on, pp. 54–60 (2014)
31. Lykkebø, O.R., Harding, S., Tufte, G., Miller, J.F.: Mecobo: A hardware and software platform for in materio evolution. In: O.H. Ibarra, L. Kari, S. Kopecki (eds.) Unconventional Computation and Natural Computation, LNCS, pp. 267–279. Springer International Publishing (2014)
32. Maass, W., Natschläger, T., Markram, H.: Real-time computing without stable states: a new framework for neural computation based on perturbations. Neural Computat. **14**(11), 2531–2560 (2002)
33. Massey, M.K., Kotsialos, A., Qaiser, F., Zeze, D.A., Pearson, C., Volpati, D., Bowen, L., Petty, M.C.: Computing with carbon nanotubes: optimization of threshold logic gates using disordered nanotube/polymer composites. J. Appl. Phys. **117**(13), 134903 (2015)
34. Miller, J.F., Downing, K.: Evolution in materio: looking beyond the silicon box. In: Proceedings of NASA/DoD Evolvable Hardware Workshop pp. 167–176 (2002)
35. Miller, J.F., Harding, S.L., Tufte, G.: Evolution-in-materio: evolving computation in materials. Evolution. Intelligen. **7**, 49–67 (2014)
36. Mohid, M., Miller, J.: Evolving robot controllers using carbon nanotubes. In: Proceedings of the 13th European Conference on Artificial Life (ECAL2015), pp. 106–113. MIT Press (2015)
37. Mohid, M., Miller, J.: Solving even parity problems using carbon nanotubes. In: Computational Intelligence (UKCI), 15th UK Workshop on. IEEE Press (2015)
38. Mohid, M., Miller, J.: Evolving solution to computational problems using carbon nanotubes. Int. J. Unconvention. Comput. **11**, 245–281 (2016)

39. Mohid, M., Miller, J., Harding, S., Tufte, G., Lykkebø, O., Massey, M., Petty, M.: Evolution-in-materio: A frequency classifier using materials. In: Proceedings of the 2014 IEEE International Conference on Evolvable Systems (ICES): From Biology to Hardware., pp. 46–53. IEEE Press (2014)

40. Mohid, M., Miller, J., Harding, S., Tufte, G., Lykkebø, O., Massey, M., Petty, M.: Evolution-in-materio: Solving bin packing problems using materials. In: Proceedings of the 2014 IEEE International Conference on Evolvable Systems (ICES): From Biology to Hardware., pp. 38–45. IEEE Press (2014)

41. Mohid, M., Miller, J., Harding, S., Tufte, G., Lykkebø, O., Massey, M., Petty, M.: Evolution-in-materio: Solving function optimization problems using materials. In: Computational Intelligence (UKCI), 2014 14th UK Workshop on, pp. 1–8. IEEE Press (2014)

42. Mohid, M., Miller, J., Harding, S., Tufte, G., Massey, M., Petty, M.: Evolution-in-materio: solving computational problems using carbon nanotube-polymer composites. Soft Comput. **20**, 3007–3022 (2016)

43. Nagel, L., Pederson, D.: Simulation program with integrated circuit emphasis. Memorandum ERL-M382, University of California, Berkeley (1973)

44. Neumann, J.V.: First draft of a report on the edvac. Tech. Rep., University of Pennsylvania (1945)

45. Nichele, S., Laketić, D., Lykkebø, O.R., , Tufte, G.: Is there chaos in blobs of carbon nanotubes used to perform computation? In: Proceedings of 7th International Conference on Future Comp. Tech. and Applications. XPS Press (2015)

46. Nichele, S., Lykkebø, O.R., Tufte, G.: An investigation of underlying physical properties exploited by evolution in nanotubes materials. In: Proceedings of 2015 IEEE International Conference on Evolvable Systems, IEEE Symposium Series on Computational Intelligence, in press. IEEE (2015)

47. Poli, R., Langdon, W.B., McPhee, N.F.: A Field Guide to Genetic Programming. Lulu Enterprises, UK Ltd (2008)

48. Rasmussen, S., Baas, N.A., Mayer, B., Nilsson, M., Olesen, M.W.: Ansatz for dynamical hierarchies. Artific. Life **7**(4), 329–353 (2001)

49. Sak, H., Senior, A.W., Beaufays, F.: Long short-term memory based recurrent neural network architectures for large vocabulary speech recognition. CoRR (2014). arXiv:1402.1128

50. Sekanina, L.: Design methods for polymorphic digital circuits. In: Proceedings of the 8th IEEE Design and Diagnostics of Electronic Circuits and Systems Workshop DDECS, pp. 145–150 (2005)

51. Sutskever, I., Vinyals, O., Le, Q.V.: Sequence to sequence learning with neural networks. In: Z. Ghahramani, M. Welling, C. Cortes, N.D. Lawrence, K.Q. Weinberger (eds.) Advances in Neural Information Processing Systems 27: Annual Conference on Neural Information Processing Systems 2014, December 8-13 2014, Montreal, Quebec, Canada, pp. 3104–3112 (2014). http://papers.nips.cc/paper/5346-sequence-to-sequence-learning-with-neural-networks

52. Turing, A.M.: On computable numbers, with an application to the entscheidungs problem. Proc. London Mathemat. Soc. **42**(2), 230–265 (1936)

53. Wasshuber, C.: Computational Single-Electronics. Springer (2001)

54. Wasshuber, C.: Single-Electronics–How it works. How it's used. How it's simulated. In: Proceedings of the International Symposium on Quality Electronic Design, pp. 502–507 (2012)

55. Wasshuber, C., Kosina, H., Selberherr, S.: A simulator for single-electron tunnel devices and circuits. IEEE Trans. Comput. Aided Des. Integ. Circ. Syst. **16**, 937–944 (1997)

56. Wolfram, S.: Universality and complexity in cellular automata. Physica D Nonlin. Phenom. **10**(1), 1–35 (1984)

Using Reed-Muller Expansions in the Synthesis and Optimization of Boolean Quantum Circuits

Ahmed Younes

Abstract There have been efforts to find an automatic way to create efficient Boolean quantum circuits, because of their wide range of applications. This chapter shows how to build efficient Boolean quantum circuits. A direct synthesis method can be used to implement any Boolean function as a quantum circuit using its truth table, where the generated circuits are more efficient than ones generated using methods proposed by others. The chapter shows, using another method, that there is a direct correspondence between Boolean quantum operations and the classical Reed-Muller expansions. This relation makes it possible for the problem of synthesis and optimization of Boolean quantum circuits to be tackled within the domain of Reed-Muller logic under manufacturing constraints, for example, the interaction between qubits of the system.

1 Introduction

The need to do computation faster and more reliably was and still is an essential aim. People are trying to find ways to achieve this goal by designing devices/machines to accelerate the process of computation, starting from using their own hands, passing on to simple devices like, for example, the abacus.

The conventional computers we know today, based on Turing Machines [17, 65], came to life after John Barden, Walter Brattain, and Will Shockley developed the *transistor* in 1947. For inventing new families of computers with more powerful capabilities, it was necessary to increase the number of transistors used in approximately the same physical space. The only way to achieve this goal is by decreasing the size of the components with a process known as *miniaturisation* [50]. According to *Moore's law*: "The number of transistors per square inch on integrated circuits has

A. Younes (✉)
Faculty of Science, Department of Mathematics and Computer Science,
Alexandria University, Alexandria, Egypt
e-mail: ayounes@alexu.edu.eg

A. Younes
School of Computer Science, University of Birmingham, Birmingham B15 2TT, UK

© Springer International Publishing AG 2018
S. Stepney and A. Adamatzky (eds.), *Inspired by Nature*, Emergence,
Complexity and Computation 28, https://doi.org/10.1007/978-3-319-67997-6_5

doubled every year since the integrated circuit was invented". This law has proved to be true since 1965 when Gordon Moore proposed it until now. However, according to this law, the size of the components will hit the quantum level by 2020 where it is not possible any more to control the states of the components by conventional means, so we must start looking for alternatives.

There have been efforts trying to find suitable alternatives to conventional computers. Charles Bennett in the early 1970s showed [9] that computation could be done in a reversible way, where the inputs can be re-generated from the outputs [33], by describing a *Universal Reversible Turing Machine*. It was shown that doing computation reversibly would help to decrease the energy dissipation of electronic devices per operation (a minimum of $\log_2 kT$ joules per bit erased, where k is Boltzman's constant and T is the temperature) [33, 39]. This was considered as one of the ways to go further in the miniaturisation process of components. Edward Fredkin and Tommaso Toffoli [34] examined how reversible computation could be done using traditional Boolean logic gates. They showed that there is a universal reversible three-bit gate for reversible computation, known now as the *Toffoli gate*, that could be used instead of the traditional Boolean logic gates like *AND*, *OR*, etc.

Quantum computers are computational devices that exploit quantum mechanical principles to do computation more powerfully than by conventional computers [10, 13, 60]. The field of quantum computation started in the early 1980s with the proposal of Paul Benioff [7, 8] for computers working according to the principles of quantum mechanics, and Richard Feynman [31, 32], who noticed that certain quantum mechanical effects cannot be simulated efficiently on conventional computers, so, he suggested that computers working according to the quantum principles might perform more efficiently. In 1985, David Deutsch introduced the *Universal Quantum Turing machine* [23]. The next question was how to use this new device. A few quantum algorithms were introduced by David Deutsch and Richard Jozsa [25] and by Daniel Simon [60]. In 1994, the field started to arise as one of the hot areas of research when Peter Shor [59] surprised the world by describing a polynomial-time quantum algorithm for factorising integers using the *phase estimation technique*. In 1996, Lov Grover [36] described another quantum algorithm for searching an unstructured list with quadratic speed-up over conventional algorithms by using the *amplitude amplification technique*. Most of the algorithms introduced since then have been based mainly on these two techniques of phase estimation and amplitude amplification.

Quantum computers have arisen as one of the strong possible alternatives to conventional computers. Most of the work done so far in the field of reversible computation is useful and suitable for the field of quantum computation [54], since quantum physics is itself reversible [28]. In addition, extra computational power may be gained from systems operating with the principles of quantum mechanics. A race started among mathematicians, computer scientists, and physicists trying to find new quantum algorithms, quantum complexity theory [12], and quantum theory of information [11, 15, 52, 68], beside the big challenge to build real quantum computers [18, 20, 35].

Fig. 1 **a** Truth table of a reversible function; **b** its circuit implementation

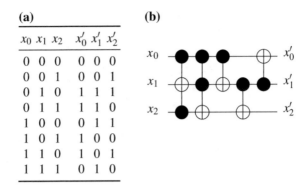

(a)

x_0 x_1 x_2	x_0' x_1' x_2'
0 0 0	0 0 0
0 0 1	0 0 1
0 1 0	1 1 1
0 1 1	1 1 0
1 0 0	0 1 1
1 0 1	1 0 0
1 1 0	1 0 1
1 1 1	0 1 0

Researchers are trying to build quantum circuits using reversible/quantum gates [5, 66]. Finding a universal quantum set of gates is essential to be able to build general-purpose quantum circuits. There are many alternatives for the universal set of gates, more than that in conventional and reversible computation. For instance, two-qubit universal quantum gates are possible [4, 29]. In fact, almost any two-qubit or n-qubit ($n \geq 2$) gates are also universal [24, 42].

Much work has been done trying to find universal sets for quantum computation. However, it is not clearly known how we can use any of these universal sets to automatically build a quantum circuit for a given problem. A few quantum circuits are known for some famous problems [19, 66]. However, it still not clear how to automatically build efficient quantum circuit for any given operation. There have been efforts to evolve quantum circuits using evolutionary methods [6, 56, 61, 62] but it is not clear if such ways can be used in general to create efficient quantum circuits for any given operation.

Reversible logic [9, 34] is one of the hot areas of research. It has many applications in quantum computation [37, 51], low-power CMOS [21, 67] and many more. Synthesis of reversible circuits cannot be done using conventional approaches [63]. A function is reversible if it maps an input vector to a unique output vector, and vice versa [58], i.e. we can re-generate the input vector from the output vector (reversibility). It is allowed for the inputs to be changed as long as the function maintains reversibility as shown in Fig. 1.

Much work has been done trying to find an efficient reversible circuit for an arbitrary reversible function. Reversible truth table can be seen as a permutation matrix of size $2^n \times 2^n$. In one research direction, it has been shown that the process of synthesizing linear reversible circuits can be reduced to a row reduction problem of $n \times n$ non-singular matrix [53]. Standard row reduction methods such as Gaussian elimination and LU-decomposition have been proposed [14]. In another research direction, search algorithms and template matching tools using reversible gates libraries have been used [30, 44–47]. These work efficiently for small circuits. Benchmarks for reversible circuits have been established [22, 57].

Table 1 (a) Reversible truth table representing a Boolean function; (b) dropping half the truth table depends on whether the output qubit is initialized to 0 or 1

(a)						(b)					
x_0	x_1	x_2	x'_0	x'_1	x'_2	x_0	x_1	x_2	x'_0	x'_1	x'_2
0	0	0	0	0	1	0	0	0	0	0	1
0	0	1	0	0	0	0	1	0	0	1	0
0	1	0	0	1	0	1	0	0	1	0	0
0	1	1	0	1	1	1	1	0	1	1	1
1	0	0	1	0	0						
1	0	1	1	0	1						
1	1	0	1	1	1						
1	1	1	1	1	0						

A Boolean function is a function that takes n Boolean inputs and generates a single Boolean output. Implementing Boolean functions as reversible functions using some of the above methods is not possible for certain classes of Boolean functions [58]. For example, representing the truth table of a Boolean function as a non-singular $n \times n$ matrix may not be possible in some cases. In case of Boolean functions, we may need to preserve the inputs unchanged after applying the circuit. Implementing Boolean function as a reversible function using search algorithms could be unnecessarily exhaustive, since we can immediately drop half the reversible truth table and keep track only for the changes to the single output, since no inputs will be changed, Table 1.

Recently, there have been some efforts to find methods to create efficient Boolean quantum circuits. One method uses a ROM-based model such that the inputs might not be changed even during an intermediate stage. Using this model requires an exponential number of ROM calls [64]. It has been suggested [49] that fixed polarity Reed-Muller expansions (FPRM) [3] can be used with binary decision diagrams (BDD) [2, 16] in an iterative algorithm to generate reversible circuits for simple incompletely specified Boolean functions with fewer than 10 variables. Another method [40] uses a modified version of *Karnaugh maps* [26] and depends on a clever choice of certain minterms to be used in the minimization process. However, this algorithm may have poor scalability because of the usage of Karnaugh maps. Another method is given in [38], where a useful set of transformations for Boolean quantum circuits is shown. In this method, extra auxiliary qubits are used in the construction that increase the hardware cost.

The chapter is organized as follows: Sect. 2 introduces the definition of the Boolean quantum logic and defines the structure of the Boolean quantum circuits to be constructed. Section 3 presents a method that can be used to convert the truth table of any Boolean function to its quantum circuit by following certain steps, then compares the efficiency of the circuits created by this method with the circuits created by others. This section ends with a general analysis on the efficiency of the created circuits. Section 4 presents a method that shows that there is a direct relation between Boolean quantum logic and certain form of classical logic known as *Reed-Muller expansions* (RM). This section reviews the principles of the classical Reed-Muller logic, then defines the Boolean quantum logic to be used by the method. This section shows the steps we may follow to construct quantum circuits directly from the corresponding classical Reed-Muller expansions. Section 5 shows the benefits we may gain from handling the problem of synthesis and optimization of Boolean quantum circuits under practical constraints within the domain of Reed-Muller. The chapter ends with a general conclusion in Sect. 6.

2 Quantum Logic

Quantum logic gates [5], or quantum transformations, are unitary operations applied on a quantum register during the process of the computation to change the state of the system from one state to another. A quantum circuit is constructed from simpler quantum gates. A quantum gate can be represented as a unitary matrix [51]. The reason that the quantum gates must be unitary is that quantum systems follow the fundamental laws of quantum physics and must be reversible [5]. To satisfy this condition, using any matrix U as a quantum gate, it must be unitary, i.e. the inverse of that matrix must be equal to its complex conjugate transpose: $U^{-1} = U^\dagger$ and $UU^\dagger = I$, where U^{-1} denotes the inverse of U, $U^\dagger$ denotes the complex conjugate transpose of U and I is the identity matrix. This agrees with the definition of reversibility [33].

A quantum register of n qubits can be represented as a vector in the 2^n—dimensional complex vector space. Any gate applied on that register can be understood by its action on the basis vectors and can be represented as a unitary matrix of size $2^n \times 2^n$.

2.1 The General n-Qubit Controlled Gate

In general, the controlled operations are the basic building blocks that are usually used to synthesize quantum Boolean circuits. The controlled operations can have multi-control/target qubits as shown in Fig. 2. It works as follows: an operation U is applied on the target qubit(s) $|x_{n+j}\rangle$ $(j = 0, \ldots, l-1)$ if and only if all the control qubit(s) $|x_k\rangle$ $(k = 0, \ldots, n-1)$ are set to $|1\rangle$. Notice that, U can be represented as a unitary matrix of size $2^l \times 2^l$, and the whole controlled operation can be represented as a unitary matrix of size $2^{n+l} \times 2^{n+l}$. The state of the qubits after applying this gate are transformed according to the following rule:

$$|x_k\rangle \to |x_k\rangle, \quad k = 0, \ldots, n-1,$$
$$\bigotimes_{j=0}^{l-1} |x_{n+j}\rangle \to U^{x_0 x_1 \cdots x_{n-1}} (\bigotimes_{j=0}^{l-1} |x_{n+j}\rangle), \tag{1}$$

Fig. 2 The general n-qubit controlled-U gate

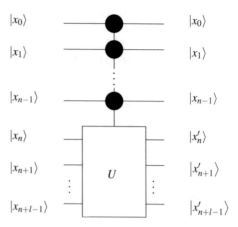

where $x_0 x_1 ... x_{n-1}$ in the exponent of U denotes the *AND* operation of the qubit values $x_0, x_1, ..., x_{n-1}$.

If U is replaced with the X gate (NOT gate), the resulting gate is called the *CNOT* gate. It works as follows: It inverts the target qubit if and only if all the control qubits are set to $|1\rangle$. Thus the system is transformed according to the following rule:

$$
\begin{aligned}
|x_k\rangle &\rightarrow |x_k\rangle \,; \;\; k = 0, ..., n-1, \\
|x_n\rangle &\rightarrow |x'_n\rangle = |x_n \oplus x_0 x_1 ... x_{n-1}\rangle .
\end{aligned}
\tag{2}
$$

To avoid ambiguity, C_{not} denotes a controlled-*NOT* operation with a single control qubit, CC_{not} denotes a controlled-*NOT* operation with two control qubits and *CNOT*, in general, denotes a controlled-*NOT* operation with $n \geq 1$ control qubits.

2.2 Generalized CNOT Gate

The operations of the *CNOT* gate [38] shown above can be generalized as follows [43]: The *CNOT* $(C|t)$ is a gate where the target qubit $|t\rangle$ is controlled by a set of qubits C such that $|t\rangle \notin C$. The state of $|t\rangle$ is flipped if and only if the conditions specified over the control qubits evaluate to true. The condition specified over certain control qubit may evaluate to true depends on whether the state of the qubit is $|0\rangle$ (cond-0; $\delta = 1$) or $|1\rangle$ (cond-1; $\delta = 0$), where δ is a Boolean parameter that is used in the Boolean algebraic expressions to indicate the condition being set on the qubit. For example, if the condition over certain qubit $|x_0\rangle$ evaluates to true if $|x_0\rangle = |1\rangle$, then we set $\delta_0 = 0$ so that $x_0 \oplus \delta_0 = x_0$ and if the condition over that qubit evaluates to true if $|x_0\rangle = |0\rangle$, then we set $\delta_0 = 1$ so that $x_0 \oplus \delta_0 = \overline{x_0}$. That is, the new state

Fig. 3 $CNOT\left(\{x_0,\overline{x_1},x_2\}\,|x_3\rangle\right)$ gate, where $\circ$ and $\bullet$ means that the condition on the qubit evaluates to true if and only if the state of that qubit is $|0\rangle$ (cond-0) and $|1\rangle$ (cond-1) respectively

of $|t\rangle$ after applying the *CNOT* gate is found by *XOR*-ing the old state of $|t\rangle$ with the *AND*-ing of the states of the control qubits in C (under the condition being specified on each control qubit).

For example, consider the *CNOT* gate shown in Fig. 3. It can be represented as $CNOT\left(\{x_0,\overline{x_1},x_2\}\,|x_3\rangle\right)$. This means that the state of the target qubit $|x_3\rangle$ is flipped if and only if $|x_0\rangle = |x_2\rangle = |1\rangle$ and $|x_1\rangle = |0\rangle$. In general, the target qubit in a four-qubit *CNOT* gate is changed according to the operation: $x_3 \rightarrow x_3 \oplus \left(x_0 \oplus \delta_0\right)\left(x_1 \oplus \delta_1\right)\left(x_2 \oplus \delta_2\right)$, so to represent the gate shown in Fig. 3, we set $\delta_0 = \delta_2 = 0$ and $\delta_1 = 1$, so the operation for this gate over $|x_3\rangle$ is: $x_3 \rightarrow x_3 \oplus x_0\overline{x_1}x_2$.

2.3 Boolean Quantum Circuits (BQC)

A Boolean function, f, is a function that takes n Boolean variables as inputs and gives one Boolean variable as output,

$$f(x_0, x_1, \ldots, x_{n-1}) \rightarrow \{0, 1\}, x_k \in \{0, 1\}. \tag{3}$$

Any Boolean function can be represented by a truth table. In order to be reversible, the circuit must have equal number of inputs and outputs. For example, consider the Boolean function $f(x_0, x_1, x_2) = \overline{x_0} + x_1 x_2$. Classically its truth table is represented as shown in Table 2a and for quantum computation purposes, the representation is as shown in Table 2b, where the initial value of the auxiliary qubit (f_{ini}) is equal to 0, which carries the value of the Boolean function at the end of the computation (f_{fin}).

A quantum circuit QC of size m (m is the number of the *CNOT* gates) representing a Boolean function f with n inputs $|x_k\rangle$, $k = 0, \ldots, n-1$, has $n+1$ qubits where the extra qubit $|x_n\rangle$ represents the target qubit to carry the output of the Boolean function at the end of the computation. QC can be represented as a sequence of *CNOT* gates as follows [38]:

$$QC = CNOT\left(C_1|t\right) \ldots CNOT\left(C_j|t\right) \ldots CNOT\left(C_m|t\right), \tag{4}$$

where $|t\rangle = |x_n\rangle$ and $C_j \subseteq \left\{|x_0\rangle, \ldots, |x_{n-1}\rangle\right\}$, $j = 1, \ldots, m$.

Table 2 Truth tables for $f(x_0,x_1,x_2) = \overline{x_0} + x_1 x_2$: (a) Classical representation; (b) modified version, representing the initial and the final states of the system

(a)				(b)							
x_0	x_1	x_2	f	x_0	x_1	x_2	f_{ini}	x_0	x_1	x_2	f_{fin}
0	0	0	1	0	0	0	0	0	0	0	1
0	0	1	1	0	0	1	0	0	0	1	1
0	1	0	1	0	1	0	0	0	1	0	1
0	1	1	1	0	1	1	0	0	1	1	1
1	0	0	0	1	0	0	0	1	0	0	0
1	0	1	0	1	0	1	0	1	0	1	0
1	1	0	0	1	1	0	0	1	1	0	0
1	1	1	1	1	1	1	0	1	1	1	1

3 Direct Synthesis of a BQC Using a Truth Table

This section shows a direct method to convert the truth table of any Boolean function to the equivalent quantum circuit by following certain steps. This method [70] can be used in the implementation of any given Boolean function as a quantum circuit to generate more efficient circuits than that generated using methods proposed by others, as is shown later.

3.1 Converting a Truth Table to a BQC

Consider a Boolean function f with n inputs. Construct the modified truth table as shown in Table 2b then apply the following steps:

1. **Initialization:** Using the modified truth table, choose $CNOT\left(C_j|t\right)$ according to the following steps:

   a. Select the input configurations from the truth table where f_{fin} is equal to 1.
   b. Add a *single CNOT* gate for *every* selected configuration taking f_{ini} as the target qubit.
   c. Set the condition over the control qubits in the corresponding *CNOT* gate such that the qubit with value 0 in the truth table has cond-0 and the qubit with value 1 has cond-1.
   d. For input configurations where f_{fin} is equal to 0, we do not add any gates, i.e. applying identity gate.

   For example, according to the truth table shown in Table 2b, we select only the configurations with $f_{fin} = 1$ as shown in Table 3 and construct the corresponding quantum circuit as shown in Fig. 4.

Table 3 Input configurations for $f(x_0, x_1, x_2) = \overline{x_0} + x_1 x_2 = 1$

	x_0	x_1	x_2	f_{fin}
G_0	0	0	0	1
G_1	0	0	1	1
G_2	0	1	0	1
G_3	0	1	1	1
G_4	1	1	1	1

Fig. 4 Initial quantum circuit for $f(x_0, x_1, x_2) = \overline{x_0} + x_1 x_2$

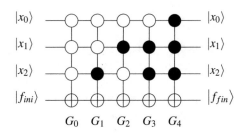

2. **Decomposition:** In the following transformations, assuming that we have classical inputs, we trace only the operations being applied on the target qubit, since no control qubits are changed during the run of the circuit. These circuit transformations are an extension and generalization of some of the equivalences between reversible circuits shown in [58]. We apply these transformations on every *CNOT* gate in the circuit we had from step-1. Applying this step eliminates cond-0 over the control qubits from the circuit. Let $|x_k\rangle$ be the control qubits, $|x_n\rangle$ be the target qubit and $\delta_k \in \{0, 1\}$ for $k = 0, 1, \ldots, n-1$. The general operation of each *CNOT* gate to be applied on the target qubit can be written as follows:

$$x_n \rightarrow x_n \oplus \left(x_0 \oplus \delta_0\right)\left(x_1 \oplus \delta_1\right) \ldots \left(x_{n-2} \oplus \delta_{n-2}\right)\left(x_{n-1} \oplus \delta_{n-1}\right). \quad (5)$$

Multiplying all terms we get the following transformation:

$$\begin{aligned} x_n \oplus \left(x_0 \oplus \delta_0\right)&\left(x_1 \oplus \delta_1\right) \ldots \left(x_{n-2} \oplus \delta_{n-2}\right)\left(x_{n-1} \oplus \delta_{n-1}\right) \\ &= x_n \oplus x_0 x_1 \ldots x_{n-1} \oplus x_0 x_1 \ldots x_{n-2} \delta_{n-1} \oplus \ldots \oplus \delta_0 x_1 \ldots x_{n-1} \\ &\oplus \ldots \oplus \delta_0 \delta_1 \ldots \delta_{n-1}. \end{aligned} \quad (6)$$

Examples

Example 1: Consider L.H.S circuit in Fig. 5 where one control qubit with cond-0. Decomposing this gate gives the R.H.S circuit in Fig. 5 by setting $\delta_0 = 1$ and $\delta_k = 0$; $k = 1, \ldots, n-1$ in L.H.S and R.H.S of Eq. 6.

Fig. 5 $n+1$–qubits *CNOT* gate with $\delta_0 = 1$ and its equivalent circuit decomposition

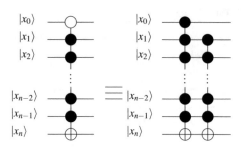

Example 2: Consider L.H.S circuit in Fig. 6 where all the control qubits with cond-0. Decomposing this gate gives the R.H.S circuit in Fig. 6 by setting $\delta_k = 1$; $k = 0, \ldots, n-1$ in L.H.S and R.H.S of Eq. 6.

Example 3: Applying this decomposition on every *CNOT* gate in the initial circuit shown in Fig. 4 for $f(x_0, x_1, x_2) = \overline{x_0} + x_1 x_2$, we get the circuit shown in Fig. 7.

3. **Optimization:** The circuit we get from step-2 contains the *CNOT* gates that have the same target qubit and all the control qubits with cond-1, so they commute with each other, and are self-inverses. Therefore pairs of identical gates can be removed from the circuit as follows:

$$CNOT(C_1|t) \ldots CNOT(C_{j'}|t) \ldots CNOT(C_m|t) \ldots CNOT(C_{j''}|t)$$
$$= CNOT(C_1|t) \ldots CNOT(C_m|t), \tag{7}$$

if and only if $C_{j'} = C_{j''}$. Applying this rule recursively on the circuit we had from step-2, we get the final quantum circuit that represents the Boolean function. For example, the final quantum circuit for $f(x_0, x_1, x_2) = \overline{x_0} + x_1 x_2$ is shown in Fig. 8.

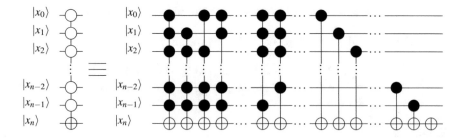

Fig. 6 $n+1$–qubits *CNOT* gate with all $\delta_k = 1$ and its equivalent circuit decomposition

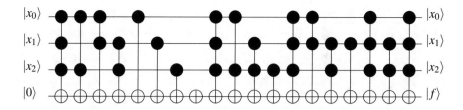

Fig. 7 Quantum circuit for $f(x_0, x_1, x_2) = \overline{x_0} + x_1 x_2$ after applying the decomposition shown in Eq. 6

Fig. 8 Final circuit for $f(x_0, x_1, x_2) = \overline{x_0} + x_1 x_2$

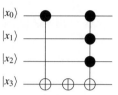

3.2 Comparison with Previous Work

The method shown in [40] uses a modified version of *Karnaugh maps* and depends on a clever choice of certain minterms to be used in the minimization process. The generated circuits may not be unique (for example, two alternatives are shown in Fig. 9). The Direct Synthesis method generates a unique form of a circuit (up to permutation) similar to that shown in Fig. 9b, which contains the smaller number of the *CNOT* gates. The generation of circuits using the method [40] is very hard for large number of inputs n because of the usage of Karnaugh maps, whereas the Direct Synthesis method can be used for arbitrary n inputs.

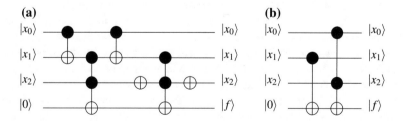

Fig. 9 Implementation of $f = \overline{x_0} x_1 + x_1 \overline{x_2} + x_0 \overline{x_1} x_2$ according to method [40] (**a**) & (**b**). Using the proposed method, we get circuit (**b**) only

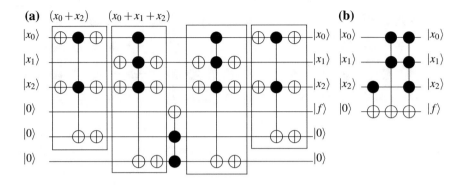

Fig. 10 Implementation of $f = (x_0 + x_2)(\overline{x_0} + x_1 + x_2)$ according to: **a** method [38], **b** the proposed method

In another work [38], a method is proposed that requires extra auxiliary qubits in the generated quantum circuits. Figure 10 compares between a circuit generated by this method (Fig. 10a) and the equivalent, but much simpler, circuit obtained by the method proposed in this chapter (Fig. 10b) where we used only one auxiliary qubit to carry the result of the Boolean function.

3.3 Analysis and Results

The circuits created by the Direct Synthesis method are based mainly on the *CNOT* gates. People sometimes refer to the *CNOT*-based circuits as reversible circuits. This construction is suitable for reversible computation in the sense that reversible operations could be done by loading the qubits by classical values. It is also suitable for quantum computation from the sense that we are not employing the reversible version of the *FAN-OUT* operation [63] in designing the Boolean circuits, which cannot be used in quantum circuits because of the *No-Cloning theory* [69].

Figure 11 shows a complete set for a three-qubit quantum circuits (two inputs) generated by this method. Using this method, after step-1, the maximum number of the *CNOT* gates we can add is up to 2^n, where all the control qubits are involved in each *CNOT* gate, with some of the control qubits with cond-0 and others with cond-1. Decomposing each *CNOT* gate in the circuit as shown in step-2 to eliminate all $\delta_k = 1$, we get a new quantum circuit where the number of the *CNOT* gates is up to 3^n. Applying the optimization step, the number of the *CNOT* gates can be up to 2^n, similar to step-1 with an important difference: Not all the control qubits have to be involved in each *CNOT* gate as in step-1, where all the control qubits have cond-1, i.e. we mainly minimize the number of control qubits in the circuit.

The worst case with 2^n *CNOT* gates exists when only a single input configuration with all the inputs with the value 0 makes the function evaluates to 1 as shown

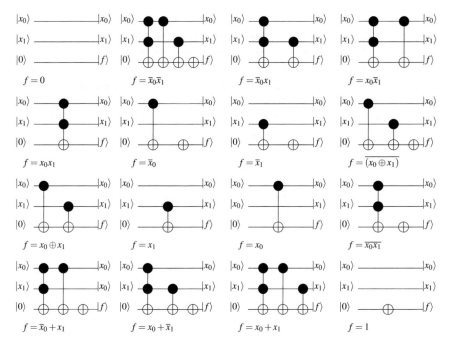

Fig. 11 A complete set of three-qubit quantum circuits

Table 4 Truth table for the worst case where the circuit generated by the method contains 2^n CNOT gates

x_0	x_1	x_2	...	x_{n-1}	f
0	0	0	...	0	1
0	0	0	...	1	0
$\vdots$	$\vdots$	$\vdots$	$\vdots$	$\vdots$	$\vdots$
0	1	1	...	1	0
1	0	0	...	0	0
1	0	0	...	1	0
$\vdots$	$\vdots$	$\vdots$	$\vdots$	$\vdots$	$\vdots$
1	1	1	...	1	0

in Table 4, where the input configuration generates an initial circuit similar to that shown in Fig. 6. In this case we can ignore the last two steps of the method and replace them with another decomposition as shown in Fig. 12, where the circuit contains $O(n)$ gates [27] instead of 2^n gates. However, this decomposition cannot be used in general to construct the BQC for any Boolean function, since the worst case we can get with this decomposition is $(n + 1)2^n$ gates subject to weak optimization based on the permutation the gates are arranged in, where adjacent *NOT* gates can be removed together, while identical *CNOT* gates cannot be removed together unless there is no *NOT* gates lying in between, which does not happen in the worst case as shown in Fig. 13a. The worst case using this decomposition where there are $O(n2^n)$ gates happens when all the input configurations can make the function evaluate to 1 (negation). Using the Direct Synthesis method in this case, we get a *single NOT* gate applied on the target qubit as shown in Fig. 13b. This problem gives the possibility of further circuit optimization that can be handled, in general, using Reed-Muller expansion, as we see in the next section.

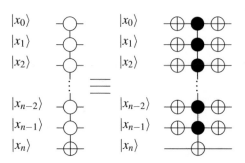

Fig. 12 Decomposition for the single input configuration case with up to $2n + 1$ gates, where all the inputs with cond-0. If any of the inputs $|x_k\rangle$ with cond-1 on the L.H.S. *CNOT* gate, when we can remove the $CNOT(x_k)$ gates for this qubit after and before the R.H.S. *CNOT* gate. This case has a special importance in designing the oracle for the original Grover's algorithm that searches for a single match (single input configuration that makes the Boolean function evaluates to TRUE, this was studied separately in [27])

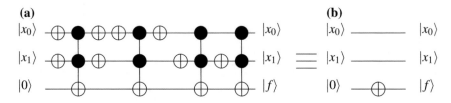

Fig. 13 **a** The worst case for the decomposition shown in Fig. 12 when applied to construct arbitrary Boolean function, **b** the equivalent circuit using the Direct Synthesis method for $f(x_0, x_1) = 1$

4 Boolean Quantum Circuits as Reed-Muller Expansions

In the previous section, we examined a method that can be used to convert the truth table of any given Boolean function to a quantum circuit. The Direct Synthesis method did not consider some of the practical constraints related to the synthesis of quantum circuits. This section presents a method that shows a direct relation between Boolean quantum logic and certain form of classical logic known as *Reed-Muller expansions* (RM). This relation makes it possible to propose that the synthesis and optimization of Boolean quantum circuits under practical constraints could be handled using RM logic to minimize the cost of the known constraint handling techniques [71].

4.1 Reed-Muller Expansions

In digital logic design, two paradigms have been studied. The first used the operations of *AND*, *OR* and *NOT* and called *canonical Boolean logic*. The second used the operations *AND*, *XOR* and *NOT* and called *Reed-Muller logic*. RM is equivalent to modulo-2 algebra. In this section, we review the properties of RM logic [3].

4.1.1 Modulo–2 Algebra

For any Boolean variable x, we can write the following *XOR* expressions:

$$x \oplus 1 = \bar{x}, \, x \oplus 0 = x,$$
$$\bar{x} \oplus 1 = x, \, \bar{x} \oplus 0 = \bar{x}.$$

Let $\dot{x}$ be a variable representing a Boolean variable in its true (x) or complemented form $(\bar{x})$, then we can write the following expressions:

$$\dot{x} \oplus 1 = \overline{\dot{x}}, \, \dot{x} \oplus 0 = \dot{x},$$
$$\dot{x} \oplus \dot{x} = 0, \, \dot{x} \oplus \overline{\dot{x}} = 1,$$
$$1 \oplus 1 = 0, \, \dot{x}_0(1 \oplus \dot{x}_1) = \dot{x}_0 \oplus \dot{x}_0 \dot{x}_1,$$

$f \oplus f\dot{x} = f\overline{\dot{x}}$, where f is any Boolean function.

For any *XOR* expression, the following properties hold:

1- $x_0 \oplus (x_1 \oplus x_2) = (x_0 \oplus x_1) \oplus x_2 = x_0 \oplus x_1 \oplus x_2$. (Associative)
2- $x_0(x_1 \oplus x_2) = x_0 x_1 \oplus x_0 x_2$. (Distributive)
3- $x_0 \oplus x_1 = x_1 \oplus x_0$. (Commutative)

4.1.2   Representation of Reed-Muller Expansions

Any Boolean function f with n variables $f : \{0, 1\}^n \rightarrow \{0, 1\}$ can be represented as a sum of products [3],

$$f(x_0, ..., x_{n-1}) = \sum_{i=0}^{2^n-1} a_i m_i, \tag{8}$$

where m_i are the minterms and $a_i = 0$ or 1 indicates the presence or absence of minterms respectively and $\sum$ means that the arguments are subject to Boolean operation inclusive-OR. This expansion can also be expressed in RM as follows [1],

$$f(\dot{x}_0, ..., \dot{x}_{n-1}) = \bigoplus_{i=0}^{2^n-1} b_i \varphi_i, \tag{9}$$

where,

$$\varphi_i = \prod_{k=0}^{n-1} \left(\dot{x}_k\right)^{i_k}, \tag{10}$$

where $\dot{x}_k = x_k$ or $\bar{x}_k$ and $x_k, b_i \in \{0, 1\}$ and i_k represent the binary digits of i.

φ_i are known as product terms and b_i determine whether a product term is present or not. $\bigoplus$ means that the arguments are subject to Boolean operation exclusive-OR (XOR) and multiplication is assumed to be the AND operation.

A RM function $f(\dot{x}_0, ..., \dot{x}_{n-1})$ is said to have fixed polarity if throughout the expansion each variable $\dot{x}_k$ is either x_k or $\bar{x}_k$ exclusively. If for some variables x_k and $\bar{x}_k$ both occur when the function is said to have mixed polarity [3].

There is a relation between a_i and b_i coefficients shown in Eqs. 8 and 9, which can be found in detail in [3].

4.1.3 π Notations

Consider the fixed polarity RM functions with $\dot{x}_k$ in its x_k form (Positive Polarity RM). The RM expansion can be expressed as a ring sum of products. For n variables expansion, there are 2^n possible combinations of variables known as the π terms [3]. 1 and 0 are used to indicate the presence or absence of a variable in the product term respectively. For example, a four variable term $x_3 x_2 x_1 x_0$ contains the four variables and is represented by $1111 = 15$, $x_3 x_2 x_1 x_0 = \pi_{15}$ and $x_3 x_1 x_0$ (x_2 is missing) $= \pi_{11}$.

Using this notation [3], the positive polarity RM expansion shown in Eq. 9 can be written as follows,

$$f(x_0, ..., x_{n-1}) = \bigoplus_{i=0}^{2^n-1} b_i \pi_i. \tag{11}$$

The conversion between φ_i and π_i used in Eqs. 9 and 11 can be done in both directions. For example, consider the three variables x_0, x_1 and x_2:

$\varphi_7 = x_0 x_1 x_2 = \pi_7,$
$\varphi_6 = x_0 x_1 \overline{x_2} = x_0 x_1 (x_2 \oplus 1)$
$\quad = x_0 x_1 x_2 \oplus x_0 x_1$
$\quad = \pi_7 \oplus \pi_6,$
$\varphi_5 = x_0 \overline{x_1} x_2 = x_0 (x_1 \oplus 1) x_2$
$\quad = x_0 x_1 x_2 \oplus x_0 x_2$
$\quad = \pi_7 \oplus \pi_5.$

Similarly we can construct the rest of conversion as follows:

$\varphi_4 = \pi_7 \oplus \pi_6 \oplus \pi_5 \oplus \pi_4,$
$\varphi_3 = \pi_7 \oplus \pi_3,$
$\varphi_2 = \pi_7 \oplus \pi_6 \oplus \pi_3 \oplus \pi_2,$
$\varphi_1 = \pi_7 \oplus \pi_5 \oplus \pi_3 \oplus \pi_1,$
$\varphi_0 = \pi_7 \oplus \pi_6 \oplus \pi_5 \oplus \pi_4 \oplus \pi_3 \oplus \pi_2 \oplus \pi_1 \oplus \pi_0.$

For the above conversion, the inverse is also true [3],

$\pi_7 = \varphi_7,$
$\pi_6 = \varphi_7 \oplus \varphi_6,$
$\pi_5 = \varphi_7 \oplus \varphi_5,$

and so on.

4.2 Boolean Quantum Logic

In the construction of Boolean quantum circuits, we initialize a single auxiliary qubit to 0, to hold the result of the Boolean function at the end of the computation.

Without loss of generality, we consider in this section the $CNOT(C|t)$ gate where only cond-1 could be set over the control qubits in C, i.e. the state of the target qubit $|t\rangle$ is flipped if and only if all the qubits in C are set to state $|1\rangle$. For example, consider the $CNOT$ gate shown in Fig. 14, it can be represented as $CNOT\left(\{x_0, x_2\} | x_3\right)$. The state of the qubit $|x_3\rangle$ is flipped if and only if $|x_0\rangle = |x_2\rangle = |1\rangle$ with whatever value in $|x_1\rangle$, i.e. $|x_3\rangle$ is changed according to the operation $x_3 \to x_3 \oplus x_0 x_2$. The case where C is an empty set is written as $CNOT(x_0)$, where $|x_0\rangle$ is the qubit that is flipped unconditionally.

Using the above definition of the $CNOT$ gate and the definition of the BQC, the quantum circuit shown in Fig. 15 can be represented as follows,

$$U = CNOT(\{x_0, x_1\} | x_2).CNOT(\{x_1\} | x_2).CNOT(x_2). \tag{12}$$

Fig. 14 $CNOT\left(\{x_0, x_2\} | x_3\right)$
gate

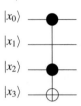

Fig. 15 Boolean quantum
circuit

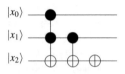

Now, to trace the operations that have been applied on the target qubit $|x_2\rangle$, we trace the operation of each of the *CNOT* gates as follows:

- $CNOT(\{x_0, x_1\}|x_2) \Rightarrow x_2 \rightarrow x_2 \oplus x_0 x_1$.
- $CNOT(\{x_1\}|x_2) \Rightarrow x_2 \rightarrow x_2 \oplus x_1$.
- $CNOT(x_2) \Rightarrow x_2 \rightarrow \bar{x}_2 = x_2 \oplus 1$.

Combining the three operations, we can see that the complete operation applied on $|x_2\rangle$ is represented as follows,

$$x_2 \rightarrow x_2 \oplus x_0 x_1 \oplus x_1 \oplus 1. \tag{13}$$

If $|x_2\rangle$ is initialized to $|0\rangle$, applying the circuit makes $|x_2\rangle$ carry the result of the operation $x_0 x_1 \oplus x_1 \oplus 1$, which is equivalent to the operation $x_0 + \bar{x}_1$.

4.3 Representation of BQC as RM

4.3.1 BQC for Positive Polarity RM

Considering the previous sections, we might notice that there is a close connection between RM and quantum circuits representing arbitrary Boolean functions. In this section, we show the steps that allow us to implement any arbitrary Boolean function f using positive polarity RM expansions as quantum circuits.

Example:

Consider the function $f\left(x_0, x_1, x_2\right) = \bar{x}_0 + x_1 x_2$. To find the quantum circuit implementation for this function, we may follow the following procedure [3]:

Fig. 16 Quantum circuit implementation for $f(x_0, x_1, x_2) = \overline{x_0} + x_1 x_2$

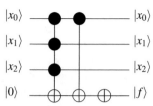

(1) The above function can be represented as a sum of products as follows,

$$f(x_0, x_1, x_2) = \overline{x_0}\,\overline{x_1}\,\overline{x_2} + \overline{x_0}\,\overline{x_1} x_2 + \overline{x_0} x_1 \overline{x_2} + \overline{x_0} x_1 x_2 + x_0 x_1 x_2. \quad (14)$$

(2) Converting to φ_i notation according to Eq. 9,

$$f(x_0, x_1, x_2) = \varphi_0 \oplus \varphi_1 \oplus \varphi_2 \oplus \varphi_3 \oplus \varphi_7. \quad (15)$$

(3) Substituting π product terms shown in Sect. 4.1.3, we get,

$$\begin{aligned} f = &\ \pi_7 \oplus \pi_6 \oplus \pi_5 \oplus \pi_4 \oplus \pi_3 \oplus \pi_2 \oplus \pi_1 \oplus \pi_0 \oplus \pi_7 \oplus \pi_5 \oplus \\ &\ \pi_3 \oplus \pi_1 \oplus \pi_7 \oplus \pi_6 \oplus \pi_3 \oplus \pi_2 \oplus \pi_7 \oplus \pi_3 \oplus \pi_7. \end{aligned} \quad (16)$$

(4) Using modulo-2 operations to simplify this expansion we get,

$$f = \pi_7 \oplus \pi_4 \oplus \pi_0 = x_0 x_1 x_2 \oplus x_0 \oplus 1. \quad (17)$$

Using the last expansion in Eq. 17, we can create the quantum circuit, which implements this function as follows:

1. Initialize the target qubit $|t\rangle$ to the state $|0\rangle$, which holds the result of the Boolean function.
2. Add a *CNOT* gate for each product term in this expansion taking the Boolean variables in this product term as control qubits and the result qubit as the target qubit $|t\rangle$.
3. For the product term, which contains 1, we add a *CNOT*(t), so the final circuit is as shown in Fig. 16.

It is important to notice here that the representation of BQC as positive polarity RM is equivalent to the Direct Synthesis method, where we get the same circuit as follows: Consider the above example for $f(x_0, x_1, x_2) = \overline{x_0} + x_1 x_2$ to be represented using the Direct Synthesis method:

i. Representing the function as a sum of products as shown in Eq. 14 is equivalent to constructing the truth table of that function with the configurations that makes the function evaluate to true.

ii. Converting to φ_i notation shown in Eq. 15 is equivalent to constructing the initial quantum circuit by step-1 of the Direct Synthesis method, where $x_k \oplus \delta_k$ is equivalent to x_k for $\delta_k = 0$ and $\overline{x_k}$ for $\delta_k = 1$ in the expressions shown in Eq. 10.

iii. Substituting π product terms shown in Eq. 16 is equivalent to the decomposition to be applied by step-2 of the Direct Synthesis method, where we remove all the occurrences of $\overline{x_k}$ from the minterm that is similar to removing all the occurrences of $\delta_k = 1$ depends on the condition being specified over each qubit.

iv. Using modulo-2 operations to simplify the expansion as shown in Eq. 17 is equivalent to the rule of optimization to be applied by step-3 of the Direct Synthesis method.

The Direct Synthesis method in the previous section provides a simpler way to demonstrate the construction of that kind of circuit without the need to understand the properties of RM.

4.3.2   BQC for Fixed Polarity RM

Consider the RM expansion shown in Eq. 9 where $\overset{\cdot}{x}_k$ can be x_k or $\overline{x}_k$ exclusively. For n variables expansion where each variable may be in its true or complemented form, but not both, then there are 2^n possible expansions. These are known as *fixed polarity generalized Reed-Muller* (GRM) *expansions* [3].

We can identify different GRM expansions by a *polarity number*, which is a number that represents the binary number calculated in the following way [3]: If a variable appears in its true form, it is represented by 0, and 1 for a variable appearing in its complemented form. For example, consider the Boolean function $f(\overset{\cdot}{x}_0, \overset{\cdot}{x}_1, \overset{\cdot}{x}_2)$: $f(x_0, x_1, x_2)$ has polarity 0 (000), $f(x_0, \overline{x_1}, x_2)$ has polarity 2 (010), $f(\overline{x_0}, x_1, \overline{x_2})$ has polarity 5 (101) and $f(\overline{x_0}, \overline{x_1}, \overline{x_2})$ has polarity 7 (111) and so on.

The RM expansion with a certain polarity can be converted to another polarity by replacing any variable x_k by $(\overline{x_k} \oplus 1)$ or any variable $\overline{x_k}$ by $(x_k \oplus 1)$. For example, consider the Boolean function $f(x_0, x_1, x_2) = \overline{x_0} + x_1 x_2$, it can be represented with different polarity RM expansions as follows,

$$f = x_0 x_1 x_2 \oplus x_0 \oplus 1 : \text{0-polarity}, \tag{18}$$

$$f = x_0 x_1 \overline{x_2} \oplus x_0 x_1 \oplus x_0 \oplus 1 : \text{1-polarity}, \tag{19}$$

$$f = \overline{x_0} x_1 \overline{x_2} \oplus x_1 \overline{x_2} \oplus \overline{x_0} x_1 \oplus x_1 \oplus \overline{x_0} : \text{5-polarity}, \tag{20}$$

$$f = \overline{x_0}\,\overline{x_1}\,\overline{x_2} \oplus \overline{x_0}\,\overline{x_2} \oplus \overline{x_1}\,\overline{x_2} \oplus \overline{x_0}\,\overline{x_1} \oplus \overline{x_1} \oplus \overline{x_2} \oplus 1 : \text{7-polarity}. \tag{21}$$

Different polarity RM expansions give different quantum circuits for the same Boolean function. For example, consider the different polarity representations for the function $f(x_0, x_1, x_2) = \overline{x_0} + x_1 x_2$ shown above. Each representation has different quantum circuit (as shown in Fig. 17) using the following procedure:

1. Initialize the target qubit $|t\rangle$ to the state $|0\rangle$, which holds the result of the Boolean function.
2. Add a *CNOT* gate for each product term in the RM expansion taking the Boolean variables in this term as control qubits and the result qubit as the target qubit $|t\rangle$.
3. For the product term, which contains 1, we add a $CNOT(t)$.
4. For control qubit $|x_k\rangle$, which appears in complemented form, we add a $CNOT(x_k)$ at the beginning of the circuit to negate its value during the run of the circuit and add another $CNOT(x_k)$ at the end of the circuit to restore its original value.

It is clear from Fig. 17 that changing polarity changes the number of the *CNOT* gates in the circuit, i.e. its efficiency. This means that there is a need to develop search algorithms for optimizing canonical Reed-Muller expansions for quantum Boolean functions similar to those found for classical digital circuit design [48, 55], taking into account that efficient expansions for classical computers may be not so efficient for quantum computers. For example, consider $f(x_0, x_1, x_2)$ defined as follows,

$$f = \overline{x_0}\,\overline{x_1}\,\overline{x_2} + \overline{x_0} x_1 \overline{x_2} + x_0 \overline{x_1} x_2 + x_0 x_1 x_2, \tag{22}$$

its 0-polarity expansion is given by $x_0 \oplus x_2 \oplus 1$ and its 3-polarity expansion is given by $\overline{x_0} \oplus x_2$. From a classical point of view, 3-polarity expansion is better than

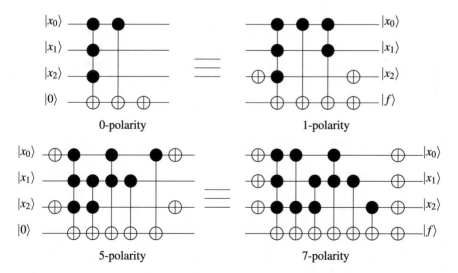

Fig. 17 Quantum circuits for the Boolean function $f(x_0, x_1, x_2) = \overline{x_0} + x_1 x_2$ with different polarities

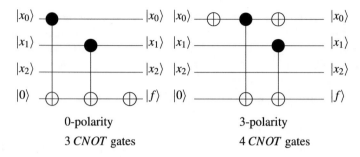

Fig. 18 Changing polarity may affect the number of the *CNOT* gates used

0-polarity expansion because it contains two product terms rather that three product terms in 0-polarity expansion. On the contrary, implementing both expansions as BQC we can see that 0-polarity expansion is better than 3-polarity expansion because of the number of the *CNOT* gates used as shown in Fig. 18.

Using the Direct Synthesis method, the worst case ocurs when the circuit contains 2^n gates and we have shown an alternative decomposition (Fig. 12) where we get $O(n)$ instead (see Sect. 3.3). Actually this is equivalent to changing the polarity of the expression in this case, i.e. the case where we get 2^n *CNOT* gates is equivalent to the 0-polarity form and the case where we get $O(n)$ *CNOT* gates is equivalent to the $2^n - 1$ polarity form where all the inputs are changed from x_k to $\overline{x_k} \oplus 1$ in the RM expression.

4.3.3 BQC for Mixed Polarity RM

Mixed polarity RM are expansions where it is allowed that some variables $\dot{x}_k$ to appear in their true form (x_k) and their complemented form $(\overline{x}_k)$ both in the same expansion [3]. To understand how this kind of expansion can be implemented as a quantum circuit, consider the three variable mixed polarity RM,

$$f = \overline{x_0}x_1x_2 \oplus x_0\overline{x_1} \oplus x_0 \oplus \overline{x_2} \oplus 1. \tag{23}$$

Using the following procedure, we get the quantum circuit as shown in Fig. 19:

1. Initialize the target qubit $|t\rangle$ to the state $|0\rangle$, which holds the result of the Boolean function.
2. Add a *CNOT* gate for each product term in the RM expansion taking the Boolean variables in this term as control qubits and the result qubit as the target qubit $|t\rangle$.
3. For the product term, which contains 1, we add a *CNOT*(t).
4. For control qubit $|x_k\rangle$, which appears in complemented form, we add a *CNOT*(x_k) directly before and after (negate/restore) the *CNOT* gate where this variable appears in its complemented form.

Fig. 19 Mixed polarity
BQC for $f =$
$\overline{x_0}x_1x_2 \oplus x_0\overline{x_1} \oplus x_0 \oplus \overline{x_2} \oplus 1$

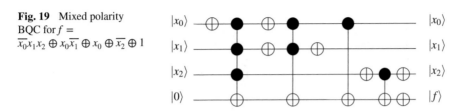

4.3.4 Calculating the Number of the *CNOT* Gates

For a **Fixed Polarity** RM expansion, the number of the *CNOT* gates in the final quantum circuit can be calculated as follows,

$$m_1 = S + 2K, \quad 0 \le S \le 2^n; 0 \le K \le n, \tag{24}$$

where m_1 is the total number of the *CNOT* gates, S is the number of product terms in the expansion, K is the number of variables in complemented form and n is the number of inputs to the Boolean function, the term $2K$ represents the number of the *CNOT* gates that are added at the beginning and the ending of the circuit (complemented form) to negate the value of the control qubit during the run of the circuit and to restore its original value respectively.

For a **Mixed Polarity** RM expansion, the number of the *CNOT* gates in the final quantum circuit can be calculated as follows,

$$m_2 = S + 2L, \quad 0 \le S \le 2^n; 1 \le L < n2^{n-1}, \tag{25}$$

where m_2 is the total number of the *CNOT* gates, S is the number of product terms in the expansion, L is the total number of occurrences of all variables in complemented form and n is the number of inputs to the Boolean function, the term $2L$ represents the number of the *CNOT* gates which may be added before and after the control qubit which appears in complemented form during the run of the circuit to negate/restore its value respectively.

5 Practical Construction of BQC

In general, the meaning of optimality for quantum circuits is connected with practical constraints, for instance, the interaction between certain control qubits [41]. Circuits depend on the physical implementation, so, it is sometimes difficult to take certain qubits as control qubits on the same *CNOT* gate (involved in the same operation) because the interaction between these qubits may be difficult to control. Another constraint is the number of control qubits per *CNOT* gate [5]. There have been efforts

Fig. 20 An abstract four-qubit system with its allowed interaction

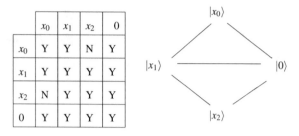

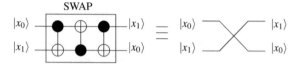

Fig. 21 The *SWAP* gate

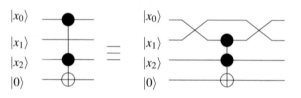

Fig. 22 Solving the interaction problem using the *SWAP* gate

to try to decrease the number of control qubits even with an increase in the total number of the *CNOT* gates in the circuit. Another constraint is the total number the *CNOT* gates in the circuit which should be kept to a minimum so we are able to maintain coherence during the operation of the circuit. In this section, we review how these constraints are being handled (one at a time) at present and how RM expansions can help in handling this problem.

Consider an abstract four-qubits system shown in Fig. 20 where Y and N in the associated table mean that the qubits can and cannot interact respectively. For simplicity, the auxiliary qubit $|0\rangle$ is able to interact with all the control qubits $|x_0\rangle$, $|x_1\rangle$ and $|x_2\rangle$. The main problem is in the interaction between $|x_0\rangle$ and $|x_2\rangle$. This problem is known to be handled using the *SWAP* gate [41] shown in Fig. 21. To illustrate this, consider the gate shown in Fig. 22, where $|x_0\rangle$ and $|x_2\rangle$ are involved in the same *CNOT* gate. By temporarily swapping the values of $|x_0\rangle$ and $|x_1\rangle$, this gate can be implemented on the above system with an increase in the total number of the *CNOT* gates because of the added *SWAP* gates.

Another constraint is the number of control qubits per *CNOT* gate. This problem has been studied in [5] and it was shown that the number of control qubits can be reduced to two control qubits/*CNOT* gate (the Toffoli gate) by using extra auxiliary qubits and an increase in the number of the *CNOT* gates in the final circuit ($4\bar{n} - 8$) Toffoli gates, $\bar{n} \geq 5$, where $\bar{n}$ is the number of control qubits/gate). Notice that we add another ($4\bar{n} - 7$) Toffoli gates to reset the auxiliary qubits (which is optional),

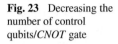

Fig. 23 Decreasing the number of control qubits/*CNOT* gate

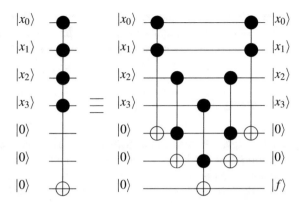

each of which can then be decomposed to two–qubit gates (Lemma 6.1 in [5]) which was proved to be universal for quantum computation [4, 29]. In Fig. 23, we show an example of an equivalent decomposition to that shown in Lemma 7.2 in [5] but more efficient using the transformation rules shown in [38], where each *CNOT* gate with $\bar{n} \geq 4$ can be decomposed to $(2\bar{n} - 7)$ Toffoli gates with another optional $(2\bar{n} - 6)$ Toffoli gates . This was proved in [27].

The above techniques deal with different constraints in a *gate level approach*. Using RM makes it possible to deal with the problem using a *circuit level approach*. To illustrate this, consider the equivalent quantum circuits shown in Fig. 24. Consider implementing these circuits on the system shown in Fig. 20. The 0-polarity form contains, in general, the interaction problem between $|x_0\rangle$ and $|x_2\rangle$ twice whereas the 2-polarity and 6-polarity contain this interaction only once. The 0-polarity contains four *CNOT* gates and 2-polarity contains two *CNOT* gates and two simple *NOT* gates and the 6-polarity contains a single *CNOT* gate with an increase in the simple *NOT* gates in favour of more complicated controlled operations. In terms of the total number of control qubits, 0-polarity contains eight control qubits, 2-polarity

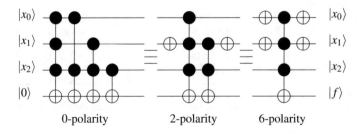

Fig. 24 Optimization using different polarities

contains five control qubits and 6-polarity contains three control qubits. In that sense, using RM can be considered as a platform for synthesis and optimization of BQC to minimize the cost of using any of the above *constraint handling techniques*.

6 Conclusion

In this chapter, we showed that there is a close connection between quantum Boolean operations and Reed-Muller expansions, which implies that a complete study of synthesis and optimization of Boolean quantum logic can be done within the domain of classical Reed-Muller logic. If we consider the positive polarity RM expansions and its corresponding BQC, then using this method we get the same circuit that we would get using the Direct Synthesis method without the explicit construction of the truth table. The worst case we get using positive polarity RM can be avoided using different polarities. The Direct Synthesis method provides a simpler way to demonstrate the construction of that kind of circuit without the need to understand the RM expansions.

We showed that the sense of optimality in quantum circuits' construction follows practical constraints, which can also be handled using the RM expansions. We showed that an efficient RM expression on classical computers may not be so efficient on quantum computers and vice versa. In that sense, we showed that the construction of Boolean quantum logic could be tackled within the domain of RM and algorithms for optimizing BQC that are required should be able to be found within that domain.

References

1. Akers, S.B.: On a theory of Boolean functions. J. SIAM **7**, 487–489 (1959)
2. Akers, S.B.: Binary decision diagrams. IEEE Trans. Comput. **C-27**(6), 509–516 (1978)
3. Almaini, A.: Electronic Logic Systems, chapter 12: Modulo-2 Logic Circuits, p. 470, 2nd edn. Prentice-Hall, Englewood Cliffs, NJ (1989)
4. Barenco, A.: A universal two-bit gate for quantum computation. Proc. R. Soc. Lond. A **449**, 679–683 (1995)
5. Barenco, A., Bennett, C., Cleve, R., Divincenzo, D.P., Margolus, N., Shor, P., Sleator, T., Smolin, J., Weinfurter, H.: Elementary gates for quantum computation. Phys. Rev. A **52**(5), 3457–3467 (1995)
6. Barnum, H., Bernstein, H., Spector, L.: Quantum circuits for OR and AND of ORs. J. Phys. A: Math. Gen. **33**(45), 8047–8057 (2000)
7. Benioff, P.: Quantum mechanical hamiltonian models of Turing machines. J. Stat. Phys. **29**, 515–546 (1982)
8. Benioff, P.: Quantum mechanical hamiltonian models of Turing machines that dissipate no energy. Phys. Rev. Lett. **48**, 1581–1585 (1982)
9. Bennett, C.: Logical reversibility of computation. IBM J. Res. Dev. **17**(6), 525–532 (1973)
10. Bennett, C., Bernstein, E., Brassard, G., Vazirani, U.: Strengths and weaknesses of quantum computing. SIAM J. Comput. **26**(5), 1510–1523 (1997)
11. Bennett, C., Bessette, F., Brassard, G., Salvail, L., Smolin, J.: Experimental quantum cryptography. J. Cryptol. **5**(1), 3–28 (1992)

12. Bernstein, E., Vazirani, U.: Quantum complexity theory. SIAM J. Comput. **26**(5), 1411–1473 (1997)
13. Berthiaume, A., Brassard, G.: The quantum challenge to complexity theory. In: Proceedings of the 7th IEEE Conference on Structure in Complexity Theory, pp. 132–137 (1992)
14. Beth, T., Roetteler, M.: Quantum algorithms: applicable algebra and quantum physics. In: Quantum Information, pp. 96–150. Springer (2001)
15. Braunstein, S., Fuchs, A.: A quantum analog of Huffman coding. IEEE Trans. Inf. Theory **46**(4), 1644–1649 (2000)
16. Bryant, R.E.: Graph-based algorithms for Boolean function manipulation. IEEE Trans. Comput. **C-35**(8), 677–691 (1986)
17. Church, A.: An unsolvable problem of elementary number theory. Am. J. Math. **58**(2), 345–363 (1936)
18. Cirac, J., Zoller, P.: Quantum computations with cold trapped ions. Phys. Rev. Lett. **74**, 4091–4094 (1995)
19. Cleve, R., Watrous, J.: Fast parallel circuit for the quantum Fourier transform. In: Proceedings of the 41th Annual Symposium on Foundations of Computer Science, p. 526 (2000)
20. Cory, D., Fahmy, A., Havel, T.: Nuclear magnetic resonance spectroscopy: an experimentally accessible paradigm for quantum computing. In: Proceedings of the 4th Workshop on Physics and Computation, pp. 87–91 (1996)
21. De Vos, A., Desoete, B., Janiak, F., Nogawski, A.: Control gates as building blocks for reversible computers. In: Proceedings of the 11th International Workshop on Power and Timing Modeling, Optimization and Simulation, pp. 9201–9210 (2001)
22. Deptartment of Computer Science, North Carolina State University. N.C.S.U. collaborative benchmarking laboratory. http://www.cbl.ncsu.edu/benchmarks/
23. Deutsch, D.: Quantum theory, the Church-Turing principle and the universal quantum computer. Proc. R. Soc. Lond. A **400**, 97–117 (1985)
24. Deutsch, D., Barenco, A., Ekert, A.: Universality in quantum computation. Proc. R. Soc. Lond. A **449**, 669–677 (1995)
25. Deutsch, D., Josza, R.: Rapid solution of problems by quantum computation. Proc. R. Soc. Lond. A **439**, 553–558 (1992)
26. Devadas, S., Ghosh, A., Keutzer, K.: Logic Synthesis. McGraw-Hill, New York (1994)
27. Diao, Z., Zubairy, M.S., Chen, G.: Quantum circuit design for Grover's algorithm. Z. Naturforschung, vol. 57a, pp. 701–708 (2002)
28. Dirac, P.: The Principles of Quantum Mechanics. Clarendon Press, Oxford (1947)
29. Divincenzo, D.: Two-bit gates are universal for quantum computation. Phys. Rev. A **51**(2), 1015–1022 (1995)
30. Dueck, G.W., Maslov, D.: Reversible function synthesis with minimum garbage outputs. In: Proceedings of the 6th International Symposium on Representations and Methodology of Future Computing Technologies, pp. 154–161 (2003)
31. Feynman, R.: Simulating physics with computers. Int. J. Theor. Phys. **21**, 467–488 (1982)
32. Feynman, R.: Quantum mechanical computers. Opt. News **11**(2), 11–20 (1985)
33. Feynman, R.: Feynman Lectures on Computation. Addison-Wesley, Reading (1996)
34. Fredkin, E., Toffoli, T.: Conservative logic. Int. J. Theor. Phys. **21**, 219–253 (1982)
35. Gershenfeld, N., Chuang, I.: Bulk spin-resonance quantum computation. Science **275**, 350–356 (1997)
36. Grover, L.: A fast quantum mechanical algorithm for database search. In: Proceedings of the 28th Annual ACM Symposium on the Theory of Computing, pp. 212–219 (1996)
37. Gruska, J.: Quantum Computing. McGraw-Hill, London (1999)
38. Iwama, K., Kambayashi, Y., Yamashita, S.: Transformation rules for designing CNOT–based quantum circuits. In: Proceedings of the 39th Conference on Design Automation, pp. 419–424. ACM Press (2002)
39. Keyes, R., Landauer, R.: Minimal energy dissipation in logic. IBM J. Res. Dev. **14**, 152–157 (1970)

40. Lee, J., Kim, J., Cheong, Y., Lee, S.: A Practical Method for Constructing Quantum Combinational Logic Circuits (1999). arXiv:9911053
41. Lloyd, S.: A potentially realizable quantum computer. Science **261**, 1569–1571 (1993)
42. Lloyd, S.: Almost any quantum logic gate is universal. Phys. Rev. Lett. **75**(2), 346–349 (1995)
43. Maslov, D., Dueck, G.W.: Complexity of reversible Toffoli cascades and EXOR PLAs. In: Proceedings of the 12th International Workshop on Post-Binary ULSI Systems, pp. 17–20 (2003)
44. Maslov, D., Dueck, G.W., Miller, D.M.: Fredkin/Toffoli templates for reversible logic synthesis. In: Proceedings of the ACM/IEEE International Conference on Computer-Aided Design, p. 256 (2003)
45. Maslov, D., Dueck, G.W., Miller, D.M.: Simplification of Toffoli networks via templates. In: Proceedings of the 16th Symposium on Integrated Circuits and Systems Design, p. 53 (2003)
46. Miller, D.M., Dueck, G.W.: Spectral techniques for reversible logic synthesis. In: Proceedings of the 6th International Symposium on Representations and Methodology of Future Computing Technologies, pp. 56–62 (2003)
47. Miller, D.M., Maslov, D., Dueck, G.W.: A transformation based algorithm for reversible logic synthesis. In: Proceedings of the 40th Conference on Design Automation, pp. 318–323 (2003)
48. Miller, J., Thomson, P.: Highly efficient exhaustive search algorithm for optimising canonical Reed-Muller expansions of Boolean functions. Int. J. Electron. **76**, 37–56 (1994)
49. Mishchenko, A., Perkowski, M.: Logic synthesis of reversible wave cascades. In: Proceedings of International Workshop on Logic and Synthesis (2002)
50. Moore, G.: Cramming more components onto integrated circuits. Electronics **38**(8), 114–117 (1965)
51. Nielsen, M., Chuang, I.: Quantum Computation and Quantum Information. Cambridge University Press, Cambridge, UK (2000)
52. Ogawa, T., Nagaoka, H.: Strong converse to the quantum channel coding theorem. IEEE Trans. Inf. Theory **45**(7), 2486–2489 (1999)
53. Patel, K.N., Markov, I.L., Hayes, J.P.: Efficient Synthesis of Linear Reversible Circuits (2003). arXiv:0302002
54. Peres, A.: Reversible logic and quantum computers. Phys. Rev. A **32**, 3266 (1985)
55. Robertson, G., Miller, J., Thomson, P.: Non-exhaustive search methods and their use in the minimisation of Reed-Muller canonical expansions. Int. J. Electron. **80**(1), 1–12 (1996)
56. Rylander, B., Soule, T., Foster, J., Alves-foss, J.: Quantum evolutionary programming. In: Proceedings of the Genetic and Evolutionary, Computation Conference, pp. 1005–1011 (2001)
57. School of Computer Science, University of Victoria. Maslov reversible logic synthesis benchmarks. http://www.cs.uvic.ca/dmaslov/
58. Shende, V., Prasad, A., Markov, I., Hayes, J.: Reversible logic circuit synthesis. In: Proceedings of ACM/IEEE International Conference on Computer-Aided Design, pp. 353–360 (2002)
59. Shor, P.: Algorithms for quantum computation: discrete logarithms and factoring. In: Proceedingsof the 35th Annual Symposium on Foundations of Computer Science, pp. 124–134. IEEE Computer Society Press (1994)
60. Simon, D.: On the power of quantum computation. In: Proceedings of the 35th Annual Symposium on Foundations of Computer Science, pp. 116–123 (1994)
61. Spector, L., Barnum, H., Bernstein, H., Swami, N.: Finding better-than-classical quantum AND/OR algorithm using genetic programming. In: Proceedings of the Congress on Evolutionary Computation, vol. 3, pp. 2239–2246. IEEE Press (1999)
62. Surkan, A., Khuskivadze, A.: Evolution of quantum algorithms for computer of reversible operators. In: The 2002 (NASA/DOD) Conference on Evolvable Hardware, IEEE Computer Society, pp. 186–187 (2001)
63. Toffoli, T.: Reversible computing. In: de Bakker, W., van Leeuwen, J. (eds.), Automata, Languages and Programming, p. 632. Springer, New York (1980). Technical Memo MIT/LCS/TM-151, MIT Lab for Computer Science (unpublished)
64. Travaglione, B.C., Nielsen, M.A., Wiseman, H.M., Ambainis, A.: ROM-based Computation: Quantum Versus Classical (2001). arXiv:0109016

65. Turing, A.: On computable numbers, with an application to the Entscheidungsproblem. In: Proceedings of the London Mathematical Society, Series 2, vol. 42, pp. 230–265 (1936)
66. Vedral, V., Barenco, A., Ekert, A.: Quantum networks for elementary arithmetic operations. Phys. Rev. A **54**(1), 147–153 (1996)
67. De Vos, A., Desoete, B., Adamski, A., Pietrzak, P., Sibinski, M., Widerski, T.: Design of reversible logic circuits by means of control gates. In: Proceedings of the 10th International Workshop on Integrated Circuit Design, Power and Timing Modeling, Optimization and Simulation, pp. 255–264 (2000)
68. Wiedemann, D.: Quantum cryptography. SIGACT News **18**(2), 48–51 (1987)
69. Wootters, W.K., Zurek, W.H.: A single quantum cannot be cloned. Nature **299**, 802 (1982)
70. Younes, A., Miller, J.: Automated method for building CNOT based quantum circuits for Boolean functions. In: Proceedings of the 1st International Computer Engineering Conference New Technologies for the Information Society, pp. 562–565, (2004)
71. Younes, A., Miller, J.: Representation of Boolean quantum circuits as Reed-Muller expansions. Int. J. Electron. **91**(7), 431–444 (2004)

Part II
Cartesian Genetic Programming Applications

Some Remarks on Code Evolution with Genetic Programming

Wolfgang Banzhaf

Abstract In this chapter we take a fresh look at the current status of evolving computer code using Genetic Programming methods. The emphasis is not so much on what has been achieved in detail in the past few years, but on the general research direction of code evolution and its ramifications for GP. We begin with a quick glance at the area of Search-based Software Engineering (SBSE), discuss the history of GP as applied to code evolution, consider various application scenarios, and speculate on techniques that might lead to a scaling-up of present-day approaches.

1 Search-Based Software Engineering

Search-based Software Engineering is a technique within the area of Software Engineering that formulates software engineering problems as search problems which can be addressed by established (meta-)heuristic search methods [20]. More generally, all kinds of Machine Learning approaches can brought to bear on these search/optimization problems, and possibly help accelerate the pace of programming, debugging and testing or repairing and repurposing of software. The ultimate goal of these approaches would be to allow computers to program computers, in other words "automatic programming".

All Engineering disciplines are pragmatic in the approach to their respective object of study. Software Engineering (SE) is no exception. When Harman and Jones wrote their manifesto on search-based approaches to SE in 2001 [20], they wondered why SE had been so slow to take up the challenges of search/optimization that other Engineering disciplines had taken up much earlier. They cite the widespread use of genetic algorithms and other metaheuristic search techniques in order to find satisfactory solutions to problems in SE often characterized by competing constraints, inconsistencies and contradictory goals, multiple solutions of varying complexity

W. Banzhaf (✉)
BEACON Center for the Study of Evolution in Action and Department
of Computer Science and Engineering, Michigan State University,
East Lansing, MI 48824, USA
e-mail: banzhafw@msu.edu

© Springer International Publishing AG 2018   145
S. Stepney and A. Adamatzky (eds.), *Inspired by Nature*, Emergence,
Complexity and Computation 28, https://doi.org/10.1007/978-3-319-67997-6_6

and reachable perfect solutions. Perhaps it was the "youth" of SE that has been one of the reasons why this engineering discipline was not yet adopting techniques that at that time had already made serious in-roads in civil or mechanical, chemical, bio-medical or electric engineering. But a decade after John Koza's seminal book [29] on Genetic Programming, the time seemed to have been right to make this bold suggestion about the possible use of metaheuristics in SE.

A review of the search-based software engineering (SBSE) literature shows the main current thrusts of this area to be:

- Software testing [43].
- Non-functional property optimization [1, 55].
- Software bug repair [14].
- Requirement analysis [17].
- Software design, development and maintenance planning [2, 3, 9].
- Software refactoring [34].

These techniques are, however, far from a complete list of search methods applied in software engineering. For example, there is model checking as a way to provide a quality measure for searching in the space of programs [27, 28], software duplication (clone) search using various methods [51], semantics-based code search [50], code completion [46], code refactoring [45], code smell detection [13] and feature summarization [42], to name a few more examples where search techniques are or can be applied.

2 History of Genetic Programming as Applied to Code Evolution

The core idea of Genetic Programming has, of course, always been to *program* computers using inspiration from the theory of natural evolution [29]. However, for the first two decades of its application, GP mostly was used to evolve algorithms that could model systems (symbolic regression) [30] or classify data (a typical machine learning task) [12].

When we wrote our textbook for GP in 1998 [6], the ambition of writing larger pieces of software with GP—and thus to automatically program—was considered to be far from being realistic. Though the vision was upheld and we claimed that we live, as far as programming computers is concerned, in an age similar to the Middle Ages before the invention of printing with movable letters, we did not see an easy path to achieving this goal. But why would one have to put code together by hand? Wouldn't an automatic method, just taking the specification of program behavior in a suitable form be enough to instruct a GP-driven process to fulfill these expectations? We also claimed that programmers are acting in a similar way as one would have to expect from a GP method [6], by writing code in pieces, copying and pasting, inserting changes, i.e. mutating, by recombining and adapting functions that were originally designed for different purposes or in different contexts.

While GP had successfully evolved small pieces of code from scratch, the method could not easily scale up to large program packages or anything beyond 100–200 lines of code. So scalability was a big problem, along with the need to specify complex behavior for a multi-faceted application with many different use cases.

Nevertheless, researchers examined the search spaces of programs, often making use of radical simplifications, like examining GP's behavior in the search space of Boolean functions. In a series of works, W.B. Langdon and R. Poli examined these search spaces and the behavior of GP in it [37]. And while this led to the recognition of important factors, like the complexity of a solution [32], or the different effects of operators when searching in this space [33], or the advantages of neutrality [25], the question of how to scale up GP to large search spaces remained unresolved [7].

In the domain of symbolic regression and classification/supervised learning, it was again John Koza who has demonstrated how to scale up to more complex problems. Using a method devised by Gruau [18] for neural networks akin to a developmental process, Koza and co-workers were able to demonstrate the ability of GP to solve more complex problems [31]. Development and the rewriting of code was also at the center of self-modifying Cartesian GP (SMCGP) [19]. In that case, rather than complexity, the issue was generalization ability.

As mentioned, these latter developments took place in the classical GP application spaces of regression/modeling and classification/learning. In the second half of the last decade, however, important steps were taken that ultimately led to the realization that we might not be so far away from using GP systematically for code evolution.

One of the most important was the demonstration of bug repair capabilities of a GP system later called GenProg [54]. A program is taken that performs well under many test cases, but fails under a few (the bug), and repaired by patching the bug. Three key innovations allowed this step, (i) operation in a high-level search space provided by the abstract syntax tree (AST) of programs; (ii) assuming that the correct program behavior is hidden somewhere in the program, and thus no new code needs to be generated, only already existing code of that program is used; (iii) variation operators focus on code executed on the failed test cases only. In a series of demonstrations, the authors showed convincingly that GP can be applied to this domain, even when the programs consist of hundreds of thousands of lines of code [40].

Another important step was what has become known as "evolutionary" or "genetic improvement" of programs. Again, existing programs were taken as input, and subjected to evolutionary/genetic optimization processes for non-functional features like energy consumption, execution time etc., functions that sometimes are under the influence of compiler flags [5]. In this case, test cases that could assure correct program behaviour were either used or co-evolved with the program. The goal was to keep the functionality of the program while optimizing for other features (non-functional). One of the main issues in these approaches is the seeding of the population. One wants to start with a semantically correct program, and move through the search space in a way that allows to keep the semantic validity of program variants, while at the same time optimizing one or more non-functional goals.

Arcuri et al. [5] solve this problem by applying variations to the program, then filtering—based on test cases that are taken from a test case population—only those program variants that behave correctly. In their study, which took 8 algorithms written as C functions of a maximum of 100 lines of code, they found that thousands of improved code versions were generated with various trade-offs. Many times, GP exploited features of the code that were on the semantic level and could not be addressed by a regular compiler.

Harman et al. [23] approach the improvement of code using a BNF grammar for decomposing the program to be improved. Again, non-functional features, like complexity or speed of execution are targeted, but also an improvement in the accuracy of the program behavior is selected for, so the process is multi-objective. The BNF grammar extracted is a representation of the program that can be subjected to variations via mutation and crossover, and the programs that can be generated by those modifications are subjected to test cases, with their behavior compared to the original program. While this works well for correct programs, the authors point out that often it is beneficial to have additional information at hand that allows to judge the functional properties of a program. The original program as well as other sources of information can be used as an oracle for testing whether the variant program is working correctly or not. In order to scale up to reasonable code examples, the authors focus the search on heavily used parts of the program and their corresponding BNF grammar parts. As for the non-functional criteria, again the pieces of code that are critical for their fulfillment are most targeted by variation operators [35].

In summary, we can see techniques being developed to improve or repair existing programs with a focus on instructions that play a role in the behavior to be repaired or improved. Methods are also developed to either co-evolve test cases, or distribute test cases such that more difficult tests are sufficiently represented in the set. More generally, all kinds of Machine Learning approaches can brought to bear on these search/optimization problems, and possibly help accelerate the pace of programming, debugging and testing or repairing and repurposing of software. I briefly indicate this research direction in the next section.

3   Machine Learning

The field of program synthesis is clearly much bigger than genetic programming. Learning programs from examples, a task frequently posed to Genetic Programming, is known in Machine Learning as "oracle-based" program synthesis [26]. Integer Logic Programming (ILP) and other classical machine learning approaches have been applied on program synthesis [39].

For example, code completion is a task that can be approached from the language perspective. By analyzing usage patterns and building probabilistic models of patterns in code, it is possible to complete code in constrained applications with high precision [46]. So instead of the programmer being forced to type out all lines of code completely, a less tedious approach could be used. While this does not allow really automatic programming, the transition between manual and automatic programming becomes continuous.

Machine learning has also been applied to predict a correct program from a set of programs on the behavior of input/output pairs. The task here is to rank programs that are correct versus those that are not, based on their behavior. Programming-by-example tasks [44] are becoming more widespread and the ranking of correct programs is at the border between machine learning and formal methods in computing [53].

While Genetic Programming is a pioneer in the field of code synthesis, other Machine Learning paradigms have taken notice. We can expect, over the next decade, that a number of interesting developments in the area of automatic programming and code synthesis will emerge.

4 The Use of Genetic Programming in Code Evolution Tasks

4.1 An Overview of the Tasks Approachable by GP

Before we discuss a few principles that shine through previous successes in applying GP to code evolution, let's first have a look at the different tasks that GP could be applied to. This list is not exhaustive as I am sure I have overlooked possibilities.

1. **Code Adaptation**
   Starting from working code for one processor type, how can we evolve code for another?
   The simple answer would be to say that if we have an abstract language construct, we can compile it to different target processors, so the compiler takes care of the adaptation. This is fine in general, but does not allow to take advantage of the functionality of the target architecture.
   A typical example would be code developed for a CPU (of any kind) needing adaptation to a GPU architecture. See Langdon and Harman [36] for an example.

2. **Code Synthesis**
   Given a behavioral description of a program, how can we synthesize it from a source code base?
   We need to search in the semantic space of programs to find relevant functions or code to use and adapt, or at least snippets that can be put together to produce the overall code looked for.
   Semantic search is an active area of research in Software Engineering and its methods could be brought to bear on this problem [50].

3. **Code Generation/Code Creation**
   Given a behavioral description of a program, how can we generate the code to produce that behavior?
   This was the original question for GP. One uses test (fitness) cases to define the input/output relation of a program, to define its behavior. It normally starts from

a population of random programs to produce the desired behavior. A description is sometimes called I/O oracles. It could also be done on a higher "specification level".

4. **Code Optimization**
   Given code, how can we optimize it with respect to certain (non-functional or functional) criteria?
   Those criteria might be efficiency, speed of execution, robustness, assurance and security criteria, resilience, or accuracy. This has been proposed under the title of Genetic Improvement (GI), with the GISMOE approach being a prime example [49].

5. **Code Repair**
   Given buggy code and a behavioral description, how can we repair the code to make sure it functions properly?
   This task sometimes overlaps with Task 4. Evolving bugs is a very successful application for GP, as exemplified by the GENPROG system [14, 54].

6. **Code Grafting**
   Given the need to add some functionality to a program that already exists in another piece of code for a different purpose, how to graft this piece?
   Examples of this task have been examined in [8, 48] under the name "program transplantation".

7. **Code Testing**
   Given a working program, how can we evolve tests that lead to its failure (in order to remove those conditions that lead to the failure)?
   The production/evolution of test cases is an active area of research in Software Engineering [11]. With GP, it could be considered as the evolution of test cases, or the co-evolution of program code and test cases, as already the work of White et al. have shown [55].

4.2 Main Aspects of Code Evolution

The above applications have a couple of overlapping aspects that are studied in connection with GP in code evolution. Those can be grouped into three categories:

(a) Population seeding
(b) Search bias
(c) Test case application

(a) refers to the question of how to initialize a population of programs at the beginning of the search. The most common procedure in GP so far, but probably not the method to be used in difficult programming problems, is to start the search with a random population. Similar to other engineering problems, it might be better to start from an existing solution and to evolve away from that solution in search of one that is better in regard to different criteria. Incorporating expert knowledge via the seeding strategy has been examined in the past [52], however, we believe there is much to be done to come up with efficient methods for programming.

(b) refers to the fact that not all search directions in a large search space are equally promising. For that reason, most researchers in SBSE adopt a strategy where they first examine which parts of the code are used (at all or frequently) under the conditions required (i.e. of input/output pairs, or for non-functional criteria). Once those code segments have been found, they are subjected to intense search modifications, while other parts which are not used at all, or used only infrequently, are left aside as probably not relevant. An inquiry into similar questions has occupied Biology since a long time under the heading of hyper-mutations [41].

(c) has been an ongoing discussion in Genetic Programming. To evaluate test cases is expensive, so the question is whether we have to run all test cases on all individuals in the population all the time. There have been various scenarios how to avoid this approach. Sampling techniques have been proposed, both based on difficulty of the test case, or on the prior frequency of its use [10, 16, 38]. The co-evolution of test cases has been a topic at least since the early work of Hillis [24] and has been applied very early in bug fixing [4, 5].

All three categories of questions aim at improving the scalability of GP search in programming spaces which for any reasonable program is of enormous size. In what follows we shall provide a few speculative thoughts of how these key questions will develop over the next few years, and where, to my mind, possible solutions could be found.

5   How to Scale Up?

The current section is not a comprehensive review of work in SBSE and what has been achieved (a good survey is [21, 22]), but an attempt at pointing to a number of key areas where progress can be expected in the future.

The reader might have noticed that we haven't discussed program representation for GP at all so far. How is the code represented such that GP can search efficiently and come up with functional code versions without having to wait until the end of the universe? We know that program spaces are notoriously large, so search efficiency is of tantamount importance. So we can safely add

(d) code representation

for GP into the list above.

The original work on evolutionary improvement of existing software used abstract syntax trees (ASTs) of the source code written in a computer language like C or Java. This representation is also commonly used in program analysis and program transformation, and a set of tools exist to extract ASTs from source code, as they are an intermediate step in compiling code. However, it turns out that working on the ASTs of program source code and manipulating it directly by genetic operations does not scale well when trying to improve programs of larger sizes.

As a result, most recent work is done with a differential representation where an input program remains the reference point for a collection of changes that comprise the GP individual. This approach scales much better than the non-differential representation. A sequence of insert/delete and replace operations on the AST—or in the case of a BNF grammar, changes to lines of code represented in the grammar—comprise the individual, again always requiring the input program as a reference point. This is in line with the idea of "improvement" of an existing program and could be termed a linear GP system for code improvement.

Clearly, this approach is only useful for improvements of working programs, or is it? I think that the very same approach can be used to evolve programs more or less from scratch. The idea is to consider a sequence of program improvements just like the evolution from a primitive cell (primitive working program) to a sophisticated and specialized cell (functional program for an arbitrary task). What would need to happen is that the reference input program is, at certain moments we can call "epoch ends", replaced by the best program so far. In other words, the reference program is replaced by a (usually) more complex one that has made progress toward the task enshrined in fitness test cases. At the beginning of a new epoch, the GP individuals of earlier epochs are set aside and replaced by newly initialized very short programs (perhaps simple mutations to the now new reference individual). As changes accrue during an evolutionary epoch, individuals grow in length until again an epoch comes to an end and the reference individual is replaced.

This way, a very simple (stem cell type?) program at the beginning of a run, could be used to evolve in different directions, guided by test cases, and end up in different behavioral spaces that represent different functionalities. This type of programming would therefore require a very general and simple input program, that will gradually be augmented and specialized into different directions, depending on the test case specification. Is code evolvable in this way? I suspect so.

At first sight, this strikes as a complicated way to produce complex code, however, there are shortcuts. In this connection it is helpful to recall that, according to a statistical examination of code bases, most small segments of code (up to a length of approximately 6 or 7 lines of code) have already been written somewhere [15]. If there is a way to harness from the environment this code already existing, this would greatly facilitate the task of assembling a complex program. This will require to traverse the feature space of existing code to find relevant segments [42].

A well-known mechanism useful for building up complexity could be put to use as well: duplication and divergence of modules. Natural evolution seems to have proceeded through a number of repeated steps of duplication and divergence [47] and there is no reason to doubt that the same method couldn't be used with benefit when building up program complexity.

Another question: Are there other programmers (or programming agents) that have faced a similar task, and how did they proceed in the source language (or in another language) to write the code necessary to achieve the goal?

Finally: Is the task decomposable into smaller parts that have been addressed elsewhere in the programmer's own work or in that of others? If that is the case, the evolution could be split into independent parts run on the simpler tasks.

In summary: Besides the tricks of complexity handling indicated in [7], we would use a differential code representation that would undergo epochs of evolution under the control of test cases. Even the test cases could be switched at the changes of epochs, for instance, to provide more detailed guidance on the desired program behavior. Much of what is described here might be categorized as a developmental approach, though there are key ingredients of real computational development still missing from the picture. However, it is reasonable to expect that evo-devo approaches could become very successful.

6 Conclusion

In this chapter we have looked at code evolution and discussed a few recent developments in this area. We saw that the scalability of code evolution approaches is an important obstacle. However, I believe that in the last decade breakthroughs have been made that have fundamentally altered the playing field for the use of GP in code evolution. This application of GP has moved from a dream in the 1990s to a realistic opportunity in the last 10 years. It is time to invest huge research efforts in this area and to harness the continuing growth in speed of our computer hardware.

To give one example where the interaction of GP code evolution and new developments in Computer Science could be very useful is in the area of "big code". Following von Neumann's lead for using the same physical representation for data and code, the term "big code" has been coined in equivalence to the earlier term "big data". Big code refers to collections of code, like on GitHub, which can be scoured or exploited for other tasks, if the appropariate tools are available to find the right entries [56]. This brings to attention the need for tools for semantic search in program spaces [50].

In this contribution we have discussed a number of research questions that might be fruitfully addressed in the application of GP to code evolution. Besides the different tasks mentioned in Sect. 4, key areas of investigation should be the topics (a)–(d) raised in the same Section later. As well, research about a good way to construct complexity, in particular in connection with an approach that makes use of the "natural" complexity of the environment—how to find the right function, how to integrate it with the existing program—are pertinent to code evolution.

Code evolution is of enormous economic importance. Genetic programming has posed new kinds of questions and opened up new ways to think about this problem. Exciting new avenues are before us. I believe that GP has a good opportunity to make serious contributions to this area.

Acknowledgements This essay was written on the occasion of the Festschrift for Julian F. Miller's 60th birthday. It is dedicated to Julian, a wonderful friend and inspiring colleague.

References

1. Afzal, W., Torkar, R., Feldt, R.: A systematic review of search-based testing for non-functional system properties. Inf. Softw. Technol. **51**, 957–976 (2009)
2. Alba, E., Chicano, J.F.: Software project management with GAs. Inf. Sci. **177**, 238002401 (2007)
3. Antoniol, G., Di Penta, M., Harman, M.: Search-based techniques applied to optimization of project planning for a massive maintenance project. In: Proceedings of the 21st IEEE International Conference on Software Maintenance, 2005. ICSM'05, pp. 2400249. IEEE Press, New Jersey (2005)
4. Arcuri, A., Yao, X.: A novel co-evolutionary approach to automatic software bug fixing. In: Proceedings of the 2008 IEEE Congress on Evolutionary Computation, pp. 162–168. IEEE Press, New York (2008)
5. Arcuri, A., White, D.R., Clark, J., Yao, X.: Multi-objective improvement of software using co-evolution and smart seeding. Proc. Int. Conf. SEAL **2008**, 61–70 (2008)
6. Banzhaf, W., Nordin, P., Keller, R., Francone, F.: Genetic Programming—An Introduction. Morgan-Kaufmann, San Francisco, CA (1998)
7. Banzhaf, W., Miller, J.: The challenge of complexity. In: Menon, A. (ed.) Frontiers of Evolutionary Computation, pp. 243–260. Springer, Berlin (2004)
8. Barr, E.T., Harman, M., Jia, Y., Marginean, A., Petke, J.: Automated software transplantation. In: Proceedigns of the 2015 International Symposium on Software Testing and Analysis, ISSTA 2015, pp. 257–269 (2015)
9. Clark, J.A., Jacob, J.L.: Protocols are programs too: the meta-heuristic search for security protocols. Inf. Softw. Technol. **43**, 8910904 (2001)
10. Curry, R., Lichodzijewski, P., Heywood, M.: Scaling genetic programming to large datasets using hierarchical dynamic subset selection. IEEE Trans. Syst. Man Cybern. Part B (Cybernetics) **37** 1065–1073 (2007)
11. Dustin, E., Rashka, J., Paul, J.: Automated Software Testing. Addison Wesley (1999)
12. Espejo, P.G., Ventura, S., Herrera, F.: A survey on the application of genetic programming to classification. IEEE Trans. Syst. Man Cybern.-C **40**, 121–144 (2010)
13. Fontana, F., Zanoni, M., Marin, A.: Code smell detection: towards a machine learning-based approach. In: Proceedings of the IEEE International Conference on Software Maintenance, pp. 396-399. IEEE Press, New Jersey (2013)
14. Forrest, S., Nguyen, T., Weimer, W., Le Goues, C.: A genetic programming approach to automated software repair. In: Proceedigns of the 11th Annual Conference on Genetic and Evolutionary Computation, pp. 947–954. ACM Press, New York (2009)
15. Gabel, M., Su, Z.: A Study of the uniqueness of source code. In: Proceedings of the Eighteenth ACM SIGSOFT International Symposium on Foundations of Software Engineering, pp. 147–156. ACM Press, New York (2010)
16. Gathercole, C., Ross, P.: Dynamic training subset selection for supervised learning in genetic programming. In: Proceedings of the International Conference on Parallel Problem Solving from Nature, pp. 312–321. Springer, Berlin (1994)
17. Greer, D., Ruhe, G.: Software release planning: an evolutionary and iterative approach. Inf. Softw. Technol. **46**, 2430253 (2004)
18. Gruau, F.: Neural network synthesis using cellular encoding and the genetic algorithm. Ph.D. Thesis. Laboratoire de l'Informatique du Parallelisme, Ecole Normale Supirieure de Lyon (1994)
19. Harding, S., Miller, J.F., Banzhaf, W.: Developments in Cartesian Genetic Programming: self-modifying CGP. Genet. Program. Evolvable Mach. **11**, 397–439 (2010)
20. Harman, M., Jones, B.F.: Search-based software engineering. Inf. Softw. Technol. **43**, 833–839 (2001)
21. Harman, M., Mansouri, S.A., Zhang, Y.: Search based software engineering: a comprehensive analysis and review of trends techniques and applications. Department of Computer Science, King's College London Technical Report TR-09-03 (2009)

22. Harman, M., Mansouri, S.A., Zhang, Y.: Search-based software engineering: trends, techniques and applications. ACM Comput. Surv. **45**, 11:1–11:61 (2012)
23. Harman, M., Langdon, W.B., Jia, Y., White, D.R., Arcuri, A., Clark, J.: The GISMOE challenge: constructing the pareto program surface using genetic programming to find better programs. In: Proceedings ot the 27th IEEE/ACM International Conference on Automated Software Engineering (ASE 12), pp. 1–14. ACM Press, New York (2012)
24. Hillis, W.D.: Co-evolving parasites improve simulated evolution as an optimization procedure. Physica D: Nonlinear Phenom. **42**, 228–234 (1990)
25. Hu, T., Payne, J.L., Banzhaf, W., Moore, J.H.: Evolutionary dynamics on multiple scales: a quantitative analysis of the interplay between genotype, phenotype, and fitness in linear genetic programming. Genet. Program. Evolvable Mach. **13**, 305–337 (2012)
26. Jha, S., Gulwani, S., Seshia, S.A., Tiwari, A.: Oracle-guided component-based program synthesis. In: Proceedings of the 32nd ACM/IEEE International Conference on Software Engineering (ICSE-2010), pp. 215–224. ACM, New York (2010)
27. Johnson, C.G.: Genetic programming with fitness based on model checking. In: European Conference on Genetic Programming, pp. 114–124. Springer, Berlin (2007)
28. Katz, G., Peled, D.: Model checking-based genetic programming with an application to mutual exclusion. In: Ramakrishnan, C.R., Rehof, J. (eds.) TACAS 2008, LNCS 4963, p. 1410156 (2008)
29. Koza, J.: Genetic Programming: on the programming of computers by means of natural selection. MIT Press, Cambridge (1992)
30. Koza, J.R.: Genetic Programming II: Automatic Discovery of Reusable Subprograms. MIT Press, Cambridge, MA (1994)
31. Koza, J.R.: Genetic Programming III: Darwinian Invention and Problem Solving. Morgan Kaufmann, San Francisco, CA (1999)
32. Langdon, W.B.: Scaling of program tree fitness spaces. Evolut. Comput. **7**, 399–428 (1999)
33. Langdon, W.B.: Size-fair and homologous tree genetic programming crossovers. Genet. Program. Evolvable Mach. **1**, 95–119 (2000)
34. Langdon, W.B., Harman, M.: Genetically improved CUDA C++ software. In: Proceedings of the European Conference on Genetic Programming, pp. 87–99. Springer, Berlin (2014)
35. Langdon, W.B., Harman, M.: Optimising existing software with genetic programming. IEEE Trans. Evol. Comput. **19**, 118–135 (2015)
36. Langdon, W.B., Harman, M.: Genetically improved CUDA C++ software. In: Proceedings of the European Conference on Genetic Programming, pp. 87–99. Springer, Berlin (2014)
37. Langdon, W.B., Poli, R.: Foundations of Genetic Programming. Springer, Berlin (2002)
38. Lasarczyk, C.W.G., Dittrich, P., Banzhaf, W.: Dynamic subset selection based on a fitness case topology. Evol. Comput. **12**, 223–242 (2004)
39. Lau, T.A., Weld, D.S.: Programming by demonstration: an inductive learning formulation. In: Proceedings of the 4th international conference on Intelligent user interfaces IUI-1999, pp. 145–152. ACM, New York (1999)
40. Le Goues, C., Nguyen, T., Forrest, S., Weimer, W.: GenProg: a generic method for automatic software repair. IEEE Trans. Softw. Eng. **38**, 54–72 (2012)
41. Martincorena, I., Luscombe, N.M.: Non-random mutation: the evolution of targeted hypermutation and hypomutation. Bioessays **35**, 123–130 (2012)
42. McBurney, P.W., Liu, C., McMillan, C.: Automated feature discovery via sentence selection and source code summarization. J. Softw.: Evol. Process **28**, 120–145 (2016)
43. McMinn, P.: Search-based software test data generation: a survey. Softw. Test. Verification Reliab. **14**, 1050156 (2004)
44. Menon, A., Tamuz, O., Gulwani, S., Lampson, B., Kalai, A.: A machine learning framework for programming by example. In: JMLR W & CP Proceedings of ICML (2013), vol 28 (2013)
45. Mens, T., Tourwe, T.: A survey of software refactoring. IEEE Trans. Softw. Eng. **30**, 126–139 (2004)

46. Nguyen, A.T., Nguyen, T.T., Nguyen, H.A., Tamrawi, A., Nguyen, H.V., Al-Kofahi, J., Nguyen, T.N.: Graph-based pattern-oriented, context-sensitive source code completion. In: Proceedigns of the 34th International Conference on Software Engineering, pp. 69–79. IEEE Press, New Jersey (2012)

47. Ohno, S.: Evolution by Gene Duplication. Springer, New York (1970)

48. Petke, J., Harman, M., Langdon, W.B., Weimer, W.: Using genetic improvement and code transplants to specialise a C++ program to a problem class. In: Genetic Programming— Proceedings of the 17th European Conference, EuroGP 2014, pp. 137–149. Springer, Berlin (2014)

49. Petke, J., Langdon, W.B., Harman, M.: Applying genetic improvement to MiniSAT. In: Proceedings of the International Symposium on Search Based Software Engineering, pp. 257–262. Springer, Berlin (2013)

50. Reiss, S.P.: Semantics-Based Code Search. ICSE09. 16–24 May 2009, pp. 243–253. IEEE Press, New Jersey (2009)

51. Roy, C.K., Cordy, J.R.: A survey on software clone detection research. Queens Sch. Comput. TR 541–2007 (2007)

52. Schmidt, M.D., Lipson, H.: Incorporating expert knowledge in evolutionary search: a study of seeding methods. In: Proceedings of the 11th Annual conference on Genetic and evolutionary computation, pp. 1091–1098. ACM Press, New York (2009)

53. Singh, R., Gulwani, S.: Predicting a correct program in programming by example. In: Kroening, D., Pasareanu, C.S. (eds.) Proceedings of the CAV 2015, pp. 398–414. Springer, Switzerland (2015)

54. Weimer, W., Nguyen, T., Le Goues, C., Forrest, S.: Automatically finding patches using genetic programming. In: Proceedings of ICSE09, 16–24 May 2009, Vancouver, Canada, pp. 364–374. IEEE Press, New Jersey (2009)

55. White, D.R., Arcuri, A., Clark, J.A.: Evolutionary improvement of programs. IEEE Trans. Evol. Comput. **15**, 515–538 (2011)

56. Yahav, E.: Analysis and synthesis of "Big Code". In: Esparza, J. et al. (eds.) Dependable Software Systems Engineering. IOS Press (2016)

Cartesian Genetic Programming for Control Engineering

Tim Clarke

Abstract Genetic programming has a proven ability to discover novel solutions to engineering problems. The author has worked with Julian F. Miller, together with some undergraduate and postgraduate students, over the last ten or so years in exploring innovation through evolution, using Cartesian Genetic Programming (CGP). Our co-supervisions and private meetings stimulated many discussions about its application to a specific problem domain: control engineering. Initially, we explored the design of a flight control system for a single rotor helicopter, where the author has considerable theoretical and practical experience. The challenge of taming helicopter dynamics (which are non-linear, highly cross-coupled and unstable) seemed ideally suited to the application of CGP. However, our combined energies drew us towards the more fundamental issues of how best to generalise the problem with the objective of freeing up the innovation process from constrictions imposed by conventional engineering thinking. This chapter provides an outline of our thoughts and hopefully may motivate a reader out there to progress this still embryonic research. The scene is set by considering a 'simple' class of problems: the single-input, single-output, linear, time-invariant system.

1 Introduction

In the context of this chapter, Control Engineering concerns the modification of the behaviour of dynamic systems in response to external stimuli through feedback. Behind this seemingly abstract and dry definition, the implications and impact of Control Engineering on our very existence are remarkable. It is well understood that the behaviours of our own biological, biochemical and physiological processes and

T. Clarke (✉)
Department of Electronic Engineering, University of York,
Heslington YO10 5DD, UK
e-mail: tim.clarke@york.ac.uk

© Springer International Publishing AG 2018
S. Stepney and A. Adamatzky (eds.), *Inspired by Nature*, Emergence,
Complexity and Computation 28, https://doi.org/10.1007/978-3-319-67997-6_7

their responses to external factors (food, climate and stress are obvious examples) are effected through mechanisms that have become well understood in an engineering context in the last 150 years or so and are, truly, the cornerstone of the infrastructure for modern living.

Consider two simple examples: the humble kettle and the home heating system. Without the temperature-based on-off switch of the kettle, how many of us would regularly experience a kitchen filled with clouds of steam? Central heating and other environmental control systems can maintain pleasant living conditions for us in extreme climates. The industrial and manufacturing processes that produce our goods and chattels, foodstuffs, pharmaceuticals, fuels, chemicals (the list is virtually endless) are subject to feedback control mechanisms which regulate the process machinery. Our automotive transport systems, from car manufacture to the urban traffic control systems that maintain safe, orderly and expeditious flows also rely on feedback control.

There is something satisfyingly symmetrical about returning to Mother Nature to seek inspiration for new ways of harnessing the undoubted power of feedback. That is the focus of this chapter. It explores how Cartesian Genetic Programming (CGP) [6] can be *harnessed to discover novel solutions* to Control Engineering problems.

Referring to Fig. 1, the human solution search space tends to be limited by several factors, such as:

- procedures
- self-imposed rules
- paradigms
- dogma
- intelligence
- skill
- past experience

Automatic, intelligent search algorithms, such as Genetic Programming, open up the potential solution search space. In the figure, the Human search space boundary

Fig. 1 The three solution search spaces: human, CGP and the virtually-unbounded space of all possible designs

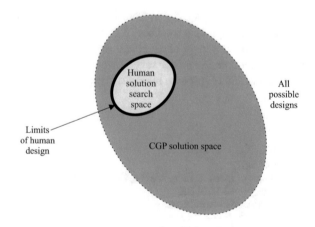

is drawn with a bold line to indicate the significant challenges of moving the state of the art in Control Engineering. The outer boundary between the CGP solution space and all possible designs is much more insubstantial, indicating that it is only limited by the choices made in how CGP is employed. These choices are discussed later.

The term Cartesian Genetic Control (CGC) is used for convenience. Rather than focus strongly on specific results, this chapter considers some of the drivers that might have a bearing upon the efficacy of an evolutionary approach. The objective is to stimulate further interest and activity in this surprisingly embryonic field.

Control Engineering is a huge field. Therefore we do two things to make the task of this chapter tractable. Firstly, we assume that the reader has some knowledge of the fundamentals of Control Theory. Alternatively, the reader will need to access the plethora of undergraduate-level Control Engineering texts (e.g. [3, 7]) and become familiar with the concepts listed below. Another approach, and one which the author highly recommends, is for a collaboration between a practising Control Engineer and an aficionado of CGP. Concepts we assume understood include the following:

- Laplace transforms; transfer functions; poles and zeros.
- frequency responses; gain and phase margins.
- time responses; system type.
- digital control systems; anti-aliasing; sample-and-hold.

2 Feedback Control

2.1 Outline Principles

In its broadest context, feedback control can be considered, algorithmically, as a recursive, three-part loop.

As shown in Fig. 2, a dynamic system is stimulated in order to perform a task whilst conforming to some behaviour. The behaviour of the system is measured using appropriate sensors. Measured outputs are compared with the externally-applied stimulus. This may define the required behaviour of the system. The difference between the demanded and achieved response is then used directly, or after predetermined modification via a controller, to change the system behaviour, which is measured using those same sensors. The loop is repeated *dum repetunt*. This feedback loop is the basis for the physical realisation of controllers. The comparison between desired and achieved responses generates an error, which is the defining characteristic of negative feedback. The idea of using feedback loops was implicitly understood and accepted as far back in time as the pre-Christian Arab and Greek civilisations. There was, at that time, a significant preoccupation with keeping accurate time using water clocks. Feedback-based water regulators were used to maintain constant water levels. The interested reader is directed to the wealth of internet-based material to enjoy the some remarkable examples of human ingenuity.

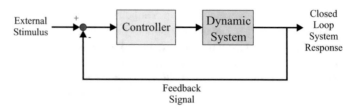

Fig. 2 A basic feedback control system

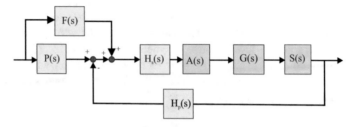

Fig. 3 Conventional control structural blocks

2.2 Control Strategies and Objectives

On the basic premise that we shall apply negative feedback, we can identify some conventional structural options open to the Control Engineer.

Beginning with the blue blocks in Fig. 3, $G(s)$[1] represents the dynamic system itself with $A(s)$ and $S(s)$ respectively representing the actuation and sensor systems. The control function is effected by the yellow blocks. $H_s(s)$, operating on the error signal, is known as a series controller whilst $H_p(s)$, acting directly on the feedback signals, is a parallel controller. $P(s)$ is a pre-filter and $F(s)$ is commonly called a feed-forward controller. Some or all of the yellow controller blocks, shown in Fig. 3, may be used to effect control action on the system (sometimes known as the plant) in order to attain a set of control objectives. Sensors and actuators elicit information and effect action, respectively, on the plant. The controller blocks act as filters, modifying the content (in amplitude and frequency content) of the signals passing into them. The Control *Strategy* is the selection of which blocks to employ.

There are two fundamental Control *Objectives*: regulation and tracking. If the objective of the controlled system is to maintain a constant output (such as the desired, uniform thickness of strip steel produced in a hot rolling mill from a large steel slab), it is a *regulation* problem. Otherwise, if the objective of the controlled system is to follow a changing input (such as the desired positioning of the read/write head of a blu-ray disc system—a 580 nm diameter laser light spot moving over a track width of 320 nm [9]), then it is a *tracking* problem. We can further elaborate these

---

[1]Here we use the common and convenient Laplace transform notation of a *transfer function* (•(s)) to represent the dynamic characteristics of a continuous dynamic system.

fundamental objectives. If the system is subjected to a disturbance, is the regulation or tracking objective still tightly maintained? Examples would be a localised change in the hardness of the steel slab, or a physical shock applied to the blu-ray player. How long does the system take to settle to its steady operation after a change of input or a disturbance? Is there an error between the desired and actual response, once any transient behaviour dies away? What happens if components within the system $G(s)$ start to wear, or drift off-specification due, for example, to a temperature rise?

2.3 Representations of Plant and Control Systems

Later, we consider how the mathematical representation of a controller can affect the behaviour of the evolutionary processes. Here we consider a variety of formats which are in common use. Beforehand, we will look at two alternative domains of representation. The behaviour of dynamic systems can be described, in the time domain, using differential equations. However, their manipulation can be cumbersome, error-prone and time-consuming. By the application of linear transforms, it is possible to generate representations that can be handled algebraically, using relatively simple polynomial forms. For continuous systems, it is usual to employ the Laplace transform [7]. For sampled data, digital or discrete systems, the z-transform or one of its variants [3] is employed.

2.4 Control Specifications

Clearly, the desired closed loop performance of the controlled system will influence how the evolutionary cost or fitness function is defined. This section discusses some alternative control specifications in common use.

2.4.1 Step Response

The input step response of a system is a commonly used as a way to define the specification of a closed loop system and therefore to assess the performance of the controller. The step response is the measured output response to a unity amplitude change in the input level. The response can be quantitatively described by the following set of characteristics.

1. Rise Time (T_r). This is usually defined as the time taken for the response to pass from 10 to 90% of the settled final value (steady state).
2. Percentage Overshoot (%OS). This is the maximum amplitude of the largest overshoot of the response (normally the first overshoot) expressed as a percentage of the final value. Some overshoot is accepted as a compromise to allow a faster rise time.

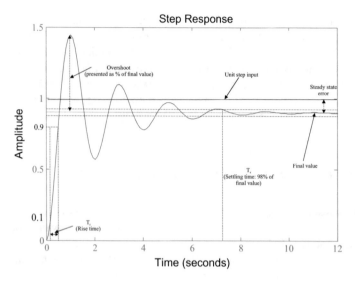

Fig. 4 Typical response of a second order system to a unit amplitude step demand

3. Settling Time (T_s). This is the time at which the response falls within $\pm 2\%$ of the final value and remains within these limits.
4. Steady State Error (SSE). For a constant input amplitude, the steady state response amplitude should ideally be the same. If not, there exists a steady state error. The reader is referred to textbook discussions of system type and steady state error performance for full elucidation of this problem.

Figure 4 defines these quantities graphically. The actual system is

$$G(s) = \frac{9}{(s^2 + s + 10)} = \frac{0.9\omega_n^2}{(s^2 + 2\zeta\omega_n s + \omega_n^2)} \tag{1}$$

There is a trade off between some of these characteristics. Resolution is challenging without resorting to pareto-optimal methods.[2] An alternative is to integrate all characteristics into one quantifiable measure by comparing the actual step response to some ideal and expressing costs as the error. Such integral-based, time-scaled error algorithms are the cornerstone of a class of optimal control design methods such as the Linear Quadratic Regulator (LQR) approach [1]. The most common measures are the integral of the time-weighted absolute error (ITAE) [4] and the integral of the time-weighted squared error (ITSE) [8].

---

[2]which would require a significant modification of CGP.

2.4.2 Dominant Dynamics

Many control problems can be specified in terms of the desired dominant time response. By this we mean the most enduring transient behaviour when a system is stimulated. We need to take a brief look at the relationship between the system poles and the transient response. In the simplest analysis, poles that lie further to the left in the left half s-plane are associated with dynamics that decay quickest. Therefore, dominant dynamics are associated with poles close to the imaginary axis in the s-plane—so-called slow poles. As a good rule of thumb, if the time constant of the slowest pole(s) is five times that of the next slowest, then the composite response is dominated by the slow dynamics and the system largely behaves as a system comprising only the slow poles. The s-plane plot of Fig. 5 should clarify.

The complex conjugate pair of poles are said to be *dominant* and the system response is largely second order if the horizontal distance between the origin and the real pole at $s = -p$ is at least $5 \times \zeta\omega_n$—the horizontal distance between origin and the second order poles at $s = -\zeta\omega_n \pm j\omega_n\sqrt{1 - \zeta^2}$. From this, it is possible to specify the real system closed loop response to be "dominantly second order with a natural frequency ω_n and a damping ratio of ζ". Even if there are dominant poles, the higher order complement of poles affects the actual system response. It is therefore often the case that controllers, once designed to meet the dominant dynamics specification, are then tested on the plant model against the specification and may then need tuning to accommodate the higher order reality. Just as a reminder, the reader is assumed to understand the concept of poles and zeros. Any good control text will provide an explanation, under the treatment of the Laplace transform.

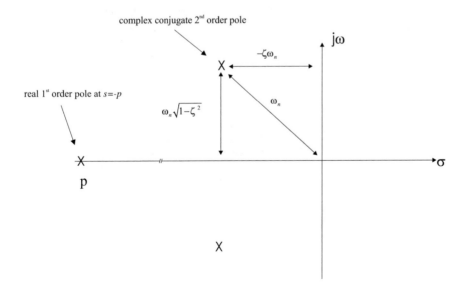

Fig. 5 Dominant dynamics—an s-plane representation

2.4.3   Frequency Specifications

There is a close relationship between positions of poles/zeros in the s-plane, time responses and frequency responses. Indeed a Control Engineer often moves between all three when tackling a suitable, challenging control problem. That said, it is less often the case that the specification of the behaviour of a closed loop system is fully represented in the frequency domain. Characteristics often considered are the gain and phase margins, the unity gain crossover frequency and, less often, the closed loop system bandwidth.

3   CGC and Its Implementation

For the purposes of brevity, the reader is assumed to have a working knowledge of CGP. In this section, we bridge the gap between Control Theory and CGP implementation for CGC. We refer to the experiences of the author and Julian F. Miller in this section. We have worked together with postgraduate students and a succession of very capable final year undergraduate students whose project focus was CGC. We had many discussions and sent our charges off to ponder, theorise and test—as did we! We firstly define some terms used in our discussions, based upon Fig. 6.

3.1   Consequences of the Design Sequence

We now summarise our position using, as a focus, the process flow for designing to meet a step response specification—as illustrated in Fig. 7. The two most important choices to be made are where to specify fitness and where to evolve the controller.

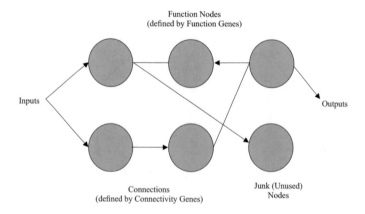

Fig. 6 CGP nodal structure

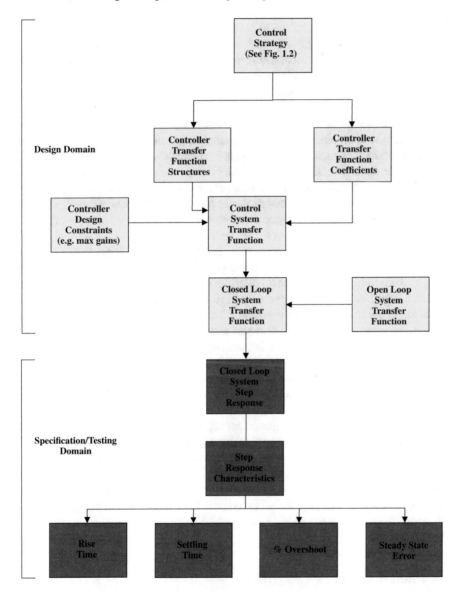

Fig. 7 Controller design flow based upon step response criteria

Looking at Fig. 7, blocks close to the top represent the typical design domain for conventional controller solutions, such as pole placement and, ultimately, individual controller transfer functions; whilst those down towards the bottom represent the characteristics of the physical system most often used in control problem specifications such as step response characteristics which, in turn, are derived from the actual continuous time-domain step response. Here comes the paradox.

Specifications for control problems are often represented by the step response characteristics, often obtained through simulation or real stimulation. However, the physical factors that cause the step response outcomes are represented *mathematically*. Those factors through which control systems are usually designed are represented by the positions of the system poles and zeroes and controller (compensator) poles and zeroes or the compensator transfer function coefficients. There is a significant degree of disjunction between simulation and mathematical representations.

If the controller is evolved as a transfer function and fitness is determined by the resulting step response characteristics, it may prove very difficult to provide a smooth fitness space. This happens because the effect of subtle variations in an arbitrary transfer function coefficient (or altering the polynomial structure of the transfer function itself) can produce potentially huge changes in the step response characteristics. There is a non-obvious mapping between minor evolutionary changes in coefficient value or polynomial structure and the corresponding change in fitness. This increases the 'randomisation' effect of evolution and reduces the potential for incremental optimisation. Thus, the full capabilities of CGP are possibly not utilised. CGP is considered to be capable of handling very abstract fitness definitions. However the broad nature of the solution space for most control problems would seem to counter this claim.

Conversely, specifying both fitness and controller design at the transfer function level *would* allow fitness to be directly related to the evolved solution, but an improvement in such a fitness value would not necessarily represent an equivalent improvement in the step response characteristics. Thus 'perfect' fitness would not necessarily represent a perfect solution. Furthermore, it would be difficult to characterise 'perfect' fitness in such an algorithm. This presents possibly the greatest challenge in producing an algorithm with which to evolve controllers.

4 Phenotypes for a Controller

Now we get to one nub of the CGC problem. How do we represent a controller? This underpins a subsidiary question which is: what are the function genes? If a change in a gene affects the eventual controller, how radical should this change be allowed to be?

4.1 Evolving Control Strategy

Should CGP change the Control Strategy or just the migration towards achieving the Control Objective(s)—or both? CGP provides us with a ready method for strategy change (Fig. 3) through changes in its connectivity genes (Fig. 6). In all of our past work, we never used the connectivity to affect strategy. Rather, we selected either series $(H_s(s))$ or parallel $(H_p(s))$ compensation, and used connectivity genes to modify the internal structure of the controller. *This use of the input and output connectivity genes of CGP to specify and alter Control Strategy is a largely unexplored problem.*

4.2 Evolving Controller Structural Blocks

4.2.1 Choice of the CGC Function Set

There are many options for the choice of function set which constitute the fundamental elements for a controller structural block ($H_s(s)$ or ($H_p(s)$) for example). So far, we have considered linear, time-invariant target plant systems and wish to utilise linear, continuous, time-invariant controllers. Therefore, the functions would generally be continuous-time operators which could be static (scaling, sum, difference, product, protected divide, etc.), basic dynamic (integration, differentiation) or classical controller functions (lead and lag terms). Alternatively, the fundamental function set could be s-plane structures (simple and complex poles and zeros along with gain terms).

A more radical approach could be to make use of sampled data. Here, we assume that sensor outputs are filtered for anti-aliasing purposes then sampled at an appropriate rate. Controller outputs would be passed through a sample-and-hold function before being presented as 'continuous' inputs to the plant. The function set, as defined by the function genes, can be any operations associated with digital processing to build up the controller. It is possible to work at the lowest levels of representations such as bit-wise operators and shifts. At a higher level, one could work with delays and scalings to form recursive realisations of digital filters (after all, this is essentially what a dynamic compensator is) or, at a still higher level, the classical controller functions could be specified as numerator/denominator polynomials of evolvable order with evolvable coefficients, or as discrete z-plane poles/zeros and gains.

We come back to the paradox expressed earlier in Sect. 3.1. The more fundamental the operators in the function set (static in the continuous case or bit-wise operators and shifts in the discrete case) the greater the remove from the physicality of the step response specification. The solution space is unlikely to be smooth and monotonically changing. Genetic operators have unpredictable effects on the progression of the evolutionary process. The upside is that the outer boundary, depicted in Fig. 1, is likely to enclose a much larger CGC solution space.

4.2.2 Interpreting Solutions

A compelling example of the use of GP for Control is that of Koza et al. [5]. They evolved a controller for a two-lag plant

$$G(s) = \frac{K}{(1 + \tau s)^2} \qquad (2)$$

K is allowed to vary between 1 and 2 whilst τ can vary between 0.5 and 1. The set of specifications is as follows:

- minimise ITAE for step input
- overshoot below 4% and settling time less than 2 s
- controller robust in the face of variations in plant gain and time constant
- no extreme plant inputs (from controller)
- controller/plant system bandwidth limited
- no controllers that cannot be physically realised.

The controller related function set was:

- gain
- inverter
- simple lead (zero)
- simple lag (pole)
- 2nd order lag
- integrator
- differentiator
- add
- subtract
- automatically defined functions, created through architecture-altering operations.

The fitness function had to encapsulate the multi-objective nature of the specification set. It did this by combining a lexical approach (instantly high-weighting pathological solutions that could not be simulated, for example, and summing other fitness measures—ITAE and bandwidth, for example). They suggest that setting a time limit penalty for the simulations of the closed loop system put indirect pressure on the process to reduce the complexity of solutions. Our own experience is that this reduces the settling time for evolved solutions. Their explicit use of bandwidth restrictions is interesting. The intention here is to limit the high frequency content of demands placed on the plant by the controller, thereby preventing damage to delicate components. Our approach has always been to limit the positioning of poles and zeroes. This is done based upon good engineering principles. In practice, there should be no controller action faster than the fastest dynamic of the plant. To allow otherwise creates rate-limiting non-linearities which, in turn, lead to destabilising dynamic lags. Often, 'toy' problems neglect this.

The evolved solution was a prefilter and a series controller ($P(s)$ and $H_s(s)$ of Fig. 3 respectively). The form of $H(s)$ is a slightly more sophisticated form of a proportional-plus-integral-plus-derivative (PID) controller—a PID-plus-squared-derivative (PIDD^2) controller. To all intents and purposes this combines a PID controller with a proportional-plus-derivative (PD) controller. The solution was compared with the results of a design produced by Dorf and Bishop [2] who originally posed the problem. To be fair to the original authors, they set a specification, met it and stopped there. Koza et al. evolved a solution that bettered it, most certainly. However, had a competent Control Engineer wished to do the same, the route of developing a supplementary PD controller would have been a sensible strategy. What is impressive is that the GP solution did the same thing.

5 Representative CGP Control Experiments

Here, we present a couple of examples of the type of experiments we conducted to test the ability of CGP to evolve useful controllers for simple test problems. Although there is no sophistication in the approach, they demonstrate the ability of CGP to realise controllers using a conventional controller function set:

- gain
- simple lead (zero) and lag (pole)
- 2nd order lead and lag
- integrator
- differentiator
- phase lead/lag

Limits were placed on controller pole locations to avoid damage to the plant due to high frequency demands by the controller. As 'toy' problems, we did not consider the effects of actuator rate limiting. Rather, we assume the system would response instantaneously to commands. The problems were chosen to represent two typical controller design challenges.

5.1 Problem 1—Speeding up Response

This is a typical issue where the open loop plant is incapable of providing a quick enough response to a change of set point under simple feedback. A compensator is necessary to achieve a quicker response than would be possible with a simple gain controller. The plant is

$$G(s) = \frac{1}{s(s+4)(s+6)} \tag{3}$$

In this case the fastest settling time with a simple gain controller is 3.88 s. We set the following requirements:

- $T_s < 2$ s
- $\%OS < 35\%$

A classically designed controller in the form of a phase advance network (PAN)

$$H_s(s) = 1423\frac{s+5}{s+42.93} \tag{4}$$

achieves the following specification:

- zero steady state error
- $T_s = 1.96$ s
- $\%OS = 30.5$

The CGC solution is

$$H_s(s) = 112.4(s+2.61) \tag{5}$$

and achieves the following specification:

- zero steady state error
- $T_s = 0.986$ s
- $\%OS = 27.5$

The unit step responses for both controllers is shown at Fig. 8.

The outcome for CGC is to create a PD controller rather than the original PAN. To achieve this the ITAE criterion was used as the test for fitness. Bounds were placed on the controller poles to lie between $s = -0.2$ and $s = -20$.

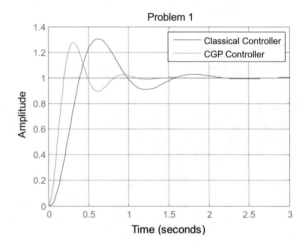

Fig. 8 The solutions to problem 1—the unit step response

5.2 Problem 2—Eliminating a Steady State Error

Here we have a Type 0 plant (no open loop plant integrators) which, without an appropriate controller, develops a 95% steady state error when a step input is applied. What is required here is integral action by the controller. The plant is:

$$G(s) = \frac{1}{(s+1)(s+2)(s+10)} \tag{6}$$

We wish to remove the steady state error and enable a settling time within 10 s. The problem is that the required integral control action creates a very slow residual response, once the faster transient dynamics have subsided.

Our classically designed solution is a proportional-plus-integral (PI) controller

$$H_s(s) = 164.565 \frac{s+0.1}{s} \tag{7}$$

which achieves the following:

- 'zero' steady state error (after a very long wait)
- $T_s = 17.3$ s
- $\%OS = 41.5\%$

The CGC controller is very different:

$$H_s(s) = 2476(s+3.37) \tag{8}$$

It achieves the following:

- a steady state error of 0.2%
- $T_s = 0.775$ s
- $\%OS = 74.1\%$

The step responses are shown at Figs. 9 and 10.

The CGC solution is approached through a compromise. A PD controller is used to speed up the step response and high gain reduces the steady state error. It is possible now to reduce the gain of 2476 which would improve the damping (less pronounced oscillations) but at the expense of increasing the settling time and the steady state error. ITSE was used to gauge the fitness of solutions.

5.3 Observations

In our two illustrative experiments, CGC is able to present alternative design strategies for the designer to consider. In Problem 1, the application of a PD controller instead of a PAN is predictable and realistic. Also, without appropriate restraint, the

Fig. 9 The solutions to problem 2—plotted over short timescale

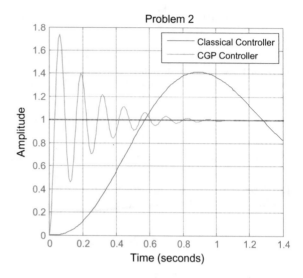

Fig. 10 The solutions to problem 2—plotted over a longer timescale

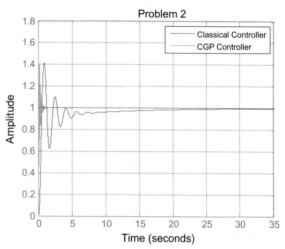

CGC solution exceeds the desired specification—here in %OS and in settling time. In this case CGC is seen to demonstrate a sensible alternative solution.

In the case of Problem 2, CGC presents a radically different approach to the steady state error problem without resorting to problematical integral action. The normal alternative approach to P + I control is a Phase Lag Network (PLN). It does not eliminate the error, but reduces it. The outcome will be a long-term residual error as in Fig. 10, but it never reaches an error-free settled state. The CGC PD solution is much faster, settling at a very small residual steady state error. It enables the designer to then adjust the gain to meet a desirable compromise between steady state error, speed of response and an acceptable level of transient oscillation.

6 Conclusions

In this paper we set out the case for studying CGC. We suggest some potentially fruitful endeavours that could explore the search space beyond the limits of human design (Fig. 1)

- representation of plant/controller—analogue (continuous sensing and analogue circuit controllers), digital (sampled sensors and computer-based controllers)
- matching control specifications to fitness (Sect. 2.4)
- concurrent exploration of Control Strategy and Structures (Sect. 3.1)
- form and choice of CGC function sets (Sect. 4.2)

In addition, one could explore the application of CGC for bootstrapping the unconventional use of conventional approaches, such as noted with Problem 2 in Sect. 5.2.

Here we have only considered linear, time-invariant (LTI), single-input, single-output (SISO) problems. Many real-world problems are non-linear (the reality is that everything in non-linear to some degree or other), time-varying and multi-input, multi-output (MIMO). Even with LTI, MIMO problems, there is a truly vast and rich field of enquiry involving sophisticated mathematics, considerable ingenuity and a large measure of either intractability or complexity. CGP would be very capable of directing the exploration process.

References

1. Clarke, T., Davies, R.: Robust eigenstructure assignment using the genetic algorithm and constrained state feedback. Proc. Inst. Mech. Eng. Part I—J. Syst. Control Eng. **211**(1), 53–61 (1997)
2. Dorf, R.C., Bishop, R.H.: Modern Control Systems, 13th edn. Pearson (2017)
3. Dutton, K., Thompson, S., Barraclough, B.: The Art of Control Engineering, 1st edn. Addison-Wesley Longman Publishing Co., Inc, Boston, MA, USA (1997)
4. Graham, D., Lathrop, R.C.: The synthesis of "optimum" transient response: criteria and standard forms. Trans. Am. Inst. Electr. Eng. Part II: Appl. Ind. **72**(5), 273–288 (1953)
5. Koza, J.R., Keane, M.A., Jessen, Y., Bennett III, F.H., Mydlowec, W.: Automatic creation of human-competitive programs and controllers by means of genetic programming. Genet. Program. Evolvable Mach. **1**(1–2), 121–164 (2000)
6. Miller, J.F., Thomson, P.: Cartesian genetic programming. In: Genetic Programming, European Conference, Edinburgh, Scotland, UK, April 15–16, 2000, Proceedings, pp. 121–132 (2000)
7. Nise, N.S.: Control Systems Engineering, 7th edn. Wiley (2015)
8. Schultz, W.C., Rideout, V.C.: Control system performance measures: past, present, and future. IRE Trans. Autom. Control AC-6 **1**:22–35 (1961)
9. Singla, N., O'Sullivan, J.A.: Influence of pit-shape variation on the decoding performance for two-dimensional optical storage (TwoDOS). In: 2006 IEEE International Conference on Communications, vol. 7, pp. 3185–3190, June 2006

Combining Local and Global Search: A Multi-objective Evolutionary Algorithm for Cartesian Genetic Programming

Paul Kaufmann and Marco Platzner

Abstract This work investigates the effects of the periodization of local and global multi-objective search algorithms. We rely on a model for periodization and define a multi-objective evolutionary algorithm adopting concepts from Evolutionary Strategies and NSGAII. We show that our method excels for the evolution of digital circuits on the Cartesian Genetic Programming model as well as on some standard benchmarks such as the ZDT6, especially when periodized with standard multi-objective genetic algorithms.

1 Introduction

Pareto-based multi-objective genetic algorithms show excellent performance when optimizing for multiple and often conflicting goals. In our work, we are interested in multi-criteria optimization of digital hardware [12, 16] using the Cartesian Genetic Programming model [20] to represent circuits. Experience shows that for this specific application domain global multi-objective genetic optimizers can be rather slow, especially when compared with local Evolutionary Strategy (ES) techniques [17]. However, in the presence of multiple objectives local search techniques typically work with fitness functions that are linear combinations of the single objectives, rather than with the Pareto-based principle. Linear weighting schemes need to be balanced by a human designer each time there is a new set of goal functions, in order to obtain the highest possible performance. The demand for an unsupervised and preference-free multi-objective method for CGP is a challenge we address in the presented work.

We describe a *periodization technique* that alternates the execution of global and local evolutionary optimizers [13]. The technique relies on a *periodized execution*

P. Kaufmann (✉) · M. Platzner
Computer Science Department, Paderborn University,
Warburger Str. 100, 33098 Paderborn, Germany
e-mail: Paul.Kaufmann@gmail.com

M. Platzner
e-mail: Platzner@uni-paderborn.de

© Springer International Publishing AG 2018
S. Stepney and A. Adamatzky (eds.), *Inspired by Nature*, Emergence,
Complexity and Computation 28, https://doi.org/10.1007/978-3-319-67997-6_8

model that blends algorithm properties, functionalities and convergence behaviors in a simple and straight-forward way. A typical example is the combination of global search for the early phase of an optimization run with local search for the final phase [7, 25].

Additionally, we present a local search algorithm termed *hybrid Evolutionary Strategy (hES)*, a synthesis of a standard ES and a Pareto set preserving technique, and investigate its performance when periodized with multi-objective genetic optimizers NSGAII [4] and SPEA2 [29]. The characteristic of hES is that it applies a $\mu + \lambda$ ES on the Pareto-dominant individuals obtained by a multi-objective genetic algorithm while keeping diversity in and avoiding deterioration of the Pareto set.

The remainder of this Chapter is structured as follows: Sect. 2 presents related work on hybrid evolutionary search techniques. Our periodization model is defined in Sect. 3, followed by a discussion of the hybrid Evolutionary Strategy (hES) in Sect. 4. Section 5 defines the fitness metrics used in our experiments and Sect. 6 shows the benchmarks and presents the results. Finally, Sect. 7 concludes the work.

2   Related Work

Early work on multi-objective optimization of CGP was presented by Kaufmann and Platzner in [14]. The authors used NSGAII and SPEA2 for the optimization of Boolean circuits for functional quality, area, and delay. In the following years the authors have refined their method in [11, 24]. In the meantime, Walker et al. have also proposed a similar approach for the optimization of digital circuits regarding multiple objective in [27].

Hernández-Díaz et al. [7] presented a two-stage multi-objective evolutionary algorithm based on Differential Evolution (DE) and Rough Sets (RS) theory. In the first stage, the authors employed a fast converging multi-objective DE scheme to compute an initial Pareto frontier approximation. In the second stage, they improve the Pareto set diversity using RS theory for detecting loosely-covered regions. The algorithm's performance is verified on the standard ZDT$\{1, \ldots, 6\}$ and DTLZ $\{1, \ldots, 4\}$ benchmarks [6, 19]. To compare the computed Pareto sets, the authors used three metrics, the unary additive epsilon indicator [32], the standard deviation of crowding distances (SDC) [5] and the space covered by a Pareto set [31]. The proposed algorithm generally outperformed NSGAII, except on the DTLZ2 and DTLZ4 benchmarks using the SDC metric.

Talbi et al. [25] proposed a similar two-stage approach and used a multi-objective genetic algorithm (GA) to calculate a first rough Pareto frontier approximation, followed by a local search technique for refining the approximation. The authors observed improved behavior to a GA-only approach as soon as the complexity of the test problems increases.

Zapotecas et al. [30] presented a hybrid approach combining the global optimizer NSGAII with the local optimizers of Nelder and Mead [21] and the golden section method. The authors enhanced the exploratory NSGAII by local search methods in order to reduce the number of fitness evaluations. The hybrid algorithm was compared to standard NSGAII on continuous benchmarks ZDT$\{1, \ldots, 4\}$, ZDT6 and DTLZ$\{1, 2\}$ using the metrics inverted generational distance [26], spacing [22] and coverage indicator [28]. With the exception of the ZDT6 and DTLZ$\{1, 2\}$ benchmarks in combination with the spacing metric, the hybrid algorithm outperformed NSGAII.

Harada et al. [8] analyzed *GA-and-LS* and *GA-then-LS* schemata in which local search is applied either after each generation or after a completed run of a genetic algorithm. The authors concluded that *GA-then-LS* is superior to *GA-and-LS* on multiple benchmarks and used generational and Pareto-optimal frontier distances [5] for comparison.

The work of Ishibuchi et al. [9, 10] is close to our approach. The authors discuss various implementations of standard multi-objective optimizers such as SPEA2 and NSGAII combined with local search. The key idea of their approach is to periodically swap between different optimizers during a run. The authors conclude that the performance of such a hybrid optimizer is sensitive to the balance between global and local search. However, by carefully weighting global and local search strategies the periodized hybrid optimizer outperformed the standard multi-objective optimizer.

In our work, which base on the work of Kaufmann et al. [13], we investigate hybrid Evolutionary Strategies (hES) and its periodization with the multi-objective optimizers NSGAII and SPEA2 in a *GA-and-LS* manner.

3   The Periodization Model

Let $A = (a_1, a_2, \ldots, a_n)$ be the set of algorithms used in the periodization. As an illustrative example, consider $A = \{GA1, GA2, LS\}$. For a hypothetical periodized algorithm that executes a single step/generation of GA1, followed by two steps of LS, then a single step of GA2 and two steps of LS, the index sequence I for the algorithm selection is given by (a_1, a_3, a_2, a_3), and the repetition sequence F is $(f_1, f_2, f_3, f_4) = (1, 2, 1, 2)$. While in this specific example, F is a vector of constants, the number of repetitions can be adaptively adjusted based on the history of the optimization run $\mathscr{H}$. In particular, global search GAs with fast convergence in the beginning of an optimization run could be repeated more often in the early search phases, while local search algorithms that excel at improving nearly optimal nondominated sets could be used more intensively in the final optimization phase.

With t as the current generation number, $\mathscr{H}$ as the history of the current optimization run, $A = (a_1, a_2, \ldots, a_n), n \in \mathbb{N}$ as the set of algorithms used in the periodization, $I = (i_1, i_2, \ldots, i_m), m \in \mathbb{N}, i_k \in (1, 2, \ldots, n)$ as the set of indices for the selected algorithms in the execution sequence, and $F = (f_1, f_2, \ldots, f_m), f_k(t, \mathscr{H}) \to \mathbb{N}$ as the number of repetitions for the algorithms in I, the complete *periodized execution model P* is defined as:

$$P := A_I^F = (d_{i_1}^{f_1(t,\mathcal{H})}, d_{i_2}^{f_2(t,\mathcal{H})}, \ldots, d_{i_m}^{f_m(t,\mathcal{H})}).$$

The history $\mathcal{H}$ can be large if considering the complete information of an optimization run, or more compact if considering, for example, only the dominated space of the current nondominated set. In our experiments, we choose $f_k(t, \mathcal{H})) := f_k(t) \equiv const$. For the general case, however, $\mathcal{H}$ changes with each generation. Accordingly, the number of algorithm repetitions $f_k(t, \mathcal{H}_t)$ computed in generation t for algorithm a_k may differ from the $f_k(t + 1, \mathcal{H}_{t+1})$ computed in the next generation. Therefore, the repetition vector F needs to be updated after each generation. An example where this effect becomes relevant is when some algorithm is iterated until local convergence occurs. That is, if for l algorithm repetitions the best individual or the nondominated area does not change, the periodization scheme proceeds with the next algorithm. However, if the population can be improved, the algorithm is executed again for at least l generations.

4 Hybrid Evolutionary Strategies

Evolutionary Strategies in their original form rely solely on a mutation operator. The $\{\mu \overset{,}{+} \lambda\}$ ES uses μ parents to create λ offspring individuals and selects μ new parents from all individuals in case of a '+' variant or from the new individuals in case of the ',' variant.

hES is a $1 + \lambda$ ES designed for periodization with multi-objective evolutionary algorithms. In particular, we include two concepts from the Elitist Nondominated Sorting GA II in hES: fast nondominated sorting and crowding distance as a diversity metric. Fast nondominated sorting calculates nondominated sets for the objective space points. The *crowding distance* for a point is defined as the volume of a hypercube bounded by the adjoining points in the same nondominated set. Consequently, the crowding distance creates an order, denoted by $\prec_n$, on the points of a nondominated set. hES uses fast nondominated sorting to decompose parents and offspring individuals into nondominated sets, and uses crowding distances to decide which of the individuals might be skipped in order to keep the nondominated set diverse. In summary, the key ideas are:

1. A local search style algorithm is executed for every element of a given set of solutions. Exactly one individual, which is nondominated, from a parent and its offspring individuals proceeds to the next population.
2. Offspring individuals that are mutually nondominated to their parent but have a different Pareto vector are skipped. This prevents unnecessary fluctuations in the nondominated set.
3. Neutral genetic drift, as presented by Miller in [20], is achieved by skipping a parent if at least one of its offspring individuals holds an equal Pareto vector.
4. Parents and offspring individuals are partitioned into nondominated sets and new parents are selected using NSGAII's crowding distance metric.

Algorithm 1 shows the pseudocode of an hES implementation, hES-step. The algorithm starts with the creation of offspring individuals in lines 1–4. To this end, for every individual in the parent population P_t, hES-step executes $1 + \lambda$ ES appending the newly created offspring individuals to Q_t. The $1 + \lambda$ ES loop is implemented by the ES-generate in Algorithm 1. After the offspring individuals are created, hES-step proceeds with the concatenation of parents and offspring individuals by calling the add-replace procedure, listed in Algorithm 3. add-replace clones the parent population and successively adds offspring individuals that have a unique Pareto vector to this population. An offspring individual with a Pareto vector identical to its parent replaces the parent. Then, hES-step partitions the concatenated set R_t in line 6 into nondominated sets $\mathcal{F}_i$ using NSGAII's fast-nondominated-sort. After that, starting with the dominant set $\mathcal{F}_1$, the algorithm partitions $\mathcal{F}_1$ by the parents into $G = \{G_1, G_2, \ldots\}$. That means all individuals of G_i have the same parent p. Additionally, if $p \in \mathcal{F}_1$, then $p \in G_i$. Should a non-empty set G_i not contain the parent p, one of the least crowded individuals of G_i is selected to proceed to the next generation. Otherwise, the parent proceeds to the next generation. Once p or one of its offspring individuals is transferred to the next generation, p and all of its offspring individuals are skipped by hES-step from further processing in the currect generation.

Algorithm 1: hES-step (λ, P_t) —perform a single hES step

    Input: λ, parent population P_t
    Output: new archive P_{t+1}
1 $Q_t \leftarrow \emptyset$
2 **foreach** $p \in P_t$ **do**
3    $\mid$ $Q_t \leftarrow Q_t \cup$ ES-generate(p, λ)
4 **end**
5 $R_t \leftarrow$ add-replace(P_t, Q_t)
6 $\mathcal{F} \leftarrow$ fast-nondominated-sort(R_t)
7 $P_{t+1} \leftarrow \emptyset$
8 **foreach** $\mathcal{F}_i \in \mathcal{F}$ **do**
9   crowding-distance-assignment$(\mathcal{F}_i)$
10     $\mathcal{G} \leftarrow$ group-ordered-by-parent$(\mathcal{F}_i)$
11   **foreach** $\mathcal{G}_j \in \mathcal{G}$ **do**
12   **if** *parent of $\mathcal{G}_j$ not already replaced* **then**
13   **if** $parent(\mathcal{G}_j) \in \mathcal{G}_j$ **then**
14            $P_{t+1} \leftarrow P_{t+1} \cup \{parent(\mathcal{G}_j)\}$
15   **else**
16   sort$(\mathcal{G}_j, \prec_n)$
17            $P_{t+1} \leftarrow P_{t+1} \cup \{\mathcal{G}_j[0]\}$
18   **end**
19   mark parent of $\mathcal{G}_j$ as replaced
20   **end**
21   **end**
22 **end**

Algorithm 2: ES-generate (p,λ) —generate λ offspring individuals

---

Input: parent p, number of offspring individuals λ
Output: offspring set Q
1　$Q \leftarrow \emptyset$
2　**for** $i \leftarrow 1$ **to** λ **do**
3　　　$p' \leftarrow$ mutate(p)
4　　　$Q \leftarrow Q \cup \{p'\}$
5　**end**

Algorithm 3: add-replace (P,Q) —return copy of P joint by Q, replace parents in P by offspring individuals in Q with equal Pareto vectors, avoid adding multiple offspring individuals with equal Pareto vectors.

Input: sets P, Q
Output: set R
1　$R \leftarrow P$
2　**foreach** $q \in Q$ **do**
3　　　**if** $\nexists r \in R : r \preceq q \wedge q \preceq r$ **then**
4　　　　$R \leftarrow R \cup \{q\}$
5　　　**end**
6　　　**if** $\exists r \in R : r \preceq q \wedge q \preceq r \wedge parent(\{q\}) == r$ **then**
7　　　　$R \leftarrow R \cup \{q\}$
8　　　　$R \leftarrow R \backslash \{r\}$
9　　　**end**
10　**end**

5　Performance Assessment

To analyze the performance of multi-objective optimizers, we need to compare the calculated Pareto sets. In this work we employ two methods: the ranking of Pareto sets by a quality indicator and the analysis of the mean Pareto set, attained during multiple runs. Both methods are described by Knowles et al. [18] and are also implemented in the PISA toolbox by Bleuler et al. [1].

5.1　Quality Indicators

To compare Pareto sets, Zitzler et al. [32] introduced the concept of a Quality Indicator (QI) as a function mapping a set of Pareto sets to a set of real numbers. Under QI, the Pareto sets define a relation on the Pareto set quality. In our work, we use the unary additive epsilon indicator $I_{\varepsilon+}^1$. It is based on the binary additive epsilon indicator $I_{\varepsilon+}$ which is defined for two Pareto sets A and B as:

$$I_{\varepsilon+}(A, B) = \inf_{\varepsilon \in \mathbb{R}} \{\forall b \in B \,\exists a \in A : a \preceq_{\varepsilon+} b\}.$$

Table 1 Interpretation of the Kruskal-Wallis test: given the Kruskal-Wallis test rejects H_0, a dot denotes a p-value higher than α

	A_1	A_2	A_3
A_1	–	0.002	0.007
A_2	.	–	.
A_3	.	0.003	–

Here, the relation $\preceq_{\varepsilon+}$ is defined as $a \preceq_{\varepsilon+} b \Leftrightarrow \forall i : a_i \le \varepsilon + b$. For a reference Pareto set R, the unary additive epsilon indicator $I_{\varepsilon+}^1$ can be now derived as

$$I_{\varepsilon+}^1(A) = I_{\varepsilon+}(A, R).$$

Following Knowles et al. [18], we use the non-parametric Kruskal-Wallis (KW) test [2] to statistically evaluate sequences of quality numbers. The Kruskal-Wallis test differentiates between the null hypothesis $H_0 = $ "The distribution functions of the sequences are identical" and the alternative hypothesis $H_A = $ "At least one sequence tends to yield better observations than another sequence". In case the test rejects H_0, we provide for all sequence pairs the one-tailed p-value. Table 1 presents an example: for an algorithm tuple $(A_{\text{row}}, A_{\text{col}})$ a p-value equal or below α indicates a lower mean for A_{row}. Thus, one can conclude for Table 1 that A_1 outperforms A_2 and A_3, and A_3 outperforms A_2. In our experiments, we configure the significance level α to 0.01.

5.2 Empirical Attainment Functions

An additional way of interpreting the results of multi-objective optimizers is to look at the Pareto points that are covered, i.e., weakly dominated, with a certain probability during the multiple repetitions of an optimization algorithm. All Pareto points that have been reached in $x\%$ of the runs are referred to as the $x\%$-attainment. The attainment allows for a direct graphical interpretation as shown in the examples of Figs. 2 and 3.

In order to statistically compare the attainments we use the two-tailed Kolmogorov-Smirnov test [23]. It distinguishes between $H_0 = $ "Sequences A and B follow the same distribution" and $H_A = $ "Sequences A and B follow different distributions". Table 2 contains exemplary results for the Kolmogorov-Smirnov (KS) test. It can be interpreted as: A_1 differs significantly from A_2 and A_3. In our experiments, we configure the significance level α to 0.05.

Table 2 Interpretation of the Kolmogorov-Smirnov test: A dot denotes an accepted H_0 hypothesis at the given α

	A_1	A_2	A_3
A_1	–	*	*
A_2	*	–	.
A_3	*	.	–

* indicates significantly different nondominated set distributions

6 Evaluation

We experimented with several benchmarks to compare hES and the periodized variants of hES, NSGAII and SPEA2. At first, we used the standard benchmarks for multi-objective algorithms DTLZ{2, 6} and ZDT6. These benchmarks are available with the PISA toolbox [1] and are described in [19]. Second, we compared our algorithms on the evolution of digital circuits, i.e., even 5- and 7-parity and $(2, 2)$ and $(3, 3)$ adders and multipliers, using Cartesian Genetic Programming (CGP) [20] as the hardware representation model. Figure 1 illustrates the CGP phenotype. Besides the functional quality of the digital circuit, which in this case is set as a constraint, we select the circuit's area and speed to define a multi-objective benchmark [15].

In our experiments we executed 20 repetitions for every combination of goal function and algorithm. For the hES, ES-generate produces 32 offspring individuals for each parent. For the other benchmarks, Algorithm 1 was configured to have one offspring individual per parent.

Table 3 presents the configuration of the benchmarks DTLZ{2, 6} and ZDT6. For these benchmarks, NSGAII and SPEA2 employ the SBX crossover operator [3]. The optimization runs were stopped after 10,000 fitness evaluations. Table 4 shows for the digital circuit benchmarks the CGP configurations, termination criteria, and

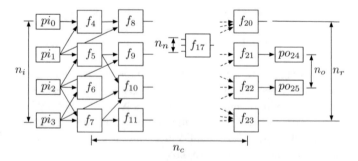

Fig. 1 Cartesian Genetic Programming (CGP) encodes a two dimensional grid of functional units connected by feed forward wires, thus forming a directed acyclic graph. The CGP model is parametrized with the number of primary inputs n_i and outputs n_o, number of rows n_r and columns n_c, number of functional block inputs n_n, the maximal length of a wire l and the functional set F that can be computed by the nodes

Table 3 DTLZ2, DTLZ6, and ZDT6 benchmark configurations

Number objectives	2
Number of decision variables	100
Individual mutation probability	1
Individual recombination probability	1
Variable mutation probability	1
Variable swap probability	0.5
Variable recombination probability	1
Eta mutation	20
Eta recombination	15
Use symmetric recombination	1

Table 4 CGP benchmark configurations. S is the termination number, measured in fitness evaluations. $|P|$ and $|A|$ denote the capacity of the parent population and the archive

	5-parity	7-parity	(2, 2) add	(2, 2) mul	(3, 3) add	(3, 3) mul
S	400, 000	800, 000	400, 000	1, 600, 000	400, 000	160, 000
n_i	5	7	4	4	6	6
n_o	1	1	4	4	6	6
n_n	2					
l	∞					
n_c	200					
n_r	1					
$|P|/|A|$	32/100 for hES, 50/100 else					
F	see Table 5					
P(mut.)	0.1					
P(rcmb.)	0.0 for hES, 0.5 else					
rcmb. type	one-point					

population sizes. We limit the functional set to the Boolean functions presented in Table 5. To simplify the nomenclature, we use the following abbreviations:

$$hES \quad \rightarrow h$$
$$SPEA2 \quad \rightarrow s$$
$$NSGAII \rightarrow n$$

Table 5 CGP configuration: functional set F

Number	Function
0	0
1	1
2	a
3	b
4	$\overline{a}$
5	$\overline{b}$
6	$a \cdot b$
7	$a \cdot \overline{b}$
8	$\overline{a} \cdot b$
9	$\overline{a} \cdot \overline{b}$
10	$a \oplus b$
11	$a \oplus \overline{b}$
12	$a + b$
13	$a + \overline{b}$
14	$\overline{a} + b$
15	$\overline{a} + \overline{b}$
16	$a \cdot \overline{c} + b \cdot c$
17	$a \cdot \overline{c} + \overline{b} \cdot c$
18	$\overline{a} \cdot \overline{c} + b \cdot c$
19	$\overline{a} \cdot \overline{c} + \overline{b} \cdot c$

6.1 Periodization of hES for DTLZ2, DTLZ6 and ZDT6

To examine the effect of local search, we first execute the standard NSGAII and SPEA2 for a given benchmark in order to determine the reference performance. Then, we increase step by step the influence of local search by periodizing NSGAII with hES until only hES is executed. In terms of our periodization model (cf. Sect. 3), we investigate the six periodization schemes: n, s, nh, nh^4, nh^{10}, and h.

Table 6 shows the results of the KW test applied to the benchmarks DTLZ{2, 6} and ZDT6 with respect to the unary additive epsilon indicator $I_{\varepsilon+}^1$ at the significance level $\alpha = 1\%$. The results of the KS test are omitted as they indicate differences significant at $\alpha = 5\%$ between the nondominated sets for almost all combinations of algorithm and benchmark.

The central observation for the DTLZ2 and DTLZ6 experiments is that the quality of the nondominated sets degrades with an increasing influence of local search. Starting with the periodization of NSGAII and hES, the KW test shows falling performance when increasing the number of hES iterations. The hES-only experiment results in the worst performance of all the algorithms. The KW test results are confirmed by the graphical interpretation of the 75% attained nondominated sets in Fig. 2a, b.

Table 6 Comparison of the nondominated sets of DTLZ2, DTLZ6, and ZDT6. A dot denotes an accepted H_0. When the KW test rejects H_0 for the algorithm pair (a_{row}, a_{col}), a one-tailed p-value lower than $\alpha = 0.01$ indicates that a_{row} evolves significantly better nondominated sets than a_{col} regarding $I_{\varepsilon+}^1$

		n	s	nh	nh^4	nh^{10}	h
KW test DTLZ2	n		·	0.0	0.0	0.0	0.0
	s	·		0.0	0.0	0.0	0.0
	nh	·	·		0.0	0.0	0.0
	nh^4	·	·	·		0.0	0.0
	nh^{10}	·	·	·	·		0.0
	h	·	·	·	·	·	
DTLZ6 KW test	n		·	0.0	0.0	0.0	0.0
	s	·		0.0	0.0	0.0	0.0
	nh	·	·		0.0	0.0	0.0
	nh^4	·	·	·		0.0	0.0
	nh^{10}	·	·	·	·		0.0
	h	·	·	·	·	·	
ZDT6 KW test	n		·	·	·	·	0.0
	s	·		·	·	·	0.0
	nh	0.0001	0.0027		·	0.0001	0.0
	nh^4	0.0	0.0	0.0014		0.0	0.0
	nh^{10}	·	·	·	·		0.0
	h	·	·	·	·	·	

For the ZDT6 benchmark, the influence of hES is not as one-sided as for the DTLZ{2,6} benchmarks. The nh periodization outperforms SPEA2 and NSGAII. Further increase in the influence of hES in the nh^4 periodization lets it dominate all other algorithms. The nh^{10} periodization is on a par with SPEA2 and NSGAII, while the execution of hES alone, as with DTLZ{2, 6}, falls behind. The 75% attainments pictured in Fig. 3 confirm this. Interestingly, despite the large gap between nh^{10} and the group of SPEA2 and NSGAII, the KW test finds no significant differences at $\alpha = 1\%$ between the corresponding indicator sequences. The KS test, similar to the results for the DTLZ{2, 6} benchmarks, reveals significant differences at $\alpha = 5\%$ for all algorithm combinations except the NSGAII and SPEA2 pair.

The DTLZ{2,6} and ZDT6 benchmarks demonstrate the various kinds of impact that local search may have when periodized with global search algorithms. To gain an insight into whether the order of the algorithms in the periodization sequence influences the results, and into how an hES-less periodizations of SPEA2 and NSGAII compares to the regular SPEA2 and NSGAII, we fixate on the ZDT6 benchmark and apply 2- and 3-tuple permutations of the NSGAII, SPEA2, and hES algorithms. All the experiments were repeated 100 times and the execution was stopped after 200 generations.

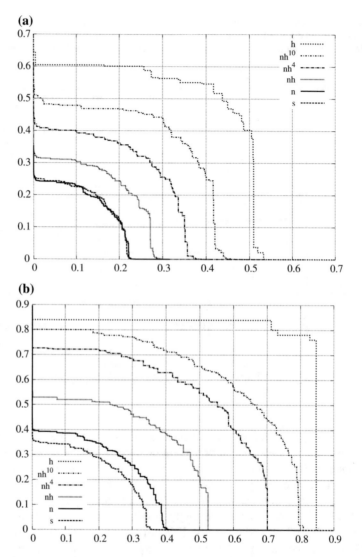

Fig. 2 75%-attainments for the 2-dimensional DTLZ2 **a** and DTLZ6 **b** benchmarks. A hypothetical optimum is located at $\vec{0}$

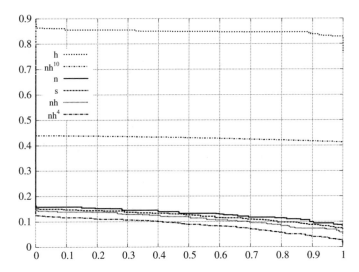

Fig. 3 75%-attainments for the 2-dimensional ZDT6 benchmarks. A hypothetical optimum is located at $\vec{0}$

Table 7 shows the results for 2-tuple combinations of NSGAII, SPEA2, and hES. The general observation taken from the KW test is that hES periodized with either NSGAII or SPEA2 outperforms standard NSGAII, SPEA2, and their combinations. Interestingly, the *hES-after-SPEA2* outperforms the *hES-after-NSGAII*, while *SPEA2-after-hES* does not. This shows that the performance of this particular periodization scheme may be sensitive to the initial order of the executed algorithms.

The KS test confirms the results observed before. There are basically two classes of algorithms, showing significantly different results: the class of algorithms periodized with hES, and the class of NSGAII, SPEA2 and their combinations. In contrast to the previous test, the differences between (hs) and (nh) are now identified as significant.

Next, Table 7 shows also the results of 3-tuple combinations of hES with NSGAII and SPEA2. Similar to the results achieved for the 2-tuple tests, all the periodized algorithms outperform (KW) and differ (KS) from NSGAII and SPEA2.

In summary, we can conclude that for the DTLZ$\{2, 6\}$ benchmarks, an increasing impact of hES reduces the quality of the nondominated sets evolved, while for the ZDT6 benchmark, the schemes periodized with hES have the dominating results.

6.2 Periodization of hES for Digital Circuit Design

In contrast to the previous benchmarks, we optimize for three objectives in this section: the functional quality, the area, and the propagation delay. In total, each

Table 7 ZDT6 nondominated sets comparison for 2- and 3-tuple combinations of NSGAII, SPEA2 and hES to NSGAII and SPEA2. A dot denotes an accepted H_0. When the KW test rejects H_0 for the algorithm pair (a_{row}, a_{col}), a one-tailed p-value lower than $\alpha = 0.01$ indicates that a_{row} evolves significantly better nondominated sets than a_{col} regarding $I^1_{\varepsilon+}$. A star indicates significantly different nondominated set distributions at $\alpha = 0.05$ according to the KS test

		n	s	nh	hn	sh	hs	ns	sn
KW test	n		·	·	·	·	·	·	·
	s	·		·	·	·	·	·	·
	nh	0.0	0.0		·	·	·	0.0	0.0
	hn	0.0	0.0	·		·	·	0.0	0.0
	sh	0.0	0.0	0.0091	·		·	0.0	0.0
	hs	0.0	0.0	·	·	·		0.0	0.0
	ns	·	·	·	·	·	·		·
	sn	·	·	·	·	·	·	·	
KS test	n		·	*	*	*	*	·	·
	s	·		*	*	*	*	·	·
	nh	*	*		·	*	*	*	*
	hn	*	*	·		·	·	*	*
	sh	*	*	*	·		·	*	*
	hs	*	*	*	·	·		*	*
	ns	·	·	*	*	*	*		·
	sn	·	·	*	*	*	*	·	

		n	s	nhs	nsh	shn	snh	hns	hsn
KW test	n		·	·	·	·	·	·	·
	s	·		·	·	·	·	·	·
	nhs	0.0	0.0		·	·	·	·	·
	nsh	0.0	0.0	·		·	·	·	·
	shn	0.0	0.0	·	·		·	·	·
	snh	0.0	0.0	·	·	·		·	·
	hns	0.0	0.0	·	·	·	·		·
	hsn	0.0	0.0	·	·	·	·	·	
KS test	n		·	*	*	*	*	*	*
	s	·		*	*	*	*	*	*
	nhs	*	*		·	·	·	·	·
	nsh	*	*	·		·	·	·	·
	shn	*	*	·	·		·	·	·
	snh	*	*	·	·	·		·	·
	hns	*	*	·	·	·	·		·
	hsn	*	*	·	·	·	·	·	

* indicates significantly different nondominated set distributions

Table 8 The number of circuits with perfect functional quality evolved during 20 runs

	$(2, 2)$ add	$(3, 3)$ add	$(2, 2)$ mul	$(3, 3)$ mul	5-parity	7-parity
n	1	0	0	0	0	0
s	6	0	7	0	0	0
nh	7	0	8	0	0	0
nh^4	9	0	11	0	0	0
nh^{10}	11	0	14	0	0	0
h	12	1	15	0	0	0

algorithm is executed 20 times for each pair of goal functions. Table 8 summarizes the number of runs that resulted in functionally correct solutions. The first observation is that for the parity, the $(3, 3)$ adder, and the $(3, 3)$ multiplier benchmarks, almost none of the algorithms managed to evolve functionally correct circuits. We focus therefore on the experiments involving the $(2, 2)$ adder and the $(2, 2)$ multiplier, when discussing the influence of local search on the evolution of correct circuits.

Despite treating equally all objectives, hES is most effective in finding functionally correct solutions. While SPEA2 outperforms NSGAII for this particular CGP configuration on the $(2, 2)$ adder and multiplier benchmarks, increasing the influence of hES in the periodization with NSGAII produces even greater success rates. nh^4 and especially nh^{10} periodization schemes reveal only a small gap with the hES-only performance. This insight is also partly confirmed by the results of the KW and the KS tests presented in Table 9. For the $(2, 2)$ adder, the KW test finds no significant differences in nondominated sets at $\alpha = 1\%$ while the KS test partitions the algorithms into groups of $\{n\}$, $\{s, nh, nh^4, nh^{10}\}$, and $\{h\}$ with significant differences in the evolved nondominated sets at $\alpha = 5\%$. For the $(2, 2)$ multiplier benchmark, NSGAII is dominated by all, and hES by SPEA2 and nh according to the KW test. The KS test splits the algorithms, similarly to what happened with the $(2, 2)$ benchmark, into groups of $\{n\}$, $\{s, nh, nh^4, nh^{10}\}$, and $\{h\}$. Additionally, in the group of $\{s, nh, nh^4, nh^{10}\}$ SPEA2 evolves different nondominated sets than it does for nh^4, nh^{10}.

The $(3, 3)$ adder and multiplier benchmarks split the algorithms into $\{n\}$ and $\{s, nh, nh^4, nh^{10}, h\}$ groups with different nondominated sets according to the KW test. The KS test reveals, similar to what happened with the $(2, 2)$ benchmarks, the same general tendency of differences between the nondominated sets evolved by the NSGAII, the hES, and the group of the remaining algorithms.

The general partitioning according to the quality of the evolved nondominated sets between NSGAII, hES, and the rest of the algorithms, is even more pronounced for the parity benchmarks, as now the KW test also confirms significant differences (Table 10). That is, SPEA2, nh, nh^4, and nh^{10} are better than NSGAII and hES for 5- and 7-parity functions and NSGAII is better than hES for the 7-parity function.

Table 9 Non-dominated sets comparison: A dot denotes an accepted H_0. When the KW test rejects H_0 for the algorithm pair (a_{row}, a_{col}), a one-tailed p-value lower than $\alpha = 0.01$ indicates that a_{row} evolves significantly better nondominated sets than a_{col} regarding $I_{\varepsilon+}^1$. A star indicates significantly different nondominated set distributions at $\alpha = 0.05$ according to the KS test

		n	s	nh	nh^4	nh^{10}	h
(2,2) add KW test	n		·	·	·	·	·
	s	·		·	·	·	·
	nh	·	·		·	·	·
	nh^4	·	·	·		·	·
	nh^{10}	·	·	·	·		·
	h	·	·	·	·	·	
(2,2) add KS test	n		*	*	*	*	*
	s	*		·	·	·	*
	nh	*	·		·	·	*
	nh^4	*	·	·		·	*
	nh^{10}	*	·	·	·		*
	h	*	*	*	*	*	
(3,3) add KW test	n		·	·	·	·	·
	s	0.0067		·	·	·	·
	nh	0.0004	·		·	·	·
	nh^4	0.0011	·	·		·	·
	nh^{10}	0.0	·	·	·		·
	h	0.0	·	·	·	·	
(3,3) add KS test	n		·	*	*	*	*
	s	·		·	·	·	*
	nh	*	·		·	·	*
	nh^4	*	·	·		·	*
	nh^{10}	*	·	·	·		*
	h	*	*	*	*	*	
(2,2) mul KW test	n		·	·	·	·	·
	s	0.0		·	·	·	0.0
	nh	0.0	·		·	·	0.0013
	nh^4	0.0	·	·		·	·
	nh^{10}	0.0	·	·	·		·
	h	0.0016	·	·	·	·	
(2,2) mul KS test	n		*	*	*	*	*
	s	*		·	*	*	*
	nh	*	·		·	·	*
	nh^4	*	·	·		·	*
	nh^{10}	*	*	·	·		*
	h	*	*	*	*	*	
(3,3) mul KW test	n		·	·	·	·	·
	s	0.0006		·	·	·	·

(continued)

Table 9   (continued)

		n	s	nh	nh^4	nh^{10}	h
	nh	0.0	·		·	·	·
	nh^4	0.0	·	·		·	·
	nh^{10}	0.0	·	·	·		·
	h	0.0002	·	·	·	·	
$(3, 3)$ mul KS test	n		*	*	*	*	*
	s	*		·	·	*	*
	nh	*	·		·	·	*
	nh^4	*	·	·		·	*
	nh^{10}	*	*	·	·		*
	h	*	*	*	*	*	

* indicates significantly different nondominated set distributions

Table 10   Non-dominated sets comparison: A dot denotes an accepted H_0. When the KW test rejects H_0 for the algorithm pair (a_{row}, a_{col}), a one-tailed p-value lower than $\alpha = 0.01$ indicates that a_{row} evolves significantly better nondominated sets than a_{col} regarding $I^1_{\varepsilon+}$. A star indicates significantly different nondominated set distributions at $\alpha = 0.05$ according to the KS test

		n	s	nh	nh^4	nh^{10}	h
5-parity KW test	n		·	·	·	·	·
	s	0.0		·	·	·	0.0
	nh	0.0	·		·	·	0.0
	nh^4	0.0	·	·		·	0.0
	nh^{10}	0.0	·	·	·		0.0
	h	·	·	·	·	·	
5-parity KS test	n		*	*	*	*	*
	s	*		·	·	·	*
	nh	*	·		·	*	*
	nh^4	*	·	·		·	*
	nh^{10}	*	·	*	·		*
	h	*	*	*	*	*	
7-parity KW test	n		·	·	·	·	0.0054
	s	0.0		·	·	·	0.0
	nh	0.002	·		·	·	0.0
	nh^4	0.0	·	·		·	0.0
	nh^{10}	0.0	·	·	·		0.0
	h	·	·	·	·	·	
7-parity KS test	n		*	*	*	*	*
	s	*		·	·	·	*
	nh	*	·		·	·	*
	nh^4	*	·	·		·	*
	nh^{10}	*	·	·	·		*
	h	*	*	*	*	*	

* indicates significantly different nondominated set distributions

The KS test finds significant differences between the three groups, also finding a significant difference between nh and nh[10] for 5-parity.

In summary, we can state that periodizations of hES with NSGAII, as well as the non-periodized SPEA2, create, for almost all benchmarks, nondominated sets which are better than those of the non-periodized hES and NSGAII. Additionally, with the increasing influence of hES in a periodization scheme, the probability for the evolution of correct CGP circuits increases.

7 Conclusion

In this work, we investigated the periodization of multi-objective local and global search algorithms. For this, we relied on a periodized execution model and on the hybrid Evolutionary Strategies as a local search technique tailored to periodization with Pareto-based genetic multi-objective optimizers such as NSGAII and SPEA2.

The results show that for the DTLZ{2, 6} benchmarks, hES and its periodization with NSGAII underperforms. For ZDT6 and, most importantly, for the evolution of digital circuit benchmarks on the CGP model, hES and its periodizations are significantly better than the reference algorithms NSGAII and SPEA2. Furthermore, the periodized execution model proved to be a simple, fast and flexible approach to combine multiple optimization algorithms for merging functional and behavior properties. Thus, blending multi- and single-objective optimizers, local and global search algorithms and differently converging methods creates a new family of optimization algorithms.

References

1. Bleuler, S., Laumanns, M., Thiele, L., Zitzler, E.: PISA—a platform and programming language independent interface for search algorithms. In: Intlernational Conference on Evolutionary Multi-Criterion Optimization (EMO) LNCS, pp. 494–508. Springer (2003)
2. Conover, W.J., Practical Nonparametric Statistics (3rd edn.). Wiley (1999)
3. Deb, K., Agrawal, R.B.: Simulated binary crossover for continuous search space. Complex Syst. **9**, 115–148 (1995)
4. Deb, K., Agrawal, S., Pratap, A., Meyarivan, T.: A fast elitist non-dominated sorting genetic algorithm for multi-objective optimisation: NSGA-II. In: Parallel Problem Solving from Nature (PPSN'00), pp. 849–858. Springer (2000)
5. Deb, K., Multi-Objective Optimization Using Evolutionary Algorithms. Wiley, Inc (2001)
6. Deb, K.,Thiele, L., Laumanns, M., Zitzler, E.: Scalable test problems for evolutionary multi-objective optimization. In: Evolutionary Multiobjective Optimization: theoretical Advances and Applications, chap. 6, pp. 105–145. Springer (2005)
7. García, A., Díaz, H., Luis, V., Quintero, S., Carlos, A., Coello, C., Caballero, R., Luque, J.M.: A new proposal for multi-objective optimization using differential evolution and rough sets theory. In: Genetic and Evolutionary Computation (GECCO), pp. 675–682. ACM (2006)

8. Harada, K., Ikeda, K., Kobayashi, S.: Hybridization of genetic algorithm and llocal search in multiobjective function optimization: recommendation of GA then LS. In: Genetic and Evolutionary Computation (GECCO), pp. 667–674. ACM (2006)

9. Ishibuchi, H., Narukawa, K.: Some issues on the implementation of local search in evolutionary multiobjective optimization. In: Genetic and Evolutionary Computation (GECCO), LNCS, pp. 1246–1258. Springer (2004)

10. Ishibuchi, H., Yoshida, T., Murata, T.: Balance between genetic search and local search in hybrid evolutionary multi-criterion optimization Algorithms. In: Genetic and Evolutionary Computation (GECCO), pp. 1301–1308. Morgan Kaufmann Publishers (2002)

11. Paul, K.: Adapting Hardware Systems by Means of Multi-Objective Evolution. Logos Verlag, Berlin (2013)

12. Knieper, T., Defo, B., Kaufmann, P., Platzner, M.: On robust evolution of digital hardware. In: Biologically Inspired Collaborative Computing (BICC), vol. 268 of IFIP International Federation for Information Processing, pp. 2313–222. Springer (2008)

13. Kaufmann, P., Knieper, T., Platzner, M.: A novel hybrid evolutionary strategy and its periodization with multi-objective genetic optimizers. In: IEEE World Congress on Computational Intelligence (WCCI), Congress on Evolutionary Computation (CEC), pp. 541–548. IEEE (2010)

14. Kaufmann, P., Platzner, M.: Multi-objective Intrinsic Hardware Evolution. In: International Conference Military Applications of Programmable Logic Devices (MAPLD) (2006)

15. Kaufmann, P., Platzner, M.: MOVES: a modular framework for hardware evolution. In: IEEE Adaptive Hardware and Systems (AHS), pp. 447–454. IEEE (2007)

16. Kaufmann, P., Platzner, M.: Toward self-adaptive embedded systems: multi-objective hardware evolution. In: Architecture of Computing Systems (ARCS), vol. 4415 of LNCS, pp. 199–208. Springer (2007)

17. Kaufmann, P., Plessl, C., Platzner, M.: EvoCaches: application-specific adaptation of cache mappings. In: IEEE Adaptive Hardware and Systems (AHS), pp. 11–18. IEEE, CS (2009)

18. Knowles, J., Thiele, L., Zitzler, E.: A Tutorial on the Performance Assessment of Stochastic Multiobjective Optimizers. Technical report, Computer Engineering and Networks Laboratory (TIK), ETH Zurich, Switzerland (2006)

19. Martello, S., Toth, P.: Knapsack Problems: Algorithms and Computer Implementations. Wiley, Inc (1990)

20. Miller, J., Thomson, P.: Cartesian genetic programming. In: European Conference on Genetic Programming (EuroGP), pp. 121–132. Springer (2000)

21. Nelder, J.A., Mead, R.: A simplex method for function minimization. Comput. J. **7**(4), 308–313 (1965)

22. Scott, J.R.: Fault Tolerant Design Using Single and Multicriteria Genetic Algorithm Optimization. Master's thesis, Department of Aeronautics and Astronautics. Massachusetts Institute of Technology (1995)

23. Shaw, K.J., Nortcliffe, A.L., Thompson, M., Love, J., Fonseca, C.M.,. Fleming, P.J.: Assessing the performance of multiobjective genetic algorithms for optimization of a batch process scheduling problem. In: Evolutionary Computation, pp. 37–45. IEEE (1999)

24. Lukas, K., Walker, J.A., Kaufmann, P., Plessl, C., Platzner, M.: Evolution of Electronic Circuits. Cartesian Genetic Programming. Natural Computing Series, pp. 125–179. Springer, Berlin (2011)

25. Talbi, El-G., Rahoual, M., Mabed, M.H., Dhaenens, M.C: A hybrid evolutionary approach for multicriteria optimization problems: application to the flow shop. In: International Conference on Evolutionary Multi-Criterion Optimization (EMO), pp. 416–428. Springer (2001)

26. David, A, Veldhuizen, V.: Multiobjective evolutionary algorithms: classifications, analyses, and new innovations. PhD thesis, Department of Electrical and Computer Engineering. Airforce Institute of Technology (1999)

27. Walker, J.A., Hilder, J.A., Tyrrell, A.M: Towards evolving industry-feasible intrinsic variability tolerant cmos designs. In: 2009 IEEE Congress on Evolutionary Computation, pp. 1591–1598 (May 2009)

28. Zitzler, E., Deb, K., Thiele, L.: Comparison of multiobjective evolutionary algorithms: empirical results. In: Evolutionary Computation, vol. 8(2), pp. 173–195. MIT Press (2000)
29. Zitzler, E., Laumanns, M., Thiele, L.: SPEA2: improving the strength pareto evolutionary algorithm. Tech. Rep. 103, ETH Zurich (2001)
30. Martínez, S.Z., Carlos, A., Coello, C.: A proposal to hybridize multi-objective evolutionary algorithms with non-gradient mathematical programming techniques. In: Parallel Problem Solving from Nature (PPSN'08), pp 837–846. Springer (2008)
31. Zitzler, E., Thiele, L.: Multiobjective evolutionary algorithms: a comparative case study and the strength pareto approach. In: IEEE Transcations on Evolutionary Computation, vol. 3(4), pp. 257–271. IEEE 1999
32. Zitzler, E., Thiele, L., Laumanns, M., Fonseca, C.M., da Fonseca, V.G.: Performance assessment of multiobjective optimizers: an analysis and review. IEEE Trans. Evol. Comput. 7(2), 117–132 (2003)

Approximate Computing: An Old Job for Cartesian Genetic Programming?

Lukas Sekanina

Abstract Miller's Cartesian genetic programming (CGP) has significantly influenced the development of evolutionary circuit design and evolvable hardware. We present key ingredients of CGP with respect to the efficient search in the space of digital circuits. We then show that approximate computing, which is currently one of the promising approaches used to reduce power consumption of computer systems, is a natural application for CGP. We briefly survey typical applications of CGP in approximate circuit design and outline new directions in approximate computing that could benefit from CGP.

1 Introduction

Julian F. Miller, a pioneer of genetic programming (GP), evolvable hardware (EHW), evolution in materio and other approaches that could be together classified as unconventional computing paradigms, is best known as the (co)inventor of Cartesian genetic programming (CGP) [25]. Genetic programming is a computational intelligence method capable of the automated designing of programs. It has been developed since the eighties, with the significant contribution of John Koza who mainly worked on a tree-based GP in which candidate solutions are syntactic trees [20]. In Miller's version of genetic programming, CGP, candidate solutions are represented using directed acyclic graphs encoded as finite-size strings of integers. This representation is especially useful for evolving electronic circuits as it naturally captures the circuit physical structure and supports multiple outputs, subgraph sharing and various types of elementary circuit components.

CGP has predominantly been developed within two research communities: genetic programming and evolvable hardware. For the genetic programming community, interesting concepts behind CGP are the specific genotype-phenotype mapping, mutation-driven search over small populations, role of neutrality and bloat free

L. Sekanina (✉)
Faculty of Information Technology, Brno University of Technology,
IT4Innovations Centre of Excellence, Brno, Czech Republic
e-mail: sekanina@fit.vutbr.cz

© Springer International Publishing AG 2018
S. Stepney and A. Adamatzky (eds.), *Inspired by Nature*, Emergence,
Complexity and Computation 28, https://doi.org/10.1007/978-3-319-67997-6_9

search. In the last decade, CGP was extended in various directions, for example, the evolution of modular code was supported and self-modifying CGP enabled the evolved programs to change themselves as a function of time. For the evolvable hardware community, CGP has represented a natural way to model and evolve digital circuits, not only in a circuit simulator but also directly on a chip. New circuit designs, including patented solutions,[1] were routinely obtained by means of CGP. Digital architectures capable of autonomous adaptation and self-repair were constructed, where CGP was responsible for autonomous changes of the circuits involved. CGP-based results have also been published outside the major evolutionary computing or evolvable hardware events, thus influencing people focusing on the mainstream circuit design approach.

In computer engineering, the following trends can be observed [7, 22]:

- Energy efficiency is strongly requested, especially for omnipresent wireless battery-powered systems (such as smart phones, consumer electronics and wireless sensor networks) and, on the other hand, for energy demanding data and supercomputer centers processing Big Data.
- Computationally-demanding applications (e.g. those based on deep neural networks) are implemented even in small battery powered embedded systems.
- Integrated circuits developed using recent fabrication technology exhibit reliability issues and uncertainties, especially when operated at low voltages.
- Massive parallelism available by implementing billions of transistors on a single die cannot be fully employed as only a fraction of them can be activated at the same time in order to prevent burning the chip.

One of the approaches capable of coping (at least partly) with the aforementioned issues is *approximate computing* [31]. It exploits the fact that many applications are inherently error resilient. If some errors can be tolerated, then the underlying hardware and software can be simplified to be faster or less energy consuming. Multimedia, datamining and machine learning applications are good candidates for the approximation because possible small imperfections in data processing usually have only a small impact on the quality of result. The approximate hardware and software system design can be viewed as an optimization problem in which a good trade-off is sought among key design variables—performance, energy and error. This optimization problem can be attacked with CGP [60, 61]. A real challenge is to employ approximate computing for the most advanced technology nodes, where transistors parameters exhibit unusual variability. One of the approaches is, by graceful performance degradation at run-time and in the presence of uncertainties or errors, to deliver the best possible trade-off between the quality of result and energy consumption for all components on a chip. It has to be noted that this new notion of approximation, which was established around 2010, is different from the approximations that have been conducted for decades in the fields of computer engineering (such as approximate signal processing by [38]) and computer science (such as approximate algorithms by [64]).

---

[1]Nonlinear image filter, Patent No. 304181, Czech Republic, 2013.

The aim of this chapter is to emphasize Miller's contribution to the evolutionary design of gate level circuits. In particular, we highlight some of the ingredients of CGP that are crucial for an efficient search, especially in the space of digital circuits (Sect. 2). We show that some of Miller's results in the area of evolutionary circuit design that were obtained over 15 years ago can be interpreted as vital contributions to the current research in approximate computing (Sect. 3). We briefly survey current applications of CGP in approximate circuit design (Sect. 4). The last part of this chapter is devoted to new directions in approximate computing that could benefit from CGP (Sect. 5). Final remarks are given in Sect. 6.

2   Cartesian Genetic Programming

According to Julian F. Miller:

Cartesian genetic programming grew from a method of evolving digital circuits developed by Miller et al. in 1997 [29]. However the term 'Cartesian genetic programming' first appeared in 1999 [23] and was proposed as a general form of genetic programming in 2000 [28].

In CGP, candidate solutions are represented in a two-dimensional array of programmable nodes. While the phenotype is represented using a directed oriented graph, the genotype is encoded in a fixed-size string of integers. In this section, key ingredients of CGP are briefly introduced. The aim is not to provide a detailed description (which is available in [25]), but to emphasize important consequences for practical evolutionary circuit design. We primarily deal with combinational gate-level circuits.

2.1   Circuit Representation

An n_i-input and n_o-output combinational circuit is modelled using an array of $n_c \cdot n_r$ programmable nodes forming a Cartesian grid. A set of available n_a-input node functions is denoted Γ, where elementary logic functions are included for gate-level circuits. The levels-back parameter l constraints which columns a node can get its inputs from. No feedbacks are allowed in the basic version of CGP. The primary inputs and programmable nodes are uniquely numbered. For each node the chromosome contains (n_a+1) values that represent the node function and n_a addresses specifying the input connections. The chromosome also contains n_o values specifying the gates connected to the primary outputs. The chromosome size is $n_c n_r (n_a + 1) + n_o$ integers.

The circuit representation used by CGP was not invented by Miller as there are papers from the early nineties utilizing the same concept [21, 37]. An important property of this representation is its extensibility to support different types of components (such as gates, 3-electrode transistors, or 6-input look-up tables), recurrent networks [51], transistor-level circuits [68], modular circuits [14, 67], decomposition strategies [49] and self-modifying code [8].

No restrictions on resulting circuits are given if $n_r = 1$ and $l = n_c$, which is the most common setting used in the literature. However, this setting is not suitable for an on-chip implementation of CGP as many options for interconnecting the nodes are allowed and must be supported by the underlying hardware platform. On the other hand, if $l = 1$ then pipeline circuits can naturally be evolved and implemented.

After decoding the genotype, each node is labelled as active or inactive according to its utilization in the phenotype. Regarding the size of the array, one of the conclusions in the literature is that more is better (i.e. the number of evaluations is minimal when the genome size far exceeds what is necessary for the problem at hand [27]). A recent work showed that there is a problem-dependent limit in the number of nodes that makes sense to employ in the array [52]. However, all results were obtained for relatively small problem instances (such as 4-bit multiplier or 8-bit parity requiring tens of gates to be implemented). It is an open question if this high redundancy is useful for the evolution of real-world circuits containing thousands of gates.

2.2 Genetic Operators

Despise several attempts to introduce a useful crossover operator to CGP [2, 47], mutation is the only genetic operator employed in almost all studies and applications conducted with CGP. The structure of fitness landscapes was studied and it has been shown that the landscapes are vastly neutral with sharply differentiated plateaus [30]. The search space is considered as very difficult, especially for circuits such as parallel multipliers. The recent work of Goldman and Punch analysed several types of mutation and their impact of the quality of search, stressing the importance of mutations on active over inactive genes [6]. Vasicek showed that there is no reason to support neutral mutations for evolutionary optimization of complex circuits (hundreds of inputs, thousands of gates) if the task is to minimize the number of gates in a fully functional circuit [53].

2.3 Seeding the Initial Population

If the initial population is randomly generated, we speak about *evolutionary circuit design*. The task for CGP is to provide a circuit structure. If conventional solutions are used in the initial population (i.e. the circuit structure is known), the goal of CGP is to optimize circuit parameters (as well as structure) and then we speak about *evolutionary circuit optimization*. While it is very time consuming to evolve non-trivial circuits from scratch, the question is whether it makes sense to evolve them at all. We believe that there are at least two reasons.

Firstly, the approach seems to be promising for adaptive embedded systems. If a new logic function has to be implemented in reconfigurable hardware, employing CGP could be a good choice in the case of simple logic circuits. The reason is that it is usually impossible to perform a standard circuit design and synthesis procedure by means of common circuit design software executed directly in the embedded system because of its high time and resources requirements. On the other hand, CGP, which is generating and testing candidate solutions directly in the reconfigurable device, can provide a suitable solution in a reasonable time, even if some circuit components are faulty [40]. This is impossible using conventional synthesis, placement and routing algorithms which expect fault-free chips.

Secondly, the circuit design problem can serve as a useful test problem for the performance evaluation and comparison of genetic programming systems. It is also occasionally possible to obtain new useful implementations of these circuits, unbiased with respect to conventional circuit designs. The most complex circuits evolved from scratch (without any decomposition) were reported in [58] and they contain less than 30 inputs and a thousand gates.

However, in the circuit design and optimization practice, there is no reason to start from scratch. A fully functional solution can always be generated from the specification by means of a basic conventional circuit design method. The most complex circuits optimized by CGP that was seeded with the best known conventional implementations were reported by [53]. They contained hundreds of inputs and thousands of gates.

It should be noted that the approach discussed so far assumes that a circuit fully compliant with the specification must be delivered. This scenario is referred to as the evolution of *completely specified circuits* and is primarily relevant for arithmetic circuits and control logic. On the other hand, *incompletely specified circuits* are used in applications (such as classification, filtering, hashing and prediction) in which the correctness can only be evaluated using a subset of all possible input vectors because verifying responses for all possible input combinations is intractable. According to the applications reported in the literature, it seems that the evolutionary circuit design is more successful for the incompletely specified circuits. We see in Sect. 3.1 that both approaches are relevant for approximate computing.

2.4   Search Algorithm

While the problem representation used in CGP is not an original invention of Julian F. Miller, the mutation-based search algorithm operating over this representation is fundamental in Miller's contribution. The search algorithm utilized by CGP is a simple $(1 + \lambda)$ search strategy. Every new population consists of the parent and its λ offspring created by a mutation operator. The parent is always the highest-scored candidate circuit. The parent from the previous generation is never selected as a new parent if there is another offspring with the same fitness value. This rule is crucial for an efficient search as it introduces new genetic material into the population through the neutral mutations. The algorithm is terminated when the maximum number of generations is exhausted or a sufficiently working solution is obtained.

It was observed in [5] that if CGP is minimizing the number of gates in an already fully functional circuit, then a modified parent selection mechanism is more efficient than the standard one. Quite unintuitively, the selection of the parent individual based solely on its functionality (i.e. all fully functional offspring can then become the new parent independent of their size) instead of compactness, led to smaller phenotypes at the end of evolution.

Contrasted to the tree-based GP, CGP employs a very small population (λ is usually between 1 and 10), but many generations are produced. Introducing a multiobjective or co-evolutionary CGP is then problematic as the populations are very small. Despite this fact, several multiobjective CGP implementations [11, 13, 16] and co-evolutionary CGP implementations [46] have been proposed.

2.5 *Fitness Evaluation and Its Acceleration*

Since its introduction, CGP has been promoted as a universal form of GP, but is especially useful for circuit design and optimization. Many authors reported digital circuits that can be evolved from scratch or optimized by CGP and at the same time these circuits show some improvements in terms of the number of gates (and delay in some cases) against conventional implementations (one of the first results in this direction was reported by [63]). The most common approach to construct the fitness function $f()$ is as follows:

$$
f = \begin{cases} b & when \quad b < n_o 2^{n_i}, \\ b + (n_c n_r - z) & otherwise, \end{cases}
\tag{1}
$$

where b is the number of correct output bits obtained as response for all possible assignments to the inputs, z denotes the number of gates utilized in a particular candidate circuit and $n_c \cdot n_r$ is the total number of available gates. It can be seen that the last term $n_c n_r - z$ is considered only if the circuit behavior is perfect (i.e. $b = b_{max} = n_o 2^{n_i}$). We can observe that the evolution has to discover a perfectly working solution firstly, while the size of circuit is not important. Then, the number of gates is optimized.

For the incompletely specified problems, the fitness is usually calculated on the basis of candidate circuit responses for a given training data set. The evolved circuit has to be validated using a test set. This approach has been widely adopted in the evolutionary design of image filters [42], classifiers [17], hash functions [15] and other circuits.

As the fitness calculation is the most time consuming procedure of CGP, various accelerators have been proposed to reduce the circuit evaluation time. The approaches include bit-level parallel circuit simulation and fitness function precompilation [62], parallel CGP [13], and FPGA [3, 54] and GPU [9] based accelerators.

2.6 Practical Aspects of Evolutionary Circuit Design

CGP for logic synthesis and optimization, as discussed so far, has not been widely accepted by the circuit design community. The reason is that this approach is far from the practical needs of the community. The main criticism addressed the issues of the (1) initial solution selection (i.e. there is usually no reason to evolve a circuit from scratch, see Sect. 2.3), (2) simplified modelling of circuit parameters, (3) poor scalability of the method, (4) non-deterministic behaviour of the method and (5) validation of the method using a few benchmark circuits only. We elaborate (2) and (3) in greater detail in the following paragraphs.

The fitness function according to Eq. 1 would be acceptable for logic synthesis, where the goal is to minimize the number of gates. It is insufficient for circuits that have to be implemented on a chip, where detailed information about the area on a chip, delay and power consumption are important. In order to estimate parameters of a given circuit, a detailed circuit analysis is requested. As many candidate circuits have to be evaluated it is very time consuming to call a professional circuit simulator for each circuit. Hence parameters of candidate circuits are estimated in the fitness function and a complete measurement is performed for the best circuits at the end of evolution. This methodology was implemented, for example, in paper [33, 59], in which the area and delay were estimated using the parameters defined in the liberty timing file available for a given semiconductor technology. Delay of a gate was modelled as a function of its input transition time and the output capacitive load. The delay of the whole circuit was then determined as a delay along the longest path. The total area was calculated as the sum of areas of all gates involved in the circuit. Finally, power consumption was estimated according to the methodology introduced for gate- and transistor-level circuits in [33]. In another transistor-level approach, power consumption and other circuit parameters were computed for all candidate designs using a circuit simulator of the Spice family [69].

Coping with the scalability problems is a more serious issue. The problem is that the circuit evaluation time (Eq. 1) grows exponentially with the number of inputs.

Hence the method is only applicable to the evolution of relatively small circuits. Because the specification is given in the form of a truth table, it is impossible to specify complex circuits in practice. Several methods have been proposed to eliminate these problems (see the overview in [43]). However, the most complex circuits evolved in this scenario are, for example, 6-bit multipliers, 9-bit adders, and 17-bit parity circuits [13, 49].

A promising solution to the fitness evaluation scalability problem is based on a completely different strategy to the fitness evaluation. In practice, a common situation is that a highly unoptimized, but fully functional circuit implementation, is always available. Such a circuit can be used in the initial population of CGP. The new fitness evaluation procedure then exploits the fact that efficient algorithms, which allow us to decide relatively quickly on whether two circuits are functionally equivalent, were developed in the field of formal verification. In our context, the task is to decide whether the parent and its offspring (created by a mutation operator) are functionally equivalent, assuming that the evolutionary algorithm is seeded by a fully functional solution and that the current parent is also fully functional. If the equivalence holds, the fitness of the offspring is given by the number of gates if the task is to minimize the number of gates. An evolutionary circuit optimization method was introduced in [55], which employs a satisfiability problem solver (SAT solver) in the fitness function in order to decide the functional equivalence. An average gate reduction of 25% was reported for benchmark circuits containing thousands of gates and having tens of inputs in comparison with state of the art academia, as well as commercial tools [56]. This result was improved in [53] by using a circuit simulator prior to a SAT solver to disprove the equivalence between a candidate solution and its parent. If the equivalence checking is based on binary decision diagrams (BDD), the circuits can also be evolved from scratch because it is possible and relatively easy to obtain the Hamming distance between the outputs of the candidate circuit and the specification [58]. However, the BDD-based approach seems to be less scalable than the SAT-based fitness evaluation.

We can summarize that CGP is currently applicable in the evolutionary design and optimization of relatively complex combinational circuits. CGP can work as a multi-objective design method in which key circuit parameters are directly estimated in the fitness evaluation procedure and optimized in the course of evolution. While improving conventionally optimized circuits is the clear advantage of CGP, its huge execution time is the main drawback.

3   Approximate Computing and Evolvable Hardware

The research dealing with approximate computing has been substantially growing since 2010. We briefly introduce this concept in this section. We emphasize that some research performed in CGP and evolvable hardware community over 10 years ago is currently very relevant for approximate computing.

3.1 Approximate Computing

According to [31]:

> Approximate computing exploits the gap between the level of accuracy required by the applications/users and that provided by the computing system, for achieving diverse optimizations.

Approximate computing was established with the goal of providing more energy efficient, faster, and less complex computer-based systems by allowing some errors in computations. One of motivations for approximate computing is that the exact computing utilizing nanometer transistors provided by recent technology nodes is extremely expensive in terms of energy requirements and reliable behavior. An open question is how to effectively and reliably compute with a huge amount of unreliable components. Another motivation is that many applications (typically in the areas of multimedia, graphics, data mining, and big data processing) are inherently *error resilient*. This resilience can be exploited in such a way that the error is exchanged for improvements in power consumption, throughput or implementation cost.

One of the approximation techniques is *functional approximation* whose principle is to implement a slightly different function with respect to the original one, provided that the error is acceptable and key system parameters are improved. The functional approximation can be conducted at the level of software as well as hardware. An approximate solution is typically obtained by a heuristic procedure that modifies the original implementation. For example, artificial neural networks were used to approximate software modules [4] and search-based methods allowed us to approximate some hardware components [39].

In addition to functional approximations, timing induced approximations (voltage over scaling) and components showing "unreliable" behavior (such as approximate memory elements) are often employed. The aim is to support approximate computing at the level of programming languages [41] and exploit new features of specialized approximate processors [1].

3.2 Approximations with CGP Before the Approximate Computing Era

Two main directions in approximate computing are (1) energy-efficient computing with unreliable components that are present in the new chips, and (2) approximation of circuits and programs on conventional platforms in order to reduce energy requirements. Interestingly, both directions can be traced in Miller's research on evolvable hardware and CGP.

3.2.1 Evolution in Materio

In the mid-1990s, Adrian Thompson evolved a tone discriminator circuit directly in the XC6216 FPGA chip. The discriminator required significantly less resources than usual conventionally designed solutions would occupy in the same FPGA [50]. Despite a huge effort, Thompson has never fully understood the evolved design. The evolved discriminator was fully functional, but its robustness was limited. For example, a higher sensitivity to fluctuations in environment (external temperature, power supply voltage) and dependence on a particular piece of FPGA were reported. This result, showing an innovative trade-off between the robustness and the amount of resources in the FPGA, can be considered as an early approach to approximate circuit design by means of evolutionary algorithms.

Thompson's work was for Julian F. Miller and his collaborators a starting point in continuing the investigation of the exploitation of physical properties of suitable substrates using artificial evolution. They have developed a new concept which is currently referred to as the *evolution in materio* (i.e. evolving useful functions in a physical system without understanding the rules of the game [26]). The idea is that there is a material to which physical signals can be applied or measured via a set of electrodes. A computer controls the application of physical inputs applied to the material, reading of physical signals from the material and the application to the material of other physical inputs known as physical configurations. Successful examples include evolved logic functions in liquid crystals [10] and carbon nanotubes materials [32].

In order to study and exploit "physics" (known from the evolution in materio) in computer simulation, Miller contributed to the development of the so-called *messy gates*. A messy gate is a gate-like component with added noise. CGP was extended to support noise modelling and then used to evolve small combinational circuits composed of messy gates. Evolved circuits exhibited implicit fault tolerance. Moreover, surprisingly efficient and robust designs were obtained for small combinational circuits.

The above-mentioned studies show that reliable computing on unknown and unreliable platforms has been one of Miller's central research topics for a long time. This topic has been further explored within an EU-funded project, NASCENCE.[2]

3.2.2 Functional Approximation and Approximation of Functions

The functional approximation is frequently used to approximate digital circuits such as adders, multipliers, filters and general logic. In 1999, Miller introduced a CGP-based method for a finite impulse response (FIR) filter design [24] that would be called functional approximation nowadays. In this method, candidate filters are composed of elementary logic gates, thus ignoring completely the well-developed techniques based on multiply–and–accumulate structures. Evolved networks of gates are extremely area-efficient (and thus potentially energy efficient) in comparison with

---

[2]NAnoSCale Engineering for Novel Computation using Evolution, http://www.nascence.eu.

conventional filters. However, only partial functionally has been obtained because of the overall simplicity of the logic networks. The evolved circuits are not, in fact, filters. In most cases, they are combinational quasi-linear circuits trained on some data. They are not able to generalize for the input signals unseen during the evolution. In order to obtain real filters, the design process must guarantee that the evolved circuits are linear, which is not the case in this method. One of the possible benefits of the method is that circuits can be evolved to perform filtering task, even if sufficient resources are not available (e.g. a part of chip is damaged). Furthermore, as Miller noted, "The origin of the quasi-linearity is at present quite mysterious … Currently there is no known mathematical way of designing filters directly at this level".

In mathematics, it is investigated how certain (usually complex) functions can best be approximated by means of basic functions that are inexpensive or suitable according to a given purpose, and with quantitatively characterizing the errors thereby introduced. CGP (and GP in general) is a method capable of performing the so-called *symbolic regression* (i.e. providing an expression which represents (with some error) an unknown function for a given data set). Miller was one of the core persons who established a *self-modifying* CGP (SMCGP), which can find general solutions to classes of problems and mathematical algorithms such as arbitrary parity, n-bit binary addition, sequences that compute π and e, etc. [8]. SMCGP is a developmental form of CGP that supports self-modification functions in addition to computational functions and enables phenotypes to vary over time. By means of the phenotype development, the degree of approximation of the target behaviour (and thus the corresponding error) can be tuned.

4 Circuit Approximation by Means of CGP

This section briefly surveys recent applications of CGP is which approximate computing is explicitly mentioned as the target application. The approximate circuit design problem is formulated as a *multi-objective* optimization problem in which the accuracy, area, delay (or performance) and power consumption are conflicting design objectives. The CGP-based approximation methods typically include the following attributes:

- Circuit parameters (delay, area, power consumption) are modelled more precisely than in the applications discussed in the previous sections.
- The search algorithm is typically constructed as multiobjective and seeded by conventional implementations showing different circuit parameters. It is assumed that the user will choose the most suitable trade-off from the Pareto front for a particular application.
- Different types of error metrics are employed. If a candidate circuit is evaluated using all possible input combinations, an arbitrary error metric can be computed. In the case of complex circuits, formal methods based on SAT solving or BDDs are currently applicable for evaluating only a few relevant error metrics [12].

In the following sections, three CGP-based approaches for circuit approximation are presented. Finally, CGP utilizing relaxed equivalence checking in the fitness function is discussed.

4.1 Resources-Oriented Method

This method exploits the fact that CGP can produce a partially working solution even if sufficient resources for constructing an accurate circuit are not available. The idea is to evolve a circuit showing a minimum error using k_i components (gates) provided that $k_i < K$ and K is the number of components (gates) required to implement a correct circuit. CGP is considered as a single-objective method which is executed several times with different k_i as the parameter. It provides a set of approximate circuits, each of which typically exhibits a different trade-off between the functionality and the number of gates. The main advantage is that the user can control the used area (and power consumption) more comfortably than by means of the error-oriented methods. The method was employed to approximate small multipliers and 9-input and 25-input median circuits operating over 8 bits [60]. Approximate software implementations of the median function were evaluated for microcontrollers by [34].

4.2 Error-Oriented Method

Vasicek and Sekanina [57] proposed a complementary design approach. The user is supposed to define the required error level e_{max} (e.g. the average error magnitude). In the first step, CGP, which is seeded by a conventional fully functional implementation, is utilized to modify the seed in order to obtain a circuit with a predefined e_{max}. After obtaining that circuit, in the second step, CGP can minimize the number of gates or other criteria providing that e_{max} is left unchanged. The method is again a single objective and multiple runs are required to construct the Pareto front. The method was evaluated in the task of approximate multipliers design. The error-oriented approach tends to be less computationally demanding than the resources oriented method.

Approximate multipliers showing specific properties were evolved for artificial neural networks implemented on a chip. Their utilization led to a significant power consumption reduction and only a very small loss in the accuracy of the image classification [35].

4.3 Multi-objective CGP

In the multi-objective method, the error and other key circuit parameters (area, delay and power consumption) are optimized together by an algorithm combining CGP with NSGA-II [36, 59]. For example, in [36], the initial population was seeded using

13 different conventional 8-bit adders and 6 different conventional 8-bit multipliers in the task of the adder and multiplier approximation. The gate set contained components of a 180 nm process library, including half and full adders. The mean relative error was used in the fitness function while the predefined worst case error and worst case relative error constrained the search space. As the task is very computationally demanding, a highly parallel implementation of CGP, exploiting a vectorised and multi-threaded code, was employed. This approach enabled us to evolve a rich library of adders and multipliers (showing different errors and parameters) that can be utilized in future applications of approximate computing.

4.4 Relaxed Equivalence Checking

So far we have discussed the approximation methods that evaluate the candidate solutions by applying a set of (all) input vectors and measuring the error of the output vectors with respect to an exact solution. This approach is not, however, applicable when approximating complex circuits. If the exact error of the approximation has to be determined, formal *relaxed* equivalence checking is requested, stressing the fact that the considered systems will be checked to be equal up to some bound w.r.t., a suitably chosen distance metric. This research area is rather unexplored as almost all formal approaches have been developed for (exact) equivalence checking [12]. Checking the worst error can be based on satisfiability (SAT) solving as demonstrated in [66]. However, while violating the worst error can be detected, no efficient method capable of establishing, for example, the average error using a SAT solver has been proposed up to now. Approaches based on binary decision diagrams (BDDs) in order to determine the average arithmetic error, worst error, and error rate were introduced [48].

In the context of CGP, a fitness function based on the average Hamming distance computed by means of BDD was introduced by [61]. The method enabled us to approximate combinational circuits consisting of hundreds of gates and up to 50 inputs (i.e. the circuits whose evolution and approximation is impossible with a direct implementation of the fitness function defined in Eq. 1).

5 New Directions

Although evolutionary approximation of arithmetic circuits [36] and general logic [61] is possible with CGP, and CGP seems to be a quite competitive method, there are other promising areas for CGP in approximate computing. If a candidate approximate circuit is evaluated using a data set (e.g. in applications such as image filtering, classification and prediction), then the scalability issues of CGP are not as burning as in the case of arithmetic circuits. Hence, we expect that CGP would be very successful in these domains.

5.1 Quality Configurable Circuits

In applications such as signal processing or image compression, it is useful when the quality of approximation (and a corresponding circuit error) can dynamically be adapted in "situ" as a response to variable requirements on the quality of results and available resources. This adaptive quality control is addressed by methods recently developed in the area of approximate computing [45, 65].

This concept has, however, been explored by the evolvable hardware community in the past. For example, a polymorphic FIR filter was proposed which can operate in two modes: a standard one and one with a reduced power budget. When the filter operates with a reduced power budget, some filter coefficients are reconfigured and some parts of the filter are disconnected. The goal is to reconfigure the filter in such a way that its original function is approximated as precisely as possible. The reconfiguration of coefficients is implemented using multiplierless polymorphic constant multipliers whose implementation was evolved using CGP [44]. In another study, Kneiper et al. investigated the robustness of EHW-based classifiers that are able to cope with changing resources at run-time. The performance and accuracy was recognized as sufficient as long as a certain amount of resources are present in the system [19]. It seems to be a natural task for CGP to automatically divide a target circuit into several sections and optimize these sections in such a way that a quality configurable solution is obtained.

5.2 Approximate Neural Networks

CGP has already been used in order to approximate neural networks, in particular, the multipliers involved in computing the product of the inputs and synaptic weights were approximated by [35]. However, there is a new opportunity for CGP in approximate neural networks. Miller has managed CGP to encode and evolve artificial neural networks. CGP is able to simultaneously evolve the networks connections weights, topology and neuron transfer functions [18]. It is also compatible with recurrent-CGP enabling the evolution of recurrent neural networks. It seems straightforward to evolve approximate neural network implementations by means of CGP and find a good trade-off between the network error and power consumption.

6 Final Remarks

In this chapter, we highlighted Julian F. Miller's contribution to the evolutionary circuit design and identified a strong connection and relevance of his research to approximate computing.

It is fair to say that this chapter could be seen as a bit biased as many of the results and studies referenced were developed by the Evolvable hardware group at Faculty of Information Technology, Brno University of Technology. However, it indicates how strongly our work was influenced by the concepts and methods proposed by Julian F. Miller. We hope that Miller's ideas were exploited by us in a good way and that we helped to disseminate these ideas to other communities, outside the evolutionary computing (Miller's home), such as mainstream circuit design conferences (DATE, ICCAD, FPL and DDECS).

Thank you, Julian!

Acknowledgements This work was supported by The Ministry of Education, Youth and Sports of the Czech Republic from the National Programme of Sustainability (NPU II); project IT4Innovations excellence in science—LQ1602.

References

1. Chippa, V., Venkataramani, S., Chakradhar, S., Roy, K., Raghunathan, A.: Approximate computing: an integrated hardware approach. In: 2013 Asilomar Conference on Signals, Systems and Computers, pp. 111–117. IEEE (2013)
2. Clegg, J., Walker, J.A., Miller, J.F.: A new crossover technique for cartesian genetic programming. In: Proceedings of GECCO, pp. 1580–1587. ACM (2007)
3. Dobai, R., Sekanina, L.: Low-level flexible architecture with hybrid reconfiguration for evolvable hardware. ACM Trans. Reconfig. Technol. Syst. **8**(3), 1–24 (2015)
4. Esmaeilzadeh, H., Sampson, A., Ceze, L., Burger, D.: Neural acceleration for general-purpose approximate programs. Commun. ACM **58**(1), 105–115 (2015)
5. Gajda, Z., Sekanina, L.: An efficient selection strategy for digital circuit evolution. In: Evolvable Systems: From Biology to Hardware, LNCS, vol. 6274, pp. 13–24. Springer (2010)
6. Goldman, B.W., Punch, W.F.: Analysis of Cartesian genetic programming's evolutionary mechanisms. IEEE Trans. Evol. Comput. **19**(3), 359–373 (2015)
7. Gupta, P., Agarwal, Y., Dolecek, L., Dutt, N., Gupta, R.K., Kumar, R., Mitra, S., Nicolau, A., Rosing, T.S., Srivastava, M.B., Swanson, S., Sylvester, D.: Underdesigned and opportunistic computing in presence of hardware variability. IEEE Trans. CAD Integr. Circuits Syst. **32**(1), 8–23 (2013)
8. Harding, S., Miller, J.F., Banzhaf, W.: Developments in cartesian genetic programming: self-modifying CGP. Genet. Program. Evolvable Mach. **11**(3–4), 397–439 (2010)
9. Harding, S.L., Banzhaf, W.: Hardware acceleration for CGP: graphics processing units. In: Cartesian Genetic Programming, pp. 231–253. Springer (2011)
10. Harding, S.L., Miller, J.F.: Evolution in materio: evolving logic gates in liquid crystal. Int. J. Unconv. Comput. **3**(4), 243–257 (2007)
11. Hilder, J., Walker, J., Tyrrell, A.: Use of a multi-objective fitness function to improve cartesian genetic programming circuits. In: NASA/ESA Conference on Adaptive Hardware and Systems, pp. 179–185. IEEE (2010)

12. Holik, L., Lengal, O., Rogalewicz, A., Sekanina, L., Vasicek, Z., Vojnar, T.: Towards formal relaxed equivalence checking in approximate computing methodology. In: 2nd Workshop on Approximate Computing (WAPCO 2016), HiPEAC, pp. 1–6 (2016)
13. Hrbacek, R., Sekanina, L.: Towards highly optimized cartesian genetic programming: from sequential via SIMD and thread to massive parallel implementation. In: Proceedings of the 2014 Conference on Genetic and Evolutionary Computation, pp. 1015–1022. ACM(2014)
14. Kaufmann, P., Platzner, M.: Advanced techniques for the creation and propagation of modules in cartesian genetic programming. In: Genetic and Evolutionary Computation (GECCO), pp. 1219–1226. ACM Press (2008)
15. Kaufmann, P., Plessl, C., Platzner, M.: EvoCaches: application-specific adaptation of cache mappings. In: Proceedings of the NASA/ESA Conference on Adaptive Hardware and Systems, pp. 11–18. IEEE Computer Society, Los Alamitos, CA, USA (2009)
16. Kaufmann, P., Knieper, T., Platzner, M.: A novel hybrid evolutionary strategy and its periodization with multi-objective genetic optimizers. In: 2010 IEEE Congress on Evolutionary Computation (CEC), pp. 1–8. IEEE (2010)
17. Kaufmann, P., Glette, K., Gruber, T., Platzner, M., Torresen, J., Sick, B.: Classification of electromyographic signals: comparing evolvable hardware to conventional classifiers. IEEE Trans. Evol. Comput. **17**(1), 46–63 (2013)
18. Khan, G.M., Miller, J.F., Halliday, D.M.: Evolution of Cartesian genetic programs for development of learning neural architecture. Evol. Comput. **19**(3), 469–523 (2011)
19. Knieper, T., Kaufmann, P., Glette, K., Platzner, M., Torresen, J.: Coping with resource fluctuations: the run-time reconfigurable functional unit row classifier architecture. In: Proceedings of the 9th International Conference on Evolvable Systems: From Biology to Hardware, LNCS, vol. 6274, pp. 250–261. Springer (2010)
20. Koza, J.R.: Genetic Programming: On The Programming of Computers by Means of Natural Selection. MIT press (1992)
21. Louis, S., Rawlins, G.J.E.: Designer genetic algorithms: genetic algorithms in structure design. In: Proceedings of the Fourth International Conference on Genetic Algorithms, pp. 53–60. Morgan Kauffman (1991)
22. Markov, I.: Limits on fundamental limits to computation. Nature **512**, 147–154 (2014)
23. Miller, J.F.: An empirical study of the efficiency of learning Boolean functions using a cartesian genetic programming approach. In: Proceedings of the 1st Annual Conference on Genetic and Evolutionary Computation, vol. 2, pp. 1135–1142. Morgan Kaufmann Publishers Inc. (1999)
24. Miller, J.F.: On the filtering properties of evolved gate arrays. In: 1st NASA-DoD Workshop on Evolvable Hardware, pp. 2–11. IEEE Computer Society (1999)
25. Miller, J.F.: Cartesian Genetic Programming. Springer (2011)
26. Miller, J.F., Downing, K.: Evolution in materio: looking beyond the silicon box. In: Proceedings of the 2002 NASA/DoD Conference on Evolvable Hardware (EH'02), pp. 167–176. IEEE Computer Society (2002)
27. Miller, J.F., Smith, S.L.: Redundancy and computational efficiency in cartesian genetic programming. IEEE Trans. Evol. Comput. **10**(2), 167–174 (2006)
28. Miller, J.F., Thomson, P.: Cartesian Genetic Programming. In: Proceedings of the 3rd European Conference on Genetic Programming EuroGP2000, LNCS, vol. 1802, pp. 121–132. Springer (2000)
29. Miller, J.F., Thomson, P., Fogarty, T.: Designing electronic circuits using evolutionary algorithms. Arithmetic circuits: A case study. In: Genetic algorithms and evolution strategy in engineering and computer science. Wiley (1998)
30. Miller, J.F., Job, D., Vassilev, V.K.: Principles in the evolutionary design of digital circuits—part II. Genet. Program. Evolvable Mach. **1**(3), 259–288 (2000)
31. Mittal, S.: A survey of techniques for approximate computing. ACM Comput. Surv. **48**(4), 62:1–62:33 (2016)
32. Mohid, M., Miller, J.F., Harding, S.L., Tufte, G., Massey, M.K., Petty, M.C.: Evolution-in-materio: solving computational problems using carbon nanotube-polymer composites. Soft Comput. **20**(8), 3007–3022 (2016)

33. Mrazek, V., Vasicek, Z.: Automatic design of low-power vlsi circuits: Accurate and approximate multipliers. In: Proceedings of 13th IEEE/IFIP International Conference on Embedded and Ubiquitous Computing, pp. 106–113. IEEE (2015)
34. Mrazek, V., Vasicek, Z., Sekanina, L.: Evolutionary approximation of software for embedded systems: median function. In: GECCO Companion '15 Proceedings of the Companion Publication of the 2015 on Genetic and Evolutionary Computation Conference, pp. 795–801. ACM (2015)
35. Mrazek, V., Sarwar, S.S., Sekanina, L., Vasicek, Z., Roy, K.: Design of power-efficient approximate multipliers for approximate artificial neural networks. In: 2016 IEEE/ACM International Conference on Computer-Aided Design (ICCAD). pp. 811–817 (2016)
36. Mrazek, V., Hrbacek, R., Vasicek, Z., Sekanina, L.: Evoapprox8b: library of approximate adders and multipliers for circuit design and benchmarking of approximation methods. In: 2017 Design, Automation & Test in Europe Conference & Exhibition (DATE), pp. 258–261 (2017)
37. Murakawa, M., Yoshizawa, S., Kajitani, I., Furuya, T., Iwata, M., Higuchi, T.: Evolvable hardware at function level. In: Parallel Problem Solving from Nature—PPSN IV, LNCS, vol. 1141, pp. 62–71. Springer (1996)
38. Nawab, S., Oppenheim, A., Chandrakasan, A., Winograd, J., Ludwig, J.: Approximate signal processing. J. VLSI Signal Process. **15**(1–2), 177–200 (1997)
39. Nepal, K., Li, Y., Bahar, R.I., Reda, S.: ABACUS: a technique for automated behavioral synthesis of approximate computing circuits. In: Proceedings of the Conference on Design, Automation and Test in Europe, EDA Consortium, DATE'14, pp. 1–6 (2014)
40. Salvador, R., Otero, A., Mora, J., la De, E.T., Riesgo, T., Sekanina, L.: Self-reconfigurable evolvable hardware system for adaptive image processing. IEEE Trans. Comput. **62**(8), 1481–1493 (2013)
41. Sampson, A., Dietl, W., Fortuna, E., Gnanapragasam, D., Ceze, L., Grossman, D.: EnerJ: approximate data types for safe and general low-power computation. In: Proceedings of the 32nd ACM SIGPLAN Conference on Programming Language Design and Implementation, pp. 164–174. ACM (2011)
42. Sekanina, L.: Evolvable components: from theory to hardware implementations. Nat. Comput. Ser. (2004)
43. Sekanina, L.: Evolvable hardware. In: Handbook of Natural Computing, pp. 1657–1705. Springer (2012)
44. Sekanina, L., Ruzicka, R., Gajda, Z.: Polymorphic fir filters with backup mode enabling power savings. In: Proceedings of the 2009 NASA/ESA Conference on Adaptive Hardware and Systems, pp. 43–50. IEEE Computer Society (2009)
45. Shubham, J., Venkataramani, S., Raghunathan, A.: Approximation through logic isolation for the design of quality configurable circuits. In: Proceedings of the 2016 Design, Automation & Test in Europe Conference and Exhibition (DATE), pp. 1–6. EDA Consortium (2016)
46. Sikulova, M., Sekanina, L.: Acceleration of evolutionary image filter design using coevolution in Cartesian GP. In: Parallel Problem Solving from Nature-PPSN XII, no. 7491 in LNCS, pp. 163–172. Springer (2012)
47. Slany, K., Sekanina, L.: Fitness landscape analysis and image filter evolution using functional-level CGP. In: Proceedings of European Conference on Genetic Programming. LNCS, vol. 4445, pp. 311–320. Springer (2007)
48. Soeken, M., Grosse, D., Chandrasekharan, A., Drechsler, R.: BDD minimization for approximate computing. In: 21st Asia and South Pacific Design Automation Conference ASP-DAC 2016, pp. 474–479. IEEE (2016)
49. Stomeo, E., Kalganova, T., Lambert, C.: Generalized disjunction decomposition for evolvable hardware. IEEE Trans. Syst. Man Cybern. Part B **36**(5), 1024–1043 (2006)
50. Thompson, A., Layzell, P., Zebulum, S.: Explorations in design space: unconventional electronics design through artificial evolution. IEEE Trans. Evol. Comput. **3**(3), 167–196 (1999)
51. Turner, A.J., Miller, J.F.: Recurrent cartesian genetic programming. In: Parallel Problem Solving from Nature—PPSN XIII, pp. 476–486. Springer (2014)

52. Turner, A.J., Miller, J.F.: Neutral genetic drift: an investigation using cartesian genetic programming. Genet. Program. Evolvable Mach. **16**(4), 531–558 (2015)
53. Vasicek, Z.: Cartesian GP in optimization of combinational circuits with hundreds of inputs and thousands of gates. In: Proceedings of the 18th European Conference on Genetic Programming—EuroGP. LCNS 9025, pp. 139–150. Springer International Publishing (2015)
54. Vasicek, Z., Sekanina, L.: An evolvable hardware system in Xilinx Virtex II Pro FPGA. Int. J. Innov. Comput. Appl. **1**(1), 63–73 (2007)
55. Vasicek, Z., Sekanina, L.: Formal verification of candidate solutions for post-synthesis evolutionary optimization in evolvable hardware. Genet. Program. Evolvable Mach. **12**(3), 305–327 (2011)
56. Vasicek, Z., Sekanina, L.: A global postsynthesis optimization method for combinational circuits. In: Proceedings of the Design, Automation and Test in Europe, DATE, pp. 1525–1528. IEEE Computer Society (2011)
57. Vasicek, Z., Sekanina, L.: Evolutionary design of approximate multipliers under different error metrics. In: IEEE International Symposium on Design and Diagnostics of Electronic Circuits and Systems 2013, pp. 135–140. IEEE (2014)
58. Vasicek, Z., Sekanina, L.: How to evolve complex combinational circuits from scratch? In: 2014 IEEE International Conference on Evolvable Systems Proceedings, pp. 133–140. IEEE (2014)
59. Vasicek, Z., Sekanina, L.: Circuit approximation using single- and multi-objective cartesian GP. In: Genetic Programming. LNCS 9025, pp. 217–229. Springer (2015)
60. Vasicek, Z., Sekanina, L.: Evolutionary approach to approximate digital circuits design. IEEE Trans. Evol. Comput. **19**(3), 432–444 (2015)
61. Vasicek, Z., Sekanina, L.: Evolutionary design of complex approximate combinational circuits. Genet. Program. Evolvable Mach. **17**(2), 169–192 (2016)
62. Vasicek, Z., Slany, K.: Efficient phenotype evaluation in cartesian genetic programming. In: Proceedings of the 15th European Conference on Genetic Programming. LNCS 7244, pp. 266–278. Springer (2012)
63. Vassilev, V., Job, D., Miller, J.F.: Towards the automatic design of more efficient digital circuits. In: Proceedings of the 2nd NASA/DoD Workshop on Evolvable Hardware, pp. 151–160. IEEE Computer Society (2000)
64. Vazirani, V.V.: Approximation Algorithms. Springer (2001)
65. Venkataramani, S., Roy, K., Raghunathan, A.: Substitute-and-simplify: a unified design paradigm for approximate and quality configurable circuits. Design, Automation and Test in Europe, DATE'13, pp. 1367–1372. EDA Consortium San Jose, CA, USA (2013)
66. Venkatesan, R., Agarwal, A., Roy, K., Raghunathan, A.: MACACO: modeling and analysis of circuits for approximate computing. In: 2011 IEEE/ACM International Conference on Computer-Aided Design (ICCAD), pp. 667–673. IEEE (2011)
67. Walker, J.A., Miller, J.F.: The automatic acquisition, evolution and reuse of modules in cartesian genetic programming. IEEE Trans. Evol. Comput. **12**(4), 397–417 (2008)
68. Walker, J.A., Hilder, J.A., Tyrrell, A.M.: Evolving Variability-Tolerant CMOS Designs. Springer, Berlin Heidelberg (2008)
69. Walker, J.A., Hilder, J.A., Reid, D., Asenov, A., Roy, S., Millar, C., Tyrrell, A.M.: The evolution of standard cell libraries for future technology nodes. Genet. Program. Evolvable Mach. **12**(3), 235–256 (2011)

Breaking the Stereotypical Dogma of Artificial Neural Networks with Cartesian Genetic Programming

Gul Muhammad Khan and Arbab Masood Ahmad

Abstract This chapter presents the work done in the field of Cartesian Genetic Programming evolved Artificial Neural Networks (CGPANN). Three types of CGPANN are presented, the Feed-forward CGPANN (FFCGPAN), Recurrent CGPANN and the CGPANN that has developmental plasticity, also called Plastic CGPANN or PCGPANN. Each of these networks is explained with the help of diagrams. Performance results obtained for a number of benchmark problems using these networks are illustrated with the help of tables. Artificial Neural Networks (ANNs) suffer from the dilemma of how to select complexity of the network for a specific task, what should be the pattern of inter-connectivity, and in case of feedback, what topology will produce the best possible results. Cartesian Genetic Programming (CGP) offers the ability to select not only the desired network complexity but also the inter-connectivity patterns, topology of feedback systems, and above all, decides which input parameters should be weighted more or less and which one to be neglected. In this chapter we discuss how CGP is used to evolve the architecture of Neural Networks for optimum network and characteristics. Don't you want a system that designs everything for you? That helps you select the optimal network, the inter-connectivity, the topology, the complexity, input parameters selection and input sensitivity? If yes, then CGP evolved Artificial Neural Network (CGPANN) and CGP evolved Recurrent Neural Network (CGPRNN) is the answer to your questions.

G.M. Khan (✉)
Department of Electrical Engineering, University of Engineering
and Technology, Peshawar, Pakistan
e-mail: gk502@uetpeshawar.edu.pk

A.M. Ahmad
Department of Computer Systems Engineering, University of Engineering
and Technology, Peshawar, Pakistan
e-mail: arbabmasood@uetpeshawar.edu.pk

© Springer International Publishing AG 2018   213
S. Stepney and A. Adamatzky (eds.), *Inspired by Nature*, Emergence,
Complexity and Computation 28, https://doi.org/10.1007/978-3-319-67997-6_10

1   Artificial Neural Networks

Artificial Neural Network (ANN) is a powerful tool for non-linear mapping between input and output. It is the ultimate solution when all the traditional algorithms fail. It provides a parametric expression for the system whose mathematical model for functionality is unknown. Despite successful application of neural networks to a variety of problems, they still have some limitations. One of the most common limitations is associated with neural network training. The back-propagation learning algorithm cannot guarantee an optimal solution. Back propagation has the tendency to get stuck-up at a sub-optimal point, and might never come out of it. Much research has been done to tackle this problem. Evolutionary methods can provide solution to this problem. They do so by producing multiple solutions at any one point, thus if one of them is stuck-up at suboptimal point the other might be in closer vicinity to the global optimum and will thus cause the network to reach the global optimum. In real-world applications, the back-propagation algorithm might converge to a set of sub-optimal weights from which it cannot escape. As a result, the neural network is often unable to find a desirable solution to a problem at hand. To get the benefits of both ANN and evolutionary methods researchers tried a hybrid of both these methods [33]. Another difficulty is related to selecting an optimal topology for the neural network. The right network architecture for a particular problem is often chosen by means of heuristics, and designing a neural network topology is still more of art than engineering. Genetic algorithm [25] and genetic programming [28] are effective optimization techniques that can guide both weight and topology selection.

2   Neuro-Evolution

The process of evolving various parameters of neural network is termed Neuro-evolution. Unlike backpropagation algorithm which is used to train only weights of the network to obtain the desired optimum characteristics, evolutionary techniques can train the network topology and even the learning rules. In case of evolving connection weights, we perform the following steps:

1. Encode the connection weights of each individual neural network into chromosomes.
2. Calculate the error function and determine the individual's fitness.
3. Reproduce children based on selection criteria.
4. Apply genetic operators.

Success or failure of an application is largely determined by the network architecture (i.e. the number of neurons and their interconnections). As the network architecture is usually decided by trial and error, a good algorithm is required to automatically design an efficient architecture for each particular application. Genetic algorithms may well be suited for this task. The basic idea behind evolving a suitable network architecture is to conduct a genetic search in a population of possible architectures. We must first choose a method of encoding a network's architecture into a chromosome. There are two types of encoding schemes:

Direct Encoding: When all the information of the network is represented directly by genes in the code, it is referred to as Direct Encoding Scheme. In this case the network has a one to one relationship between genotype and phenotype.

Indirect Encoding: In this type of encoding the genes don't represent the network directly, and show only the indirect function responsible for generation of network parameters [5]. This is biologically more plausible, as according to the findings of neuroscience it is impossible for genetic information to be encoded in humans to specify the whole nervous system directly. It is computationally more expensive, since it doesn't have any clue of the targeted application for which the network is developed.

3 CGP Evolved Artificial Neural Network (CGPANN)

We have used CGP to introduce four different ways of evolving neural networks that are as follows:

- Feed-forward CGP evolved ANN (FCGPANN)
- Recurrent CGP evolved ANN (RCGPANN)
- Plastic CGP evolved ANN (PCGPANN)
- Plastic Recurrent CGPANN (PRCGPANN).

3.1 Feed-Forward CGP Evolved ANN (FCGPANN)

In the first case, CGP is transformed to a feed-forward neural network by considering each node as a neuron, and providing each connection with a weight. The neurons of such a network are arranged in Cartesian format with rows and columns inspired by original CGP architecture, and later on restricted to a single row mostly giving the network an ability to create infinite graphs/topologies. Each neuron in the network can take connection from either a previous neuron or from the system input. Not all neurons are necessarily connected with each other or with system inputs, this provides the network with an ability to continuously evolve its complexity along with the weights. All the network parameters are represented by a string of numbers called genotype. The number of active neurons (connected from inputs to outputs) varies from generation to generation subject to the genotype selection. Output of any neuron or a system input can be a candidate for the systems output selection. The ultimate system functionality is identified by interconnecting neurons from output to input. Since CGP works best with mutation, thus only mutation operator is

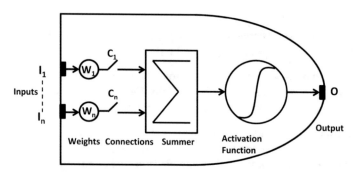

Fig. 1 A CGPANN node containing the inputs, weights, switches, summer and activation function

explored in all the variants of neural networks introduced in this work. Mutation operator specifies the percentage of genes to be mutated during the process of evolution. A gene can be an input connection to a node, a weight, a switch or a neuron function. These genes for a single node are shown in Fig. 1. An FCGPANN genotype for arity 2 (two inputs per node) is represented by the following expression: $FI_1 W_1 C_1 I_2 W_2 C_2, FI_1 W_1 C_1 I_2 W_2 C_2 \ldots, O_1, O_2 \ldots O_n$. Where $\mathbf{F}$ is the activation function chosen randomly from a list of different type of nonlinear functions. The two most popular functions, also used in this work are the Log-sigmoid and the tangent hyperbolic function, $\mathbf{I}$ represents output of a node or a system input, connected to the node under consideration, $\mathbf{W}$ is the weight being multiplied with a node input and $\mathbf{C}$ is an optional ON/OFF switch. All these genes can get only certain allowed values depending on the type of network. The network representation of a genotype is called a phenotype. Figure 2 shows a typical FCGPANN phenotype with its associated genotype. We have the option to evolve, all the parameters, or can fix one or two and evolve others.

FCGPANN was initially tested for its speed of learning, and evaluated against the previously introduced neuro-evolutionary techniques on benchmarks such as single and double Pole balancing [19]. Table 1 shows the superior performance of FCG-PANN and RCGPANN (see next section) in comparison to the other neuroevolutionary techniques evaluated for speed of learning on single pole balancing problem. Table 2 shows the performance of FCGPANN and RCGPANN compared to other techniques, for Markovian and non-Markovian cases of double pole balancing task, where a Markovian process can be defined as the one in which the conditional probability distribution of the future state depends only on the present state and not on the past history. The figures show average number of evaluations needed to achieve the target objective of balancing the poles for a specific bench marked time interval. FCGPANN is explored in a range of applications including: breast cancer detection, prediction of foreign currency exchange rates, Load forecasting, Internet multimedia traffic management, cloud resource estimation, solar irradiance prediction, wind power forecasting and arrhythmia detection [2, 13, 14, 16, 19, 24, 30]. FCGPANN outperformed all the previously introduced techniques as highlighted in

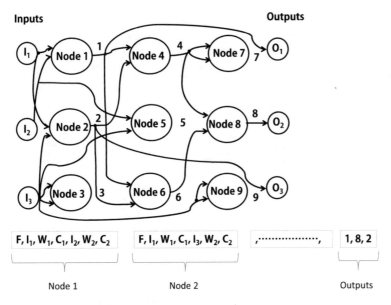

Fig. 2 A CGPANN phenotype with its corresponding genotype

Table 1 Comparison of CGPANN with other neuro-evolutionary algorithms in terms of average number of evaluations required to solve the single pole balancing task

Method	Markovian	Non-Markovian
Conventional Neuro-Evolution (CNE) [36]	352	724
Symbiotic, Adaptive Neural Evolution (SANE) [23]	302	1212
Enforced sub-population (ESP) [7]	289	589
Neuro-Evolution of Augmenting Topologies (NEAT) [35]	743	1523
Cooperative synapse neuroevolution (CoSyNE) [6]	98	127
FCGPANN [19]	21	–
RCGPANN [19]	17	55

the literature. Table 3 shows the comparative results for the mean accuracy in breast cancer detection with fine needle aspiration (FNA), using different algorithms. It can be seen that FCGPANN outperformed the other methods [19]. Another important area in which FCGPANN was successfully applied is the prediction of foreign currency exchange rates [24]. Table 4 shows the comparative results for prediction of foreign currency exchange rates using FCGPANN and other methods.

Marketing a product requires good knowledge about the demands of customers, especially in the case of food products. FCGPANN provides efficient method to predict market trends [1].

Table 2 Comparison of CGPANN with other neuro-evolutionary algorithms in terms of average number of evaluations required to solve the double pole balancing task

Method	Markovian	Non-Markovian	
	Standard fitness	Standard fitness	Damping fitness
Conventional Neuro-Evolution (CNE) [36]	22 100	76 906	87 623
Symbiotic, Adaptive Neural Evolution (SANE) [23]	12 600	262 700	451 612
Enforced sub-population (ESP) [7]	3 800	7 374	26 342
Neuro-Evolution of Augmenting Topologies (NEAT) [35]	3 600	–	6 929
Cooperative synapse neuroevolution (CoSyNE) [6]	954	1 294	3416
FCGPANN [19]	77	–	–
RCGPANN [19]	129	163	387

Table 3 Comparison of Mean Absolute Percentage Errors (MAPE) obtained using various classification methods using the processed FNA data from WDBC that contains 30 features

No.	Method	Mean (MAPE)
1	Multi-Layer Perceptron (MLP) [8]	95.56
2	Fisher Linear Discriminant Analysis (FLDA)/MLP [8]	90.92
3	Principle Component Analysis (PCA)/MLP [8]	92.02
4	Genetic Programming (GP/MDC) [8]	96.58
5	Evolutionary Neural Network (ENN) [10]	95.6
6	FCGPANN [19]	97 for Type-I and 98.5 for Type-II

Table 4 Comparison between the MAPE of other well known methods and that of FCGPANN, for predicting foreign exchange rates

Network	MAPE (%)
Hidden Markov model (HMM) [29]	1.928
ARIMA [3]	1.6108
Regression model [29]	1.9
CART model	1.62
Neural network model [9]	1.61
FCGPANN [24]	1.148

3.2 Recurrent CGPANN (RCGPANN)

The second type of CGPANN is the Recurrent CGPANN (RCGPANN). These networks are more suitable for modeling systems that are dynamic and nonlinear. This network is a modification to one of the earliest networks, the Jordan's network [11].

In the Jordan's network there are state inputs that are equal in number to the outputs. These inputs are fed by the outputs through unit weights. The state inputs are present only at the input layer. Learning of these networks take place by changing the weights of connections between input layer and the hidden layer and, the hidden and the output layer. In RCGPANN unlike the Jordan's network the state inputs can be connected, not necessarily to the first layer but to any layer. These additional inputs also have the activation functions. Figures 3 and 4 show a typical RCGPANN neuron, and the RCGPANN genotype and phenotype respectively. Here I_1 and I_2 are the normal inputs while R is a state input. Initial value of the R input to the system is considered zero. Output is taken from Node 6 as evident from genotype and the corresponding phenotype, and node 6 takes input from node 3 and input I_2 only. Node 4 and 5 do not contribute to the output and are termed inactive nodes, while 3 and 6 are active nodes as they contribute to the output. Following are the outputs of active nodes:

$$\psi_3 = tanh(I_1 \cdot W_{13} + I_2 \cdot W_{23} + R \cdot W_{R3})$$

$$\psi_6 = tanh(\psi_3 \cdot W_{36} + I_2 \cdot W_{26} + R \cdot W_{R6})$$

where I_1 and I_2 are the normal inputs to the system, R is the state input, W_{mn} is the weight of connection between system-input/node m and n, ψ_m is the output of node m. Figure 5 shows the case when all the outputs are presented as feedback

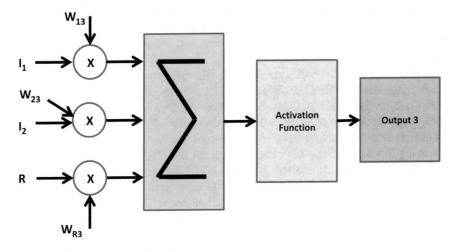

Fig. 3 A typical RCGPANN neuron

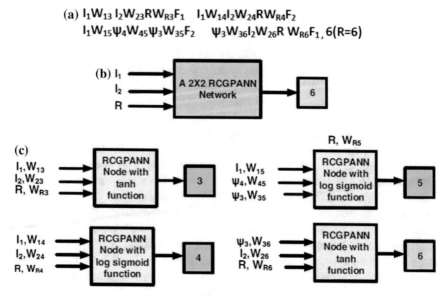

Fig. 4 Figure showing **a** RCGPANN genotype, **b** 2×2 RCGPANN Network and **c** corresponding to the phenotype

inputs to the neurons for selection. The output is averaged to find the desired output of the system. RCGPANN was also tested initially for its speed of learning similar to FCGPANN on both single and double pole balancing for both markovian and nonmorkovian cases. Its performance relative to other neuroevolutionary techniques for Markovian and non-Markovian cases explored on single pole balancing task is shown in Table 1 and that on double pole is shown in Table 2. The numbers in the tables represent the required average number of evaluations to find the desired pole balancing behaviour. The results in these two tables clearly show the superiority of FCGPANN and RCGPANN in Markovian case and that of RCGPANN in the non-Markovian case.

FCGPANN and RCGPANN are also tested for their generalization ability. Table 5 shows generalization of FCGPANN and RCGPANN: average number of random cart initializations (out of 625) that can be balanced for a desired number (benchmark) of time-steps. Table 6 presents the generalization ability of various neuroevolutionary algorithms for the double pole balancing scenario for non-Markovian case using the damping fitness function. It is observed that the RCGPANN scored 335.84 out of 625 for 50 independently evolved genotypes exhibiting greater generalization ability as compared to other techniques presented to date. Recurrent CGPANN has been successfully applied to a number of applications including: Load forecasting, foreign currency exchange rates, bandwidth management and estimation [16, 17, 31]. RCGPANN has been successfully applied to electrical load prediction for a complete year and also for different seasons of the year, enabling efficient utilization of

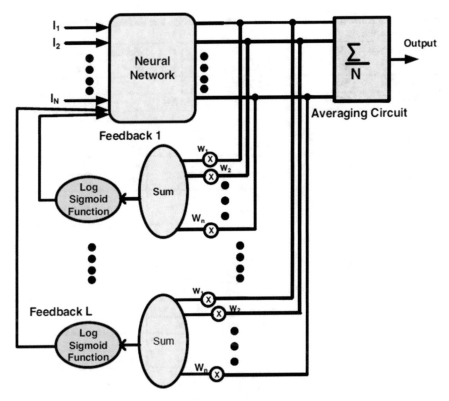

Fig. 5 RCGPANN phenotype with full feedback having all outputs available for feedback

electricity management [14]. Table 7 compares the performance of RCGPANN with other contemporary methods in terms of Mean Absolute Percentage Error (MAPE), for the load forecasting task. Bandwidth allocation for communication channels has always been a challenge for the engineers. Recurrent CGPANN has been successfully applied to predict the size of next MPEG4 video frame based on the estimate of the last ten frames [16]. Table 8 shows the comparative results of next frame prediction in terms of overall error for RCGPANN and other models. The electric load prediction discussed earlier was improved to predict very short term (about half an hour) load [18]. Table 9 shows the performance in terms of MAPE values, for the proposed RCGPANN in comparison to other methods, for predicting very short term electric load. Table 10 shows the comparative results for predicting the foreign currency exchange rate using RCGPANN and other models [31].

Table 5 Generalization of CGPANN: average number of random cart initializations (out of 625) that can be solved

Algorithm type	Markovian		Non-Markovian	
	Single pole	Double pole	Single pole	Double pole
FCGPANN	590	277.38	–	–
RCGPANN	363.94	471.92	294	335.84

Table 6 Comparison of generalization of CGPANN with other neuroevolutionary algorithms for the double pole balancing scenario for Non-Markovian case

Method	Value
CE	300
ESP	289
NEAT	286
RCGPANN	335.84

Table 7 Comparison of RCGPANN with other methods for the load forecasting task

Method	MAPE (%)
Local linear model tree	1.98
Support vector machine	1.93
Autonomous ANN	1.75
Floating search + SVM	1.70
CGPANN	1.71
ANN-back propagation	2.41
GA based adaptive ANN	1.94
RCGPANN	1.56

Table 8 Best overall error comparison in MPEG4 frame size prediction

S.no.	Scheme	Error (%)
1	Recurrent ANN	RMSE = 3.0
2	F-CGPANN	RMSE = 16
3	Laetitia et al. model	RPE = 7.30
4	SARIMA	MARE = 1.37
5	Kalman filter	MARE = 1.4
6	Proposed (RCGPANN)	RMSE = 2.7 MAPE = 1.2

Table 9 Comparison of RCGPANN with other methods in terms of MAPE for very short time electric load forecasting

S.no	Model	MAPE (%)
1	Self-supervised adaptive ANN	0.91
2	FNN for RTLF	0.88
3	ANNSTLF	2
4	RBF forecaster	1.3393
5	Model in [34]	0.66
6	Multiplicative decomposition model	0.7601
7	Seasonal ARIMA model	1.6108
8	Model 1 [22]	1.792
9	Model 2 [22]	1.813
10	RCGPANN (proposed model)	0.43

Table 10 Comparison between the accuracy distribution rates and MAPE of RCGPANN and other models for the foreign currency exchange rate prediction

Network	Accuracy (%)
Multi layer perceptron [21]	72
HFERFM [27]	69.9
AFERFM [27]	81.2
Backpropagation with Bayesian regularization [12]	93.93
RCGPANN (implemented)	98.872

3.3 Plastic CGPANN (PCGPANN)

Plasticity in neural networks has been the characteristic of choice when it comes to applications in dynamic systems due to its comparatively better performance [4, 26, 32]. The improved performance in Plastic neural networks can be attributed to the adaptability of its morphology to environmental stimuli. This is similar to the natural neural system. In this developmental form of CGPANN an additional output gene provides extra features to the system. PCGPANN has the same basic structure as that of CGPANN presented previously. With the addition of an extra output gene, that causes developmental decisions during evaluation process, the CGPANN achieves its plasticity. The decision is made on the basis of output values. The mutation of the genotype is invoked according to a decision function that decides either to invoke mutation or not. The decision function in this case is a threshold function which invokes mutation when the value of output of the CGPANN is higher than the threshold value. The threshold value is selected based on the performance of the model. In this way an unlimited number of phenotypes might be generated from a single genotype, depending on the system requirement. The network is tested for its performance in diverse learning domains. The genotypes are modified based on the defined set of constraints. The plasticity invokes mutation of genotype at runtime, modifying its genes, producing complex network structures.

3.3.1 PCGPANN Methodology

The PCGPANN generalized approach is demonstrated in Fig. 6. The figure depicts network topology, its parameters i.e. inputs, connections, outputs, the additional output gene and the algorithm of PCGPANN. The system consists of the original CGPANN as explained before. Output from this network is fed into the running sum block. Based on the evolutionary requirements, the number of outputs fed into the summation block can vary. Either half or full number of CGPANN outputs are fed into the summation block. The network is thus named Full Feedback (FFB) or Half Feedback (HFB) network, based on its architecture. Output of the summation block is fed to a decision function that generates either a 0 or a 1 so as to invoke the mutation or not. Depending on the value of the summation block the decision function either invokes mutation or leaves the genotype unchanged. The network is unique in its ability to invoke mutation during run time. The mutation may take place randomly in any of the following network genes: node inputs, weights, outputs or activation functions. The genes are mutated under the given set of constraints.

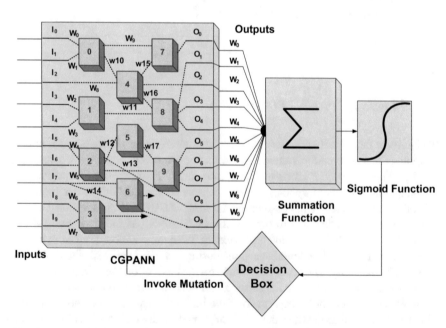

Fig. 6 A Generalized approach of PCGPANN depicting the network topology, attributes: inputs, connections, outputs, the additional output gene and the algorithm

Constraints:

1. If the gene that has to be mutated is the function of a node, then a new function is randomly selected from a list of functions and assigned to the node.
2. If a weight gene has to be mutated, then it is assigned a random value between −1 and +1.
3. If the gene to be mutated happens to be a node input, then depending on the levels-back parameter, output of any randomly selected node on the left of the node under consideration or a network input is connected to it.
4. If an output gene has to be mutated, then it is connected to the output of a randomly selected node or a system input.

In each iteration genes are mutated, the number of mutations depending on runtime mutation rate (Developmental index). The possibility of change in genotype at runtime is dependent on the output from activation function. The decision function is maintained in accordance to the expected range of output from the activation function.

3.3.2 Development in PCGPANN

The plasticity in PCGPANN is based on the decision function. Once developmental index is invoked by the function, the network initiate a step by step process of development under the given set of aforementioned constraints. An example of various possible steps of development in PCGPANN are illustrated in Fig. 7. A change may be invoked in either the input of a node, a function or a weight gene as shown in the figure. The process is highlighted both in the genotype and the phenotype. Figure 7a shows the initial genotype and phenotype. There are two node functions, a sigmoid and a tangent hyperbolic, in the initial genotype. These are the available functions that can be randomly assigned to a node when mutation takes place. The ψ_0s and ψ_1s are network inputs and ψ_2 is the output of the first node and ψ_3 is the system output, while the weights have values in the range $[-1, 1]$. The mathematical representation of the initial network is given in Eq. 1.

$$\psi3 = Sig\left[0.6\psi1 + 0.2Tanh(0.2\psi0 + 0.7\psi1)\right] \tag{1}$$

In the first case, the function of first node i.e. Tanh is transformed to sigmoid function. The updated genotype and phenotype are presented in Fig. 7b. The mathematical expression for the updated genotype is given in Eq. 2.

$$\psi3 = Sig\left[0.6\psi1 + 0.2Sig(0.2\psi0 + 0.7\psi1)\right] \tag{2}$$

In the second case, the weight changes from 0.1 to 0.7 i.e. at the first input of the second node as shown in Fig. 7c. The mathematical expression for the updated genotype is given by Eq. 3.

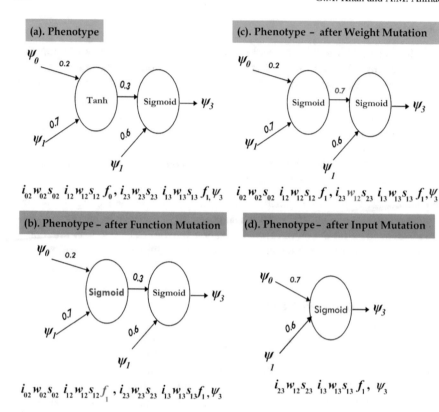

Fig. 7 Demonstration of real time development in PCGPANN

$$\psi 3 = Sig\left[0.6\psi 1 + 0.7 Sig(0.2\psi 0 + 0.7\psi 1)\right] \qquad (3)$$

In Fig. 7d, the structure is modified when an input to the neuron is altered i.e. the first input to the second node is disconnected from node 1 and connected to a system input. The network attains final expression as given by Eq. 4.

$$\psi 3 = Sig\left[0.6\psi 1 + 0.7\psi 0\right] \qquad (4)$$

Similar to FCGPANN and RCGPANN, PCGPANN is also first evaluated for its learning ability and then the generalization of performing in an unknown environment. Table 11 shows the comparative results for PCGPANN and other algorithms, used for single and double pole balancing tasks. The performance is shown in terms

Table 11 Comparison of PCGPANN with other neuroevolutionary algorithms applied on single and double pole balancing task: average number of network evaluations

Method	Single pole	Double pole
Conventional Neuro-Evolution (CNE) [36]	352	22 100
Symbiotic, Adaptive Neural Evolution (SANE) [23]	302	12 600
Enforced sub-population (ESP) [7]	289	3 800
Neuro-Evolution of Augmenting Topologies (NEAT) [35]	743	3 600
Cooperative synapse neuroevolution (CoSyNE) [6]	98	954
FCGPANN	21	77
PCGPANN	104	1 169

Table 12 Average number of random cart-pole initializations (out of 625) that can be solved

Type	Single pole	Double pole
FCGPANN	590	277.38
PCGPANN	456	349

Table 13 Comparison of PCGPANN with other ANNS for the prediction of foreign currency exchange rates

Network	Accuracy
AFERFM	81.2
HFERFM	69.9
Multi layer perceptron	72
Volterra network	76
Back propagation network	62.27
Multi neural network	66.82
CGPANN	98.85
PCGPANN [15]	98.8516

of average number of network evaluations [20]. Table 12 shows generalization of the PCGPANN genotypes in comparison to FCGPANN for both the single and double pole balancing scenarios. Plastic CGPANN has also been successfully applied to evolve a dynamic and robust computational model for efficiently predicting daily foreign currency exchange rates in advance based on past data [15]. Table 13 shows the comparative results of foreign currency exchange rate prediction using PCGPANN and other contemporary methods.

3.4   Plastic Recurrent Cartesian Genetic Programming Evolved Artificial Neural Network (PRCGPANN)

Plastic Recurrent Cartesian Genetic Programming Evolved Artificial Neural Network is an online learning approach that incorporates developmental plasticity in Recurrent Neural Networks. Recurrent Neural Networks can process arbitrary sequences of inputs due to their ability to access internal memory. In a Plastic RCG-PANN the output gene not only forms the system output but also plays a role in the developmental decision. Output of the system is applied to a decision function to invoke development of the network. Development in the phenotype takes place with the mutation of the genotype in runtime. The recurrent CGPANN has a feedback mechanism in the network, that feeds one or more outputs back to the system. The general approach of PCGPRNN is depicted in Fig. 8. The figure shows the inputs, outputs, connections, recurrent inputs and the output gene that invokes development in the network. The initial network is the original representation of the genotype that changes in response to the output of the system with the passage of time. The decision regarding the development is the reflection of output of the system fed into the sum block and the recurrent paths taken from the CGPANN block associated with the weights, summation and sigmoid function as shown in the figure. If the value obtained from the function is less than defined decision value, development is invoked in the network or otherwise the outputs are monitored without any modification. The uniqueness aspect of the approach is that network changes take place in real time according to the data flow in the network and this provides the plasticity feature to the network. PRCGPANNs invoke changes in the network by changing (or mutating) a node function, an input, a weight or by switching an input in the real time. The learning rules for development in the PCGPRNN are achieved during the process of evolution. A special case is presented here to describe the developmental mechanism in PRCGPANN at run time as shown in Figs. 9, 10 and 11. The system has three inputs (I_0, I_1 and I_2), two recurrent inputs (R_0 and R_1), five outputs (Y_0, Y_1, Y_2, Y_3 and Y_4), weights ($w_0, w_1, w_2, w_3, \ldots, w_{11}$) and the plastic feedback (PF).

Figure 9 shows the original phenotype having the following genotype:

$$\text{Genotype} = f_0, I_0, w_0, I_1, w_1, I_2, w_2, f_1, I_3, w_5, R_1, w_4, I_2, w_6,$$
$$f_2, I_3, w_7, R_1, w_4, R_0, w_8, f_3, I_3, w_{10}, I_4, w_{11}, I_0, w_9, I_3, I_1, I_4, I_5, I_6$$

Figure 10 illustrates the change in function as a result of change in function gene as highlighted in the genotype below:

$$\text{Genotype} = f_0, I_0, w_0, I_1, w_1, I_2, w_2, f_1, I_3, w_5, R_1, w_4, I_2, w_6,$$
$$f_0, I_3, w_7, R_1, w_4, R_0, w_8, f_3, I_3, w_{10}, I_4, w_{11}, I_0, w_9, I_3, I_1, I_4, I_5, I_6$$

Figure 11 shows the change in the output connectivity at runtime respectively with corresponding change in gene highlighted below:

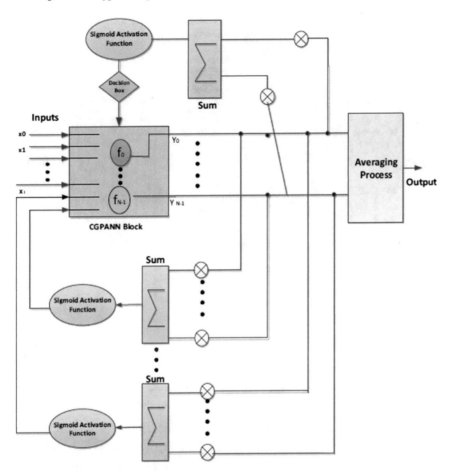

Fig. 8 Structural view of PRCGPANN

$$\text{Genotype} = f_0, I_0, w_0, I_1, w_1, I_2, w_2, f_1, I_3, w_5, R_1, w_4, I_2, w_6,$$
$$f_0, I_3, w_7, R_1, w_4, R_0, w_8, f_3, I_3, w_{10}, I_4, w_{11}, I_0, w_9, I_0, I_1, I_4, I_5, I_6$$

The various cases of genotype and phenotype diagrams above demonstrate the possible run time modification to the recurrent network, thus adding not only the signal feedback but also structural feedback through modification in the system architecture and topology at runtime. This is a very interesting system, because it can transform to feedforward and feedback structure from time to time during the processing, thus giving a unique ability to the system. This architecture is yet to be explored on various applications.

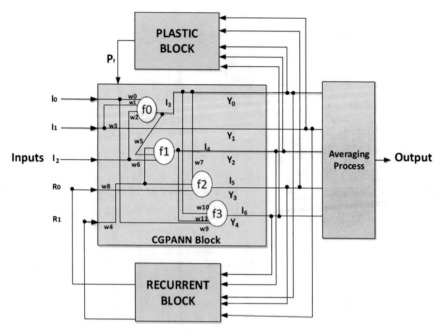

Fig. 9 Original PRCGPANN phenotype

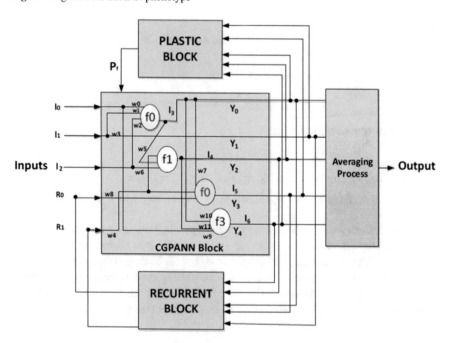

Fig. 10 Mutation in the function of the network

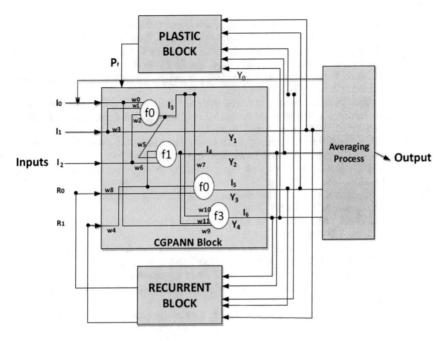

Fig. 11 Mutation in the output gene

4 Concluding Remarks

This chapter provides a detailed overview of how CGP is used to evolve artificial neural network by finding the proper set of weights and topology for the network. CGP based ANN provides an ideal platform to all Markovian and non-Markovian, Linear and non-linear problems that are static or dynamic/plastic. They can help finding the unknown mathematical model for the problem at hand. The CGPANN model not only helps in selection of topology and optimum weights for ANNs, but also helps in identifying the best possible features to be selected amongst many provided to the network and ignoring the unwanted noise. Various models of CGPANNs are tested in diverse fields of application for its speed of learning, robustness, and accuracy. Comparison with other algorithms on same set of problems show encouraging results.

References

1. Ali, J., Khan, G.M., Mahmud, S.A.: Enhancing growth curve approach using CGPANN for predicting the sustainability of new food products. In: IFIP International Conference on Artificial Intelligence Applications and Innovations, pp. 286–297. Springer (2014)

2. Arbab, M.A., Khan, G.M., Sahibzada, A.M.: Cardiac arrhythmia classification using cartesian genetic programming evolved artificial neural network. Exp. Clin. Cardiol. **20**(9) (2014)
3. Bidlo, M.: Evolutionary design of generic combinational multipliers using development. In: International Conference on Evolvable Systems, pp. 77–88. Springer (2007)
4. Carpenter, G.A., Grossberg, S.: The art of adaptive pattern recognition by a self-organizing neural network. Computer **21**(3), 77–88 (1988)
5. Fekiač, J., Zelinka, I., Burguillo, J.C.: A review of methods for encoding neural network topologies in evolutionary computation. In: Proceedings of 25th European Conference on Modeling and Simulation ECMS 2011, pp. 410–416 (2011)
6. Gomez, F., Schmidhuber, J., Miikkulainen, R.: Accelerated neural evolution through cooperatively coevolved synapses. J. Mach. Learn. Res. **9**(May), 937–965 (2008)
7. Gomez, F.J., Miikkulainen, R.: Solving non-markovian control tasks with neuroevolution. IJCAI **99**, 1356–1361 (1999)
8. Guo, H., Nandi, A.K.: Breast cancer diagnosis using genetic programming generated feature. Pattern Recognit. **39**(5), 980–987 (2006)
9. Haider, A., Hanif, M.N.: Inflation forecasting in Pakistan using artificial neural networks. Pak. Econ. Soc. Rev. 123–138 (2009)
10. Iranpour, M., Almassi, S., Analoui, M.: Breast cancer detection from fna using svm and rbf classifier. In: 1st Joint Congress on Fuzzy and Intelligent Systems (2007)
11. Jordan, M.I.: Attractor dynamics and parallellism in a connectionist sequential machine. In: Proceedings of the 8th Confererence of the Cognitive Science Society, pp. 531–546. Lawrence Erlbaum Associates (1986)
12. Kamruzzaman, J., Sarker, R.A.: Forecasting of currency exchange rates using ANN: a case study. In: Neural Networks and Signal Processing, 2003. Proceedings of the 2003 International Conference on, vol. 1, pp. 793–797. IEEE (2003, December)
13. Khan, G.M., Ali, J., Mahmud, S.A.: Wind power forecastingan application of machine learning in renewable energy. In: 2014 International Joint Conference on Neural Networks (IJCNN), pp. 1130–1137. IEEE (2014)
14. Khan, G.M., Khattak, A.R., Zafari, F., Mahmud, S.A.: Electrical load forecasting using fast learning recurrent neural networks. In: The 2013 International Joint Conference on Neural Networks (IJCNN), pp. 1–6. IEEE (2013)
15. Khan, G.M., Nayab, D., Mahmud, S.A., Zafar, H.: Evolving dynamic forecasting model for foreign currency exchange rates using plastic neural networks. In: 2013 12th International Conference on Machine Learning and Applications (ICMLA), vol. 2, pp. 15–20. IEEE (2013)
16. Khan, G.M., Ullah, F., Mahmud, S.A.: MPEG-4 internet traffic estimation using recurrent CGPANN. In: International Conference on Engineering Applications of Neural Networks, pp. 22–31. Springer (2013)
17. Khan, G.M., Zafari, F.: Dynamic feedback neuro-evolutionary networks for forecasting the highly fluctuating electrical loads. Genet. Program. Evolvable Mach. **17**(4), 391–408 (2016)
18. Khan, G.M., Zafari, F., Mahmud, S.A.: Very short term load forecasting using cartesian genetic programming evolved recurrent neural networks (CGPRNN). In: 2013 12th International Conference on Machine Learning and Applications (ICMLA), vol. 2, pp. 152–155. IEEE (2013)
19. Khan, M.M., Ahmad, A.M., Khan, G.M., Miller, J.F.: Fast learning neural networks using cartesian genetic programming. Neurocomputing **121**, 274–289 (2013)
20. Khan, M.M., Khan, G.M., Miller, J.F.: Developmental plasticity in cartesian genetic programming artificial neural networks. In: Proceedings of the 8th International Conference on Informatics in Control, Automation and Robotics (ICINCO 2011), pp. 449–458. SciTePress (2011)
21. Kryuchin, O.V., Arzamastsev, A.A., Troitzsch, K.G.: The prediction of currency exchange rates using artificial neural networks. Exch. Organ. Behav. Teach. J. **4** (2007)
22. Liu, K., Subbarayan, S., Shoults, R., Manry, M., Kwan, C., Lewis, F., Naccarino, J.: Comparison of very short-term load forecasting techniques. IEEE Trans. Power Syst. **11**(2), 877–882 (1996)
23. Moriarty, D.E.: Symbiotic evolution of neural networks in sequential decision tasks. Ph.D. thesis, University of Texas at Austin USA (1997)

24. Nayab, D., Khan, G.M., Mahmud, S.A.: Prediction of foreign currency exchange rates using cgpann. In: International Conference on Engineering Applications of Neural Networks, pp. 91–101. Springer (2013)
25. Pan, Z., Nie, L.: Evolving both the topology and weights of neural networks. Parallel Algorithms Appl. **9**(3–4), 299–307 (1996)
26. Papadrakakis, M., Papadopoulos, V., Lagaros, N.D.: Structural reliability analyis of elastic-plastic structures using neural networks and monte carlo simulation. Comput. Methods Appl. Mech. Eng. **136**(1–2), 145–163 (1996)
27. Philip, A.A., Taofiki, A.A., Bidemi, A.A.: Artificial neural network model for forecasting foreign exchange rate. World Comput. Sci. Inf. Technol. J. (WCSIT) **1** (3) 110–118 (2011)
28. Pujol, J.C.F., Poli, R.: Evolving the topology and the weights of neural networks using a dual representation. Appl. Intell. **8**(1), 73–84 (1998)
29. Refenes, A.N., Azema-Barac, M., Chen, L., Karoussos, S.: Currency exchange rate prediction and neural network design strategies. Neural Comput. Appl. **1**(1), 46–58 (1993)
30. Rehman, M., Ali, J., Khan, G.M., Mahmud, S.A.: Extracting trends ensembles in solar irradiance for green energy generation using neuro-evolution. In: IFIP International Conference on Artificial Intelligence Applications and Innovations, pp. 456–465. Springer (2014)
31. Rehman, M., Khan, G.M., Mahmud, S.A.: Foreign currency exchange rates prediction using cgp and recurrent neural network. IERI Procedia **10**, 239–244 (2014)
32. Sadeghi, B.: A bp-neural network predictor model for plastic injection molding process. J. Mater. Process. Technol. **103**(3), 411–416 (2000)
33. Sarangi, P.P., Sahu, A., Panda, M.: A hybrid differential evolution and back-propagation algorithm for feedforward neural network training. Int. J. Comput. Appl. **84**(14) (2013)
34. Singh, D., Singh, S.: A self-selecting neural network for short-term load forecasting. Electr. Power Compon. Syst. **29**(2), 117–130 (2001)
35. Stanley, K.O., Miikkulainen, R.: Efficient reinforcement learning through evolving neural network topologies. In: Proceedings of the 4th Annual Conference on Genetic and Evolutionary Computation, pp. 569–577. Morgan Kaufmann Publishers Inc. (2002)
36. Wieland, A.P.: Evolving neural network controllers for unstable systems. In: IJCNN-91-Seattle International Joint Conference on Neural Networks, 1991, vol. 2, pp. 667–673. IEEE (1991)

Multi-step Ahead Forecasting Using Cartesian Genetic Programming

Ivars Dzalbs and Tatiana Kalganova

Abstract This paper describes a forecasting method that is suitable for long range predictions. Forecasts are made by a calculating machine of which inputs are the actual data and the outputs are the forecasted values. The Cartesian Genetic Programming (CGP) algorithm finds the best performing machine out of a huge abundance of candidates via evolutionary strategy. The algorithm can cope with non-stationary highly multivariate data series, and can reveal hidden relationships among the input variables. Multiple experiments were devised by looking at several time series from different industries. Forecast results were analysed and compared using average Symmetric Mean Absolute Percentage Error (SMAPE) across all datasets. Overall, CGP achieved comparable to Support Vector Machine algorithm and performed better than Neural Networks.

1 Introduction

Predicting the future based on past data has been of interest for a long time. The further ahead the forecast can be made, the more time there is to act on tactical decisions, such as planning production resources or creating strategic contracts. Furthermore, forecasts have many applications, such as finance and telecommunication [1], economics [2, 3], meteorology [4], agricultural [5].

Multi-step ahead forecasting problem is more involved since, compared to one-step ahead forecasting, it has to deal with increased complications, like reduced accuracy and accumulated errors [6].

Rapid technological growth has made it possible to develop and test sophisticated machine learning algorithms in forecasting domain. These models use historical data to learn the underlying dependencies in the past to predict the future. For instance,

I. Dzalbs · T. Kalganova (✉)
Brunel University London, Kingston Lane, Uxbridge UB8 2PX, UK
e-mail: Tatiana.Kalganova@brunel.ac.uk

© Springer International Publishing AG 2018
S. Stepney and A. Adamatzky (eds.), *Inspired by Nature*, Emergence,
Complexity and Computation 28, https://doi.org/10.1007/978-3-319-67997-6_11

235

[7] shows how Neural Networks (NN) can be used to accurately predict long-term energy requirements of an electric utility. Furthermore, Cortez [5] shows that NN and Support Vector Machine (SVM) outperform the classical Holt-Winters (HW) forecasting method. The aim of this chapter is to introduce an alternative forecasting model based on CGP that is comparable to both NN and SVM.

2 Proposed Algorithm Applied to Data Forecasting Problem

Cartesian Genetic Programming (CGP) was first introduced as one of the evolutionary techniques to design logic circuits [8]. Due to its specific nature of the chromosome representation, CGP has been applied to a number of automatic design and optimisation problems [9, 10]. The proposed approach is based on classical CGP using a $(1 + \lambda)$ evolutionary strategy [11]. During each generation, the number of parents (P) are selected to produce a number of children (C) using mutation operator only. The mutation rate (M) is expressed as percentage of the number of genes that are being mutated. The quality of the chromosome is evaluated using fitness function. The best chromosomes in the population are defined as fittest members. Furthermore, the fittest member(s) of the population now becomes the parent(s) and the process repeats till G_n generations are reached.

Each chromosome represents one approximation equation. The equation is assembled using arithmetic operators defined by primary arithmetic equations, also called *gates*. Each gate can have a maximum of number of inputs, J. Each gate type defines one primary arithmetic operation. Variations of primary arithmetic equations used in evolutionary process are defined in the *function library* in advance. Therefore, each gate is represented using collection of the following genes:

- The number of inputs to the gate
- Gate type
- Collection of inputs for a specific gate
- Gate constant

  Each chromosome is defined as following (Fig. 1):

- Collection of gates, G
- Collection of chromosome outputs

  The number of chromosome outputs is defined by the number of outputs to be forecasted. The evolutionary process is driven by mutation that is focused on:

- Changing the number of inputs in gate
- Changing the gate function out of the *function library*
- Changing the gate inputs
- Changing the gate constant

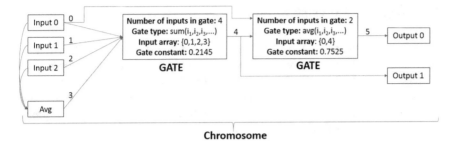

Fig. 1 Graphical representation of gate and chromosome. In this example, 3 inputs are used to generate 2 outputs. Only 2 gates are used and one additional input (avg). Output 1 uses first gate (output label 4), whereas output 0 uses output of the second gate (output label 5)

Table 1 Datasets used [5]

Series	Dataset title	The total number of data points	Duration of seasonal period (N) [5]	The number of validation data points (N_v)
Cars	Monthly car sales in Quebec 1960–1968	108	12	12
Pigs	Monthly total number of pigs slaughtered in Victoria. Jan 1980–August 1995	188	12	12
Pass	International airline passengers: monthly totals in thousands. Jan 49–Dec 60	144	12	19
Gas	Monthly gasoline demand Ontario gallon millions 1960–1975	192	12	19
Houses	Monthly sales of U.S. houses (thousands) 1965–1975	132	12	12
Cradfq	Monthly critical radio frequencies in Washington, D.C., May 1934–April 1954	240	12	24
Suns	Annual Wolfer sunspot numbers. 1770–1889	289	10	25

Furthermore, mutation of chromosome outputs is also allowed and the initial population is randomly generated.

In this chapter, the data have been forecast using two different proposed multi-step ahead forecasting methods. Depending on the method used, historical data entries are used to predict future values up to seasonal period N. Each dataset is divided into a training dataset and a validation dataset (N_v), as shown in Table 1, where validation data points, due to the sequential nature of the datasets, are the last N_v entries from all data points. The algorithm uses the training dataset to obtain an estimate function. The estimate function is then used to predict the values of non-trained validation data, also called *blind data*.

2.1 Lagging Forecast Method

Let us consider the series X be the actual time series and Y be the prediction based time series. Given the actual values in period x_t we make a forecast N periods ahead to predict value of y_{t+N}, where N is any integer value that represents the duration of seasonal period given in Table 1. Furthermore, predicted value of y_t will be based on the actual value of x_{t-N}. Therefore, this method takes one input value to predict one output value, as shown in Fig. 2.

2.2 Full Period Forecast Method

This method takes an input vector $(x_{t-N} \ldots x_{t-1})$ of N data points defined as particular seasonal period to predict some future output vector $(y_t \ldots y_{t+N-1})$. Therefore, this method can take advantages of relationships between points in the same seasonal period to predict either single or multiple future forecasted points, as shown in Fig. 3.

2.3 Evaluation of the Forecasting Performance

The forecasting performance is evaluated by an accuracy measure, such as Symmetric Mean Absolute Percentage Error (SMAPE) [12]:

$$SMAPE = \frac{1}{N} \sum_{t=1}^{N} \frac{|Y_t - X_t|}{(|X_t| + |Y_t|)/2} \qquad (1)$$

Fig. 2 Scheme of lagging forecast method, where series x is the actual time series and y is the prediction based time series

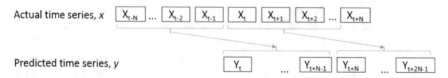

Fig. 3 Scheme of full period forecast method, where series x is the actual time series and y is the prediction based time series

where Y_t is the forecast value, X_t is the actual value and N is the number of forecasts points (seasonal period). SMAPE is particularly useful for measuring scale-independent performance, having values from 0 to 200%, with 0% indicating no error in the forecast and 200% being the worst forecast. It allows us to compare average forecasting error across multiple datasets which then is comparable to other forecasting methods.

3 Experimental Results

In order to evaluate the proposed algorithm's performance, results must be compared. Using [5] Neural Networks (NN) and Support Vector Machine's (SVM) results as the baseline, several experiments were run using both, Lagging forecast and Full period forecast. Furthermore, 10 runs were applied to the selected model and the average performance value obtained. The chosen sets of arithmetic operators in the Function library are shown in Table 2. These sets are grouped by their complexity.

Experiment 1: Feasibility study
In this experiment, the developed algorithm was tested across all 7 datasets summarized in Table 1 in order to confirm that developed CGP algorithm can be used on real data. The initial algorithm configuration was derived from the complexity of the datasets and available hardware, as shown in Table 3. Results are then compared to NN and SVM solutions as in Table 4, where CGP-L and CGP-FP is Cartesian Genetic Programming with Lagging method and with Full Period method respectively.

$$Absolute\ error = \sum_{t=1}^{N} |Y_t - X_t| \tag{2}$$

where Y_t is the forecast value, X_t is the actual value and N is the number of forecasts points (seasonal period).

As can be seen from the results in Table 4, both CGP-L and CGP-FP obtain very competitive results. CGP-L performed better than NN in 6 out of 7 forecasts. Similarly, CGP-FP performed better results compared to NN in 4 out of 7 datasets. When compared to SVM, both CGP-L and CGP-FP showed competitive results, performing worse by 1.96% and 3.01% respectively.

This experiment confirmed that CGP is able to forecast time series and produce competitive results.

Experiment 2: Optimum fitness function
The fitness function, also referred as objective function, determines how the performance of produced results are measured. It can be based on the forecasting error

Table 2 Function library combinations, where i_1 is the first input of the gate, const is the gate constant

Function name (name of the primitive arithmetic functions)	Combination				
	1—default	2—simple	3—trigonometric	4—statistic	5—logical
sum(i_1,i_2,i_3,...)	√	√	√	√	√
multiply(i_1,i_2,i_3,...)		√	√	√	√
const*sum(i_1,i_2,i_3,...)	√	√	√	√	√
avg(i_1,i_2,i_3,...)	√			√	√
substract(i_1,i_2,i_3,...)	√	√	√	√	√
division(i_1,i_2)		√	√	√	√
naturalLog(i_1)			√	√	√
logBase10(i_1)			√	√	√
exp(i_1)			√	√	√
sin(const*i_1)	√		√	√	√
cos(const*i_1)	√		√	√	√
tan(const*i_1)			√	√	√
sqrt(i_1)			√	√	√
const*0.1			√	√	√
mult(i_1,i_2,const)	√	√	√	√	√
const*sin(i_1)	√				
const*cos(i_1)	√				
const*tan(i_1)	√				
const*i_1^2	√	√	√	√	√
const*i_1^3	√	√	√	√	√
i_1^const	√	√	√	√	√
standardDev(i_1,i_2,i_3,...)				√	√
skewNess(i_1,i_2,i_3,...)				√	√
kurtosis(i_1,i_2,i_3,...)				√	√
invert(i_1)					√
min(i_1,i_2,i_3,...)					√
max(i_1,i_2,i_3,...)					√

Table 3 CGP initial configuration

The number of gates (G)	50
Mutation rate (M)	5%
The number of children (C)	2
The number of parents (P)	8
The number of generations (G_n)	100 000
The maximum number of inputs in a gate (J)	20
Fitness function	Absolute error
Function library combination	1 (default value)

Table 4 Comparison of the forecasting errors (SMAPE, %)

Series	NN [5]	SVM [5]	Proposed CGP-L	Proposed CGP-FP
Cars	9.25	9.72	9.02	12.56
Pigs	6.3	7.2	6.12	13.72
Pass	4.88	9.08	3.6	3.64
Gas	7.67	4.08	6.06	3.16
Houses	15.49	10.68	17.56	15.56
Cradfq	22.21	10.71	15.3	14.58
Suns	59.8	42.33	49.84	50.67
Average value	17.94	**13.4**	15.36	16.41

or on other metrics, such as service level or profit. Due to the nature of datasets, various representations of forecasting error have been explored.

In this experiment, two variations of initial algorithm in Experiment 1 were developed based on the configuration in Table 3. The fitness function was changed from Absolute Error (AE) to Mean Square Error (MSE) and Symmetric Mean Absolute Percentage Error (SMAPE).

Results in Table 5 clearly indicates that evaluating the error within CGP's fitness function using Absolute Error (AE) performs better than using more complex MSE or SMAPE evaluations.

Experiment 3: Most suitable function library

CGP uses a combination of primary arithmetic equations (gate functions) to estimate the time series. These arithmetic operators can be as simple as summation of two input numbers, or as complex as calculating standard deviation across multiple inputs.

In this experiment, the impact on different gate functions was investigated. Each gate can contain only one of the functions from the function library (in Table 2). Five different combinations were generated:

1. default—combination that was used in Experiment 1: Feasibility study and Experiment 2: Optimum fitness function;

Table 5 Comparison of the fitness function (SMAPE, %)

Fitness function series	AE		MSE		SMAPE	
	CGP-L	CGP-FP	CGP-L	CGP-FP	CGP-L	CGP-FP
Cars	9.02	12.56	8.08	11.03	11.61	11.41
Pigs	6.12	13.72	7.6	13.67	6.99	8.64
Pass	3.6	3.64	4.01	4.02	4.17	4.04
Gas	6.06	3.16	5.34	6.88	5.17	3.05
Houses	17.56	15.56	17.94	34.47	15.97	37.09
Cradfq	15.3	14.58	19.88	17.07	15.53	11.71
Suns	49.84	50.67	52.98	52.66	50.92	45.85
Average value	**15.36**	**16.41**	16.55	19.97	15.76	17.4

2. simple—only using basic functions such as addition, subtraction and multiplication/division;
3. trigonometric—additional trigonometric functions such as sin, cos and tan as well as exponential components;
4. statistical—additional statistical components such as average, standard deviation, skewness and kurtosis;
5. logical—additional logical expressions such as inverter, min, max and comparison functions.

CGP was set based on the parameters in Table 3 where variable "Function Library combination" was altered. The results are compared in Table 6.

This experiment shows that CGP performs the best with very simplistic function library. Although all combinations produce very comparable results, the "simple" function library combination performs on average approximately 1% better than any other combination on both CGP-L and CGP-FP. During 10 runs all models produced stable and repeatable results with variation of around 0.2%.

Experiment 4: Additional statistical inputs

Although CGP can estimate complex mathematical equations using simple mathematical operators, it can be beneficial to provide additional statistical inputs to speed up the process.

Additional 4 statistical inputs where added to the model's inputs in this experiment. Statistical inputs are calculated based on previous period's vectors values. For instance, if current input vector is $(x_{t-N} \ldots x_{t-1})$ in case of CGP-FP, then additional inputs would be calculated based on $(x_{t-2N} \ldots x_{t-N-1})$. Similarly, if input is x_{t-N} in case of CGP-L, then additional statistical inputs would be calculated from $(x_{t-N} \ldots x_{t-1})$ vector. We looked at the four common statistical expressions: mean, standard deviation, skewness [13] and kurtosis [13]:

$$skewness = \frac{\sum_{i=1}^{N} \left(X_i - \overline{X}\right)^3 / N}{s^3}$$

$$kurtosis = \frac{\sum_{i=1}^{N} \left(X_i - \overline{X}\right)^4 / N}{s^4}$$

where $\overline{X}$ is the mean and s is the standard deviation and N is the number of data points in the period. CGP was set based on the parameters in Table 7 and averages of 10 simulations can be seen in Table 8. Table 8 shows that CGP-L average SMAPE error increases by 1.47% when additional statistical inputs are used. However, CGP-FP model benefits from statistical information, decreasing the forecasting error further by 3.01% and is comparable to the forecasting error of SVM [5]. Compared to SVM results, CGP-FP is performing better on 4 out of 7 datasets, while average SMAPE results are the same (13.4%).

Experiment 5: Models

To further investigate the dynamics of Cartesian Genetic Programming, the impact of the number of generations was explored. The algorithm was set up as defined in

Table 6 Comparison of the function libraries (SMAPE, %), where CGP-L and CGP-FP is Cartesian genetic programming algorithm with lagging forecast and with full period forecast respectively

Library combination	Default (1)		Simple (2)		Trigonometric (3)		Statistical (4)		Logical (5)	
Series	CGP-L	CGP-FP	CGP-L	CGP-FP	CGP-L	CGP-FP	CGP-L	CGP-FP	CGP-L	CGP-FP
Cars	9.02	12.56	8.76	12.24	10.46	17.52	9.5	10.88	9.6	15.26
Pigs	6.12	13.72	6.73	9.42	6.66	11.98	6.62	22.45	7.25	9.93
Pass	3.6	3.64	3.43	9.76	4.45	6.57	3.46	4.41	6.61	3.59
Gas	6.06	3.16	5.9	2.84	8.91	9.98	5.51	2.91	7.3	3.61
Houses	17.56	15.56	16.2	17.65	16.52	24.29	17.61	33.12	17.27	28.02
Cradfq	15.3	14.58	15.18	14.33	15.77	17.11	15.73	13.18	16.69	11.31
Suns	49.84	50.67	45.52	40.33	49.34	46.96	51.02	40.75	52.61	41.82
Average	15.36	16.41	**14.53**	**15.22**	16.02	19.2	15.64	18.24	16.76	15.83

Table 7 CGP configuration, experiment 4

The number of gates (G)	50
Mutation rate (M)	5%
The number of children (C)	2
The number of parents (P)	8
The number of generations (G_n)	100 000
The maximum number of inputs in a gate (J)	20
Fitness function	Absolute error
Function Library combination (See Table 2)	2

Table 8 Additional statistical input comparison (SMAPE, %)

Series	NN [5]	SVM [5]	CGP-L (without inputs)	CGP-FP (without inputs)	CGP-L (with inputs)	CGP-FP (with inputs)
Cars	9.25	9.72	9.02	12.56	12.99	11.65
Pigs	6.3	7.2	6.12	13.72	8.36	8.89
Pass	4.88	9.08	3.6	3.64	4.09	5.4
Gas	7.67	4.08	6.06	3.16	5.58	4
Houses	15.49	10.68	17.56	15.56	27.88	15.72
Cradfq	22.21	10.71	15.3	14.58	10.66	8.42
Suns	59.8	42.33	49.84	50.67	48.24	39.75
Average	17.94	**13.4**	15.36	16.41	16.83	**13.4**

Table 7, with additional statistical inputs described in Experiment 4: and termination criteria "Generations (G_n)" was altered. Only the overall SMAPE averages for all datasets are shown in Table 9. It can be clearly seen from Table 9 that forecasting error decreases when generations (G_n) is increased from 1 to 1000, then forecasting error stays around the same value from 1 000 to 100 000 generations and increases when iterations reach 1 million. At 1 million generations algorithm is over-trained which leads to over-fitting the datasets. Moreover, the algorithm starts to remember the datasets instead of finding the general expression, which is very

Table 9 Generation (G_n) forecasting performance, SMAPE (%)

G_n	CGP-L	CGP-FP
1	135.9	147.5
10	138.3	130.2
100	83.7	45.3
1 000	16.0	14.0
10 000	17.1	**13.0**
50 000	16.9	14.3
100 000	16.8	13.4
1 000 000	21.9	15.3

common problem in machine learning for forecasting [14]. Furthermore, CGP-FP model at 10 thousand generations is performing the best, with average of 13.0% SMAPE value which is better than SVM [5] by 0.4% on average.

4 Conclusions and Further Work

Multi-step ahead forecasting is very useful tool for various industries. Predicting a time series many steps ahead in the future is not trivial. Our results supplement the work discussed in [15], where it was shown that Multiple-Output (in this paper, "Full Period") approach are invariably better than Single-Output (in this paper, "Lagging method") approach.

Moreover, proposed model achieved competitive performance to the SVM and NN results in [5]. On average (SMAPE), proposed CGP-FP model performed 0.4% better than SVM and 4.94% better than NN.

Furthermore, work in [16] describes a technique called "dropout" that prevents over-fitting on Neural Networks by randomly dropping the nodes along with their connections from the neural network during training. Similar approach could be adopted to CGP model, where best chromosome's currently active gates are randomly dropped in order to avoid overfitting.

Acknowledgements The authors would like to thank the supporter of this work: Intel Corporation.

References

1. Palit, A.K., Popovic, D.: Computational intelligence in time series forecasting: theory and engineering applications (advances in industrial control). Springer (2005)
2. Xiong, T., Bao, Y., Hu, Z.: Beyond one-step-ahead forecasting: evaluation of alternative multi-step-ahead forecasting models for crude oil prices. Energy Econ. **40**, 405–415 (2013)
3. Andalib, A., Atrya, F.: Multi-step ahead forecasts for electricity prices using NARX: a new approach, a critical analysis of one-step ahead forecasts. Energy Convers. Manage. **50**(3), 739–747 (2009)
4. Chang, F.J., Chiang Y.M., Chang, L.C.: Multi-step-ahead neural networks for flood forecasting. Hydrol. Sci. J. **52**(1), 114–130 (2007)
5. Cortez, P.: Sensitivity analysis for time lag selection to forecast seasonal time series using neural networks and support vector machines. In: The 2010 International Joint Conference on Neural Networks (IJCNN) (2010)
6. Sorjamaa, A., Hao, J., Reyhani, N., Ji, Y., Lendasse, A.: Methodology for long-term prediction of time series. Neurocomputing **70**(16–18), 2861–2869 (2007)
7. Al-Saba, T., El-Amin, I.: Artificial neural networks as applied to long-term demand forecasting. Artif. Intell. Eng. **13**(2), 189–197 (1999)
8. Miller, J.F., Job, D., Vassilev, V.K.: Principles in the evolutionary design of digital circuits—part I. Genet. Program Evolvable Mach. **1**(1), 7–35 (2000)

9. Gajda, Z., Sekanina, L.: Gate-level optimization of polymorphic circuits using Cartesian genetic programming. In: Evolutionary Computation (2009)
10. Walker, J.A., Liu, Y., Tempesti, G., Tyrrell, A.M.: Automatic code generation on a MOVE processor using Cartesian genetic programming. In: Evolvable Systems: From Biology to Hardware, pp. 238–249 (2010)
11. Miller, J.F.: Cartesian Genetic Programming. Springer, Berlin Heidelberg (2011)
12. Armstrong, J.S.: Long-Range Forecasting: From Crystal Ball to Computer. Wiley-Interscience (1985)
13. NIST, 1.3.5.11. Measures of Skewness and Kurtosis. http://www.itl.nist.gov/div898/handbook/eda/section3/eda35b.htm
14. Hawkins, D.M., The problem of overfitting. J. Chem. Inf. Comput. Sci. pp. 1–12 (2004)
15. Taieba, S.B., Bontempia, G., Atiyac, A.F., Sorjamaab, A.: A review and comparison of strategies for multi-step ahead time series forecasting based on the NN5 forecasting competition. Expert Syst. Appl. **39**(8), 7067–7083 (2012)
16. Srivastava, N., Hinton, G., Krizhevsky, A., Sutskever I., Salakhutdinov, R.: Dropout: A simple way to prevent neural networks from overfitting, J Mach Learn Res **15**, pp. 1929–1958, (2014)

Medical Applications of Cartesian Genetic Programming

Stephen L. Smith and Michael A. Lones

Abstract The application of machine learning techniques to problems in medicine are now becoming widespread, but the rational and advantages of using a particular approach is not always clear or justified. This chapter describes the application of a version of Cartesian Genetic Programming (CGP), termed *Implicit Context Representation CGP*, to two very different medical applications: diagnosis and monitoring of Parkinson's disease, and the differential diagnosis of thyroid cancer. Importantly, the use of CGP brings two major benefits: one is the generation of high performing classifiers, and the second, an understanding of how the patient measurements are used to form these classifiers. The latter is typically difficult to achieve using alternative machine learning methods and also provides a unique understanding of the underlying clinical conditions.

1 Introduction

Medical problems are a specific and challenging example of real-world applications of evolutionary computation which require special consideration in design and implementation to ensure a meaningful and effective outcome. Any investigation that goes beyond a simple benchmarking exercise using existing datasets will likely require substantial resources including the time and willingness of patients and clinicians which often incurs a certain level of anxiety, inconvenience and cost. It is therefore essential from the outset to clearly understand the clinical need and how the application of computational intelligence methods can meet this need through,

S.L. Smith (✉)
Department of Electronic Engineering, University of York, York, UK
e-mail: stephen.smith@york.ac.uk

M.A. Lones
School of Mathematical and Computer Sciences, Heriot-Watt University, Edinburgh, UK

for example, a quicker test, a more accurate or objective diagnosis, or providing a better understanding of the clinical condition itself. For clinical diagnosis and monitoring, it is also important to establish how the method can be easily incorporated into the exiting clinical protocol to encourage uptake and minimise errors in use. Similarly, the health economics of introducing such methods into a health system should be one of the first considerations, not the last, and will ultimately determine whether the technology will be adopted in routine clinical use.

It is clear that the application of evolutionary computation (EC) to problems in medicine has much to offer and has previously been demonstrated in areas such as: medical imaging and signal processing, data mining medical data and patient records, modelling and simulation of medical processes, clinical expert systems and knowledge-based systems, and clinical diagnosis and therapy [1]. Much previous work has focused on the use of genetic algorithms (GAs) as an optimiser in combination with feature extractors and other methods that are applied to the raw data [2–7]. This is understandable as the size, nature and type of data can vary hugely depending on the problem at hand; for example, a common set of vital signs for a patient (such as pulse rate, temperature, respiration rate, and blood pressure) will generate a few bytes of data for each reading, whereas full-field digital mammography can generate in the order of 100 MB of image data per patient and a typical MRI scan several gigabytes. The modality of the data can encompass a very wide range from discrete readings, images, sound, video and text to hugely complex multidimensional datasets. There is also potential merit in applying EC directly to the raw data with the evolved algorithm effectively providing a feature extractor as well as a classifier.

Considering the proliferation of EC methods developed over the past 10 years, it can be difficult to justify the choice of a particular algorithm, but previous work and experience has shown that some are better suited to particular applications than others [8–12]. Genetic Programming (GP) is a powerful and flexible paradigm but has drawbacks, the most common being bloat of the parse tree representation (an uncontrolled growth of the network over subsequent generations) [13, 14] and lack of built-in modularity that leads to an ungainly and inefficient solutions. In the case of medical applications, both of these features are important to deliver solutions that are both effective and have the capacity to be interrogated by, for example, providing an informed understanding of how the provided clinical data relates to the resulting classification. The directed graph representation of Cartesian Genetic Programming (CGP) addresses these drawbacks of GP and has a number of additional advantages that makes implementation of medical applications both easier and more effective, such as a variable geometry, simple encoding and adaption of an implicit context representation. These properties and benefits are considered in greater detail in the following section, after which several clinical example applications are presented that demonstrate both the flexibility and power of CGP to deliver effective clinical solutions.

2   CGP for Medical Applications

As discussed above, CGP possess a number of properties that make it an effective choice of algorithm for application to a wide range of medical problems. The directed graph representation is both flexible and powerful, and can be used as an optimiser, feature extractor and classifier.

2.1   CGP Geometry

Although CGP, as its name indicates, can adopt a Cartesian arrangement of rows and columns, it is commonly implemented with only one row and a number of columns equal to the number of processing nodes required. This provides maximum flexibility (within a directed graph representation) for any one node to provide the input to any subsequent node to its right in the network, as show in Fig. 1.

Alternatively, a Cartesian arrangement with multiple rows and columns can be implemented, such as a network of 10 rows and 3 columns, as shown in Fig. 2.

The choice of geometric arrangement will have an effect on the processing of the data as with the single row representation, the second processing node onwards may be supplied with data from an input node or any proceeding processing node. However, in the multi-row representation, there is a greater opportunity for all input node values to be processed in a similar way, as the first 10 processing nodes in the first column of the network may only take data from the input nodes. It can be argued that such an arrangement will be beneficial in scenarios such as that depicted in Fig. 3 where a steam of data is being sampled in a moving window operation before being presented to the network. In this case, it would be logical to suggest that data elements from the window should be processed in a similar way.

Reducing the number of processing nodes specified in the geometric parameters of the CGP network can be used to force simpler solutions to be evolved, although potentially with reduced fitness. Alternatively, the number of active nodes within

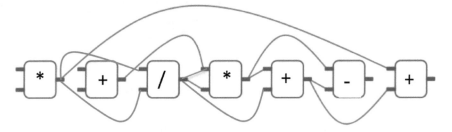

Fig. 1 Commonly adopted single row CGP geometry

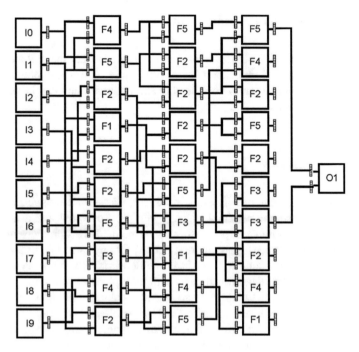

Fig. 2 Multiple row CGP geometry

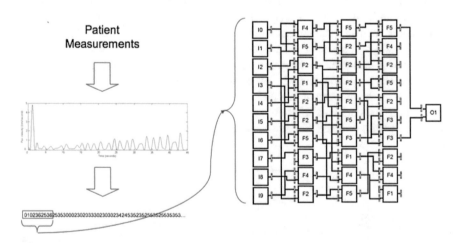

Fig. 3 CGP processing of streamed data using a moving window operation

the network can be used as a factor in the fitness function to encourage simpler networks to be evolved. In either case this is a useful property where the complexity of the evolved solution and execution time is critical, as for example, where the algorithm is to be implemented in firmware, or hardware, for real time operation.

2.2 *Implicit Context Representation CGP*

One of the benefits of CGP over standard GP is the implicit reuse of subexpressions and its use of functional redundancy provided by its graphical representation [15]. However, CGP can be said to be positionally dependent, since the inputs to, and output from processing nodes are dependent upon its Cartesian co-ordinates which are encoded within the program's representation. This positional dependence can be disruptive during recombination and, consequently, the crossover operator is often ineffective [16] (unless the operator is well matched to the problem [17]. Lones [18] and Smith [19] have proposed *Implicit context* as a means of providing positional independence to CGP by specifying that interconnections between input nodes, processing nodes and output nodes are specified in terms of each node's functional context within the network rather than its Cartesian co-ordinates. Termed Implicit Context Representation CGP (IRCGP), a node's inputs and output are augmented by functionality profiles—*binding sites* at the input and a *shape* at the output, as shown in Fig. 4. The shape functionality profile is simply a vector that describes the functional behaviour of the node within the network, and the binding site functionality profiles, vectors that describe the functionality profile of the node to which that input is best matched. Networks are therefore formed, not through stochastic operation as in conventional CGP, but by interconnecting nodes which most closely match their respective functionality profiles. This is achieved using a bottom-up process in which each node's binding site is connected to the input or processing node shape (if available) that provides the closest match, as illustrated in Fig. 5. Following selection, recombination operators are applied to the binding site functionality profiles and node function (as in conventional CGP), and the shape functionality profile calculated as a function of the two. (Further details of how functionality profiles are described and defined can be found in [19].)

Fig. 4 Functionality profiles used to describe a CGP processing node (taken from [16])

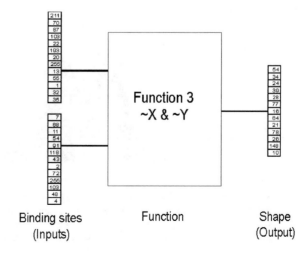

| Binding sites | Function | Shape |
| (Inputs) | | (Output) |

2.3   Fitness Function

Although not limited to CGP, the representation facilitates the use of simple fitness function, preferably a single figure. In medical applications there is often an overlap between the two classes under investigation which will mean an unavoidable misclassification of data inputs dependent on the threshold adopted. This is shown graphically in Fig. 6 where the proportion of True Positives (TP), True Negatives (TN), False Positives (FP) and False Negatives (FN) will depend on where the threshold is placed.

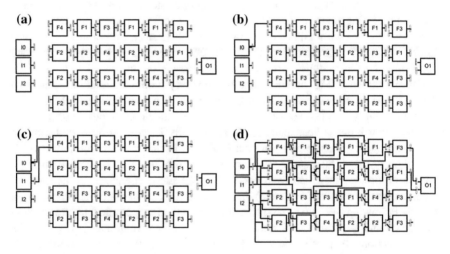

Fig. 5 The interconnection of processing nodes within a CGP network using functionality profiles (taken from [20])

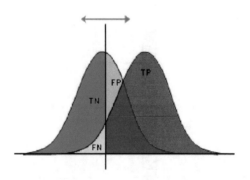

Fig. 6 Depiction of True Positives (TP), True Negatives (TN), False Positives (FP) and False Negative (FN) values dependent on the threshold chosen in an overlapped two-class problem (kakau (https://commons.wikimedia.org/wiki/File:Receiver_Operating_Characteristic.png), Receiver Operating Characteristic, https://creativecommons.org/licenses/by-sa/3.0/legalcode)

A convenient means of establishing the performance of an evolved classifier over a range of thresholds can be achieved using a receiver operating characteristic (ROC) curve [21], which can be construction as shown in Fig. 7. The area under the ROC curve provides a summary of the classifier performance over a range of thresholds and can therefore be used as an effective fitness function.

2.4 CGP as an Optimiser and a Feature Extractor

It is common for EAs to be used an optimiser in conjunction with a feature extractor. The Cartesian representation of CGP and the predefined number of inputs to the network makes it particularly appropriate to act on either previously extracted features or the raw data values. Figure 8 provides an example of how features extracted from pixel values of an image can then be presented to the inputs of a CGP network, acting as an optimiser and classifier. Figure 9 portrays the same image, but in this example, the pixel values are passed directly to the inputs of the CGP network. It is reliant on CGP to effectively evolve features that will subsequently classify.

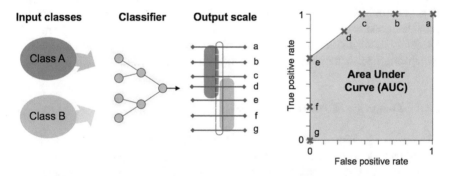

Fig. 7 Summary of how the area under a receiver operating characteristic (ROC) curve can be used as an effective fitness function for evolving classifiers

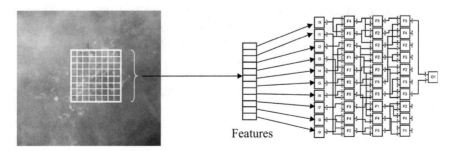

Fig. 8 CGP as an optimising classifier

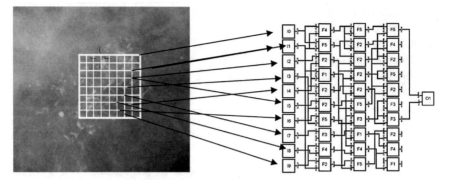

Fig. 9 CGP as a feature extractor and classifier

3 Example Applications

3.1 Parkinson's Disease

Parkinson's disease (PD) affects approximately 120,000 people in the UK alone and is projected to increase dramatically over the next decade as people live longer [22]. It is a chronic debilitating disease which has no cure, but effective medication provides good relief from symptoms in most patients for a number of years. The two main clinical challenges of the condition are accurate diagnosis and effective monitoring and management of side effects from medication.

3.1.1 Diagnosis of Parkinson's Disease

Diagnosis of Parkinson's disease can be difficult to confirm by conventional clinical assessment alone particularly in the early stages when symptoms are easily confused with other conditions such as essential tremor, progressive supranuclear palsy and dystonia. It is widely accepted that misdiagnosis of Parkinson's disease is as high as 25% [23, 24] and yet early confirmation is important for effective long term management of the condition [25].

A novel, non-invasive test has been developed by Smith et al. [26] which digitizes a conventional clinical assessment for Parkinson's disease—a simple finger-tapping task—where the patient is asked to tap their forefinger and thumb together for a period of 10–30 s. To digitize these tapping actions in real-time two small electromagnetic tracking sensors are attached to the forefinger and thumb as shown in Fig. 10. A recording from the sensors showing the distance between the finger and thumb over the course of a single tap is shown in Fig. 11 along with the associated acceleration profiles. The normalized acceleration for the entire tapping sequence is fed to the inputs of an IRCGP network via an overlapping moving window, as summarized in Fig. 12. The fitness function comprises the area under the receiver

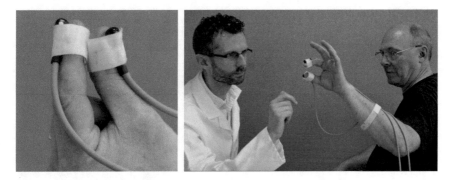

Fig. 10 Electromagnetic tracking sensors used to digitize a conventional clinical tapping task for Parkinson's disease

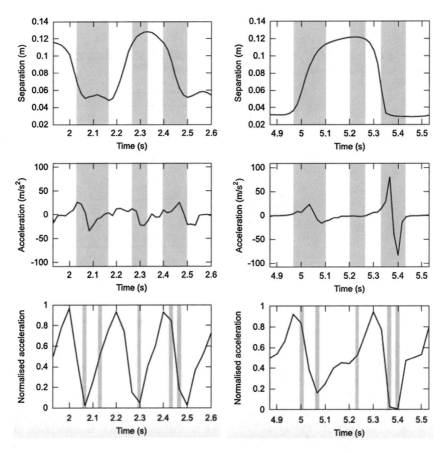

Fig. 11 Extract of data for one finger tap showing separation between finger and thumb (top), acceleration (middle) and normalised acceleration (bottom); superimposed grey bands relate to input data used in evolved classifier (taken from [28])

Fig. 12 Overview of moving
window operation feeding
tapping data to IRCGP (taken
from [28])

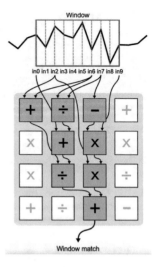

Fig. 13 ROC curves for the
best evolved classifier (taken
from [28])

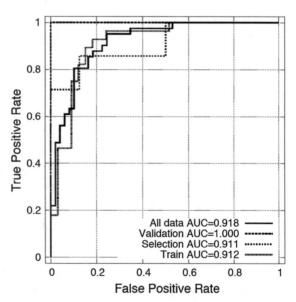

operating characteristic curve (AUC) and classifier results exceeding 0.9 AUC have
been achieved, as shown in Fig. 13. For full details see Lones et al. [27].

Figure 14 gives a schematic representation of the most discriminating classifier
evolved. This relatively simple network provides an important indication of those
aspects of the finger tapping activity that contributed to the classification of
Parkinson's disease; in this case, inputs 5, 7, 12, 16 and 17 have been mapped on to
the data in Fig. 11 as a set of grey lines. Interestingly, the evolved network has

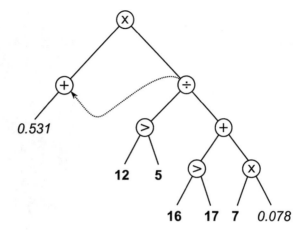

Fig. 14 Schematic for most discriminative classifier. Window offsets are in bold, constants are in italics, and > indicates the max function. The dashed line indicates sub-expression reuse, which is possible because we are using graph-based CGP, rather than a standard tree-based GP (taken from [28])

selected the deceleration of the fingers in both the opening and closing phases of the tapping activity.

3.1.2 Monitoring the Side Effects to Medication for Parkinson's Disease

The most effective form of treatment for PD symptoms is a drug called levodopa, but approximately 90% patients who take it for ten years or more develop involuntary movements called dyskinesia. These movements are a major source of disability, severely affecting the patient's quality of life and have significant financial implications for the healthcare system. Management of dyskinesia is particularly difficult as it may occur many times per day and an accurate method of monitoring is not available. Currently, physicians rely on patients' own descriptions, or in severe cases, patients are admitted to hospital for several days to monitor symptoms. However, because it is difficult to measure the dyskinesia accurately, the changes made to the patient's medication can often be ineffective. Providing an objective and accurate record of the patient's dyskinesia will allow physicians to make effective changes to medication, improve the patient's quality of life and reduce costs to the healthcare system.

The solution adopted was to monitor the patient's movements over 24–48 h by attaching small matchbox-sized sensors to the patient's arms, legs, head and torso as shown in Fig. 15. The sensors contain accelerometers and gyroscopes that measure the patient's movements in six degrees of freedom and can transmit the data via Bluetooth to a computer or smart phone, or record to an internal microSD card (Fig. 16). Once uploaded, the data is presented to an IRCGP network using a moving window in a similar way to that described above for the finger tapping task [29].

Fig. 15 Positioning of
sensors on patient's arms,
legs, torso and head

Fig. 16 Six degree of
freedom
accelerator/gyroscope sensor

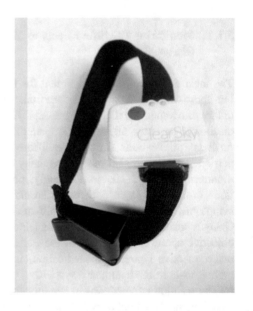

Clinically, Dyskinesia is measured on a five-point scale where 0 signifies no
occurrence of the side-effect and 4, severe occurrence. The results presented in
Fig. 17 show that the best evolved IRCGP network can predict Dyskinesia (at level
4 and level 3) with an area under RoC curve (AUC) of 0.9. Figure 18 shows how
tremor, a symptom of Parkinson's disease itself, can also be predicted with an
accuracy of 0.98 AUC. The results for dyskinesia are plotted on an easy-to-read
time chart that allow the clinicians to see how the dyskinesia varies throughout the
measurement period with respect to when medication is taken. Figure 19 is an

Fig. 17 ROC curves for dyskinesia levels 1–4 (taken from [29])

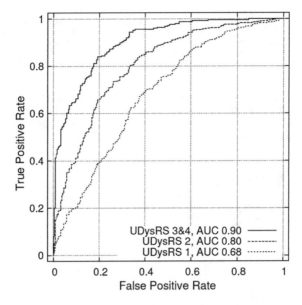

Fig. 18 ROC curves for tremor levels 1–4 (taken from [29])

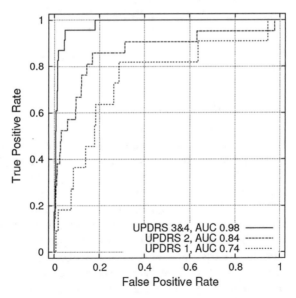

example of a summary chart for a patient measured over a 24 h period. The red dots represent the time at which medication was taken, the dark green lines the occurrence of severe Dyskinesia (level 4) and the light green lines, significant dyskinesia (level 3). The purple line indicates when the patient is asleep. Figure 20 shows a similar chart for each of the individual sensors worn.

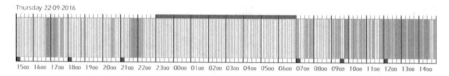

Fig. 19 Example of a summary chart for a patient measured over a 24 h period. The red dots represent the time at which medication was taken, the dark green lines the occurrence of severe dyskinesia (level 4) and the light green lines, significant dyskinesia (level 3). The purple line indicates when the patient is asleep

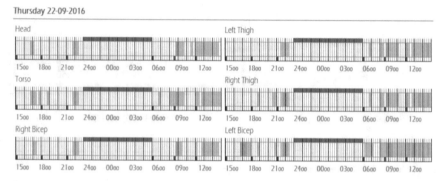

Fig. 20 Individual charts for each of the six sensors worn by the patient. See Fig. 19 for a full explanation

3.2 Thyroid Cancer

Cancer of the thyroid can be broadly categorised into four main types: Papillary, Follicular, Medullary and Anaplastic. Although overall incidence is relatively low with 3,404 new cases being diagnosed in the UK in 2014 (accounting for approximately 1% of all cancers), thyroid cancer incidence rates have increased by 149% since the 1970s. Incidence is dominated by Papillary cancer which accounts for some 75–85% of cases and has an excellent prognosis if detected early; those with cancer stages 1 and 2 have a 100% 5-year survival rate, but this decreases to 51% for cancer stage 4. Follicular thyroid cancer has a similar prognosis and accounts for 10–20% of cases, whereas Medullary thyroid cancer (5–8% of cases) has a poorer prognosis for cancer stage 4. Anaplastic thyroid cancer, accounting for less than 5% of cases, has a 7% 5-year survival rate and is usually classed as cancer stage 4 when diagnosed [30]. It is clear that a reliable and rapid means of diagnosing and differentiating between different types of thyroid cancer is needed to improve prognosis. Currently confirmation of diagnosis requires a fine needle biopsy, an invasive procedure usually taking 2 to 3 weeks.

A novel approach to confirming diagnosis is the use of Raman spectroscopy, an optical methodology that can provide detailed biochemical information from a

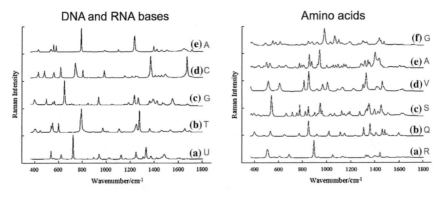

Fig. 21 Example Raman spectra for DNA and RNA bases and amino acids

Table 1 Cell lines used in the training of an IRCGP classifier

Class	Name	Cell type	Samples
1	Nthy-ori 3-1	Normal follicular epithelial	30
2	K1	Papillary thyroid carcinoma	25
3	RO82-W-1	Follicular thyroid carcinoma	25
4	TT	Medullary thyroid carcinoma	24
5	8305C	Anaplastic thyroid carcinoma	25

blood sample alone—a procedure that can be completed in a matter of minutes [31]. However, as can be seen in the example spectra shown in Fig. 21 the different energy level patterns associated with DNA and RNA bases and Amino acids are complex and cannot be easily distinguished.

Work undertaken by Lones et al. [32] used implicit context representation CGP (IRCGP) to discriminate between the Raman spectra of thyroid cancers and normal follicular epithelial, as listed in Table 1. Normalisation was applied to each cell line to compensate for experimental variability and consisted of linear scaling of the total energies of the mean spectra for each of the cell lines, as show in Fig. 22.

Selected Raman shift values for the spectra were then presented to the inputs of an IRCGP network configured to evolve a classifier as summarised in Fig. 23.

The results for a multi-class solution is presented in Fig. 24 with the fittest classier obtaining an AUC score of 0.75. Evolving classifiers for classes as a set of pair-wise evaluations is presented in Fig. 25 it can be seen that higher performing classifiers have been generated that, importantly, discriminate well between Anaplastic thyroid carcinoma (the most aggressive form of the cancer) and both Papillary thyroid carcinoma (the most commonly occurring cancer) and normal follicular epithelial. This demonstrates how the technique has the potential to be provide a clinical useful and important tool. The expressions for the top performing

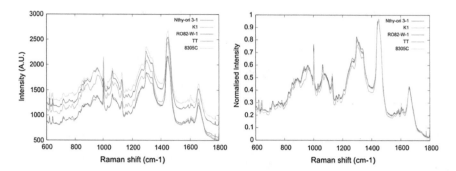

Fig. 22 Original and normalised Raman spectra for each cell line (taken from [32])

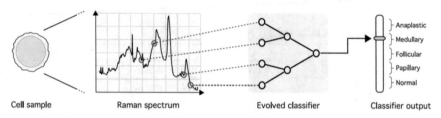

Fig. 23 Selected Raman shift values from spectra of blood samples are presented to the inputs of an IRCGP network train to classify the cancer type (taken from [32])

Fig. 24 Best multi-class classifier with AUC score of 0.75

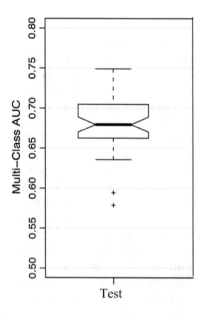

Fig. 25 Evolved classifiers
for classes as a set of
pair-wise evaluations

Classes	Train	Validate	Test
1/2/3/4/5	0.75	0.78	0.75
1/2	0.69	0.64	0.61
1/3	0.80	0.83	0.74
1/4	0.85	0.94	0.88
1/5	0.94	0.98	0.91
2/3	0.65	0.69	0.62
2/4	0.70	0.82	0.79
2/5	0.84	0.89	0.83
3/4	0.59	0.65	0.71
3/5	0.76	0.76	0.78
4/5	0.68	0.62	0.60

Table 2 Expressions for the best performing classifiers

Expression	Test AUC
$out = \overline{\{885, 1001\}} - 1607 - \max\{1656, 1117\}$	0.75
$out = (841 + 1002) - (836 + 1600) + (\max\{-622, 1657*1655\} - 1657*1655)$	0.75
$out = 780 - 1606 + (\max\{1002, 1084\} - (636 + 1664))$	0.74
$out = (1002 - 1661) - ((1607 - 1379)(625 - 1783)) - \max\{625, 636\}$	0.73
$out = -\left(\overline{\{-1001, 1270 + 1741\}} + \{-999, 1606\}\right)$	0.72
$out = -\left(-\overline{\{1002, 999\}} + \overline{\{-1473, 1173\}} - -1473*(1599 + 723)\right)$	0.72
$out = (636*1661 - \max\{878, 893\})(636*1661 - (1002 - 833))$	0.72
$out = -\left\{sub_1, \min\{833, \{\overline{1463, 636}\}\}, sub_1, 1603 + 1661\right\}$ $sub_1 = -\max\{1685, 1002\}$	0.71
$out = (1001 - 713) + (1715*1751 - 636)$	0.71
$out = \left\{\left(1439 + \overline{\{1533,\ 1759\}}\right)^2, sub_1 - sub_2\right\} + (sub_1 - sub_2)$ $sub_1 = 1603 + 1757 + (833 - 1001)$ $sub_2 = \overline{\{1566,\ 964\}}$	0.71

classifiers are shown in Table 2 These can be related back to the respective Raman spectra, as shown in Fig. 26 and provide an insight to those DNA and RNA bases and Amino acids associate with predicting the particular type of cancer.

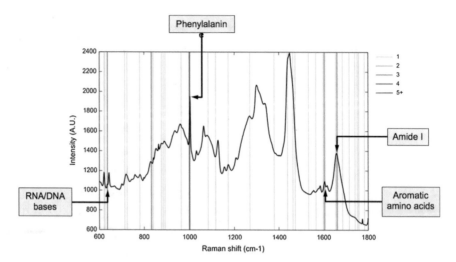

Fig. 26 RNA/DNA bases and amino acids indicated by Raman offsets in best evolved classifiers

4 Summary

The aim of this chapter has been to illustrate how properties of CGP make it particularly suitable for medical applications. Its simple and adaptable directed graph structure supports implicit modularity and avoids bloat, providing a distinct advantage over conventional GP. The representation is also amenable to customisation to an implicit context representation that overcomes the positional dependence of CGP's encoding. The ability to configure the geometry of the network allows for solutions that can be made compact which is of particular benefit if a real-time or hardware implementation is required. As well as generating high performance classifiers, the ability to decode the network as a mathematical expression provides useful insight to the functioning of the measured system.

References

1. Smith, S.L., Cagnoni, S.: Genetic and Evolutionary Computation: Medical Applications. Wiley (2011)
2. Benamrane, N., Aribi, A., Kraoula, L.: Fuzzy neural networks and genetic algorithms for medical images interpretation. In: Geometric Modeling and Imaging–New Trends (GMAI'06), pp. 259–264. IEEE (2006)
3. Delibasis, K., Undrill, P.E., Cameron, G.G.: Designing texture filters with genetic algorithms: an application to medical images. Sig. Process. **57**, 19–33 (1997)
4. Delibasis, K., Undrill, P.E., Cameron, G.G.: Designing Fourier descriptor-based geometric models for object interpretation in medical images using genetic algorithms. Comput. Vis. Image Underst. **66**, 286–300 (1997)

5. Gudmundsson, M., El-Kwae, E.A., Kabuka, M.R.: Edge detection in medical images using a genetic algorithm. IEEE Trans. Med. Imaging **17**, 469–474 (1998)
6. Maulik, U.: Medical image segmentation using genetic algorithms. IEEE Trans. Inf. Technol. Biomed. **13**, 166–173 (2009)
7. Shih, F.Y., Wu, Y.-T.: Robust watermarking and compression for medical images based on genetic algorithms. Inf. Sci. **175**, 200–216 (2005)
8. Ding, S., Li, H., Su, C., Yu, J., Jin, F.: Evolutionary artificial neural networks: a review. Artif. Intell. Rev. **39**, 251–260 (2013)
9. Fister Jr, I., Yang, X.-S., Fister, I., Brest, J., Fister, D.: A brief review of nature-inspired algorithms for optimization. arXiv:1307.4186 (2013)
10. Larrañaga, P., Karshenas, H., Bielza, C., Santana, R.: A review on evolutionary algorithms in Bayesian network learning and inference tasks. Inf. Sci. **233**, 109–125 (2013)
11. Tahmasian, M., Bettray, L.M., van Eimeren, T., Drzezga, A., Timmermann, L., Eickhoff, C. R., Eickhoff, S.B., Eggers, C.: A systematic review on the applications of resting-state fMRI in Parkinson's disease: does dopamine replacement therapy play a role? Cortex **73**, 80–105 (2015)
12. Zang, H., Zhang, S., Hapeshi, K.: A review of nature-inspired algorithms. J. Bionic Eng. **7**, 232–237 (2010)
13. Langdon, W.B.: Quadratic bloat in genetic programming. In: Proceedings of the 2nd Annual Conference on Genetic and Evolutionary Computation, pp. 451–458. Morgan Kaufmann Publishers Inc. (2000)
14. Langdon, W.B., Poli, R.: Fitness causes bloat. Soft Computing in Engineering Design and Manufacturing, pp. 13–22. Springer (1998)
15. Miller, J.F., Smith, S.L.: Redundancy and computational efficiency in Cartesian genetic programming. IEEE Trans. Evol. Comput. **10**, 167–174 (2006)
16. Cai, X., Smith, S.L., Tyrrell, A.M.: Positional independence and recombination in Cartesian genetic programming. In: European Conference on Genetic Programming, pp. 351–360. Springer (2006)
17. Clegg, J., Walker, J.A., Miller, J.F.: A new crossover technique for cartesian genetic programming. In: Proceedings of the 9th Annual Conference on Genetic and Evolutionary Computation, pp. 1580–1587. ACM (2017)
18. Lones, M.A., Tyrrell, A.M.: Modelling biological evolvability: implicit context and variation filtering in enzyme genetic programming. BioSystems **76**, 229–238 (2004)
19. Smith, S.L., Leggett, S., Tyrrell, A.M.: An implicit context representation for evolving image processing filters. In: Workshops on Applications of Evolutionary Computation, pp. 407–416. Springer (2015)
20. Smith, S.L., Lones, M.A.: Implicit context representation Cartesian genetic programming for the assessment of visuo-spatial ability. In: IEEE Congress on Evolutionary Computation, 2009. CEC'09, pp. 1072–1078. IEEE (2009)
21. Fawcett, T.: An introduction to ROC analysis. Pattern Recogn. Lett. **27**, 861–874 (2006)
22. Parkinson's, U.: Parkinson's prevalence in the United Kingdom. 2009. London, Parkinson's UK. 1–13 (2012)
23. Bajaj, N.P., Gontu, V., Birchall, J., Patterson, J., Grosset, D.G., Lees, A.J.: Accuracy of clinical diagnosis in tremulous Parkinsonian patients: a blinded video study. J. Neurol. Neurosurg. Psychiatry **81**, 1223–1228 (2010)
24. Das, R.: A comparison of multiple classification methods for diagnosis of Parkinson disease. Expert Syst. Appl. **37**, 1568–1572 (2010)
25. NICE: Parkinson's disease: national clinical guideline for diagnosis and management in primary and secondary care. Royal College of Physicians (2006)
26. Smith, S.L., Lones, M.A., Bedder, M., Alty, J.E., Cosgrove, J., Maguire, R.J., Pownall, M.E., Ivanoiu, D., Lyle, C., Cording, A.: Computational approaches for understanding the diagnosis and treatment of Parkinson's disease. IET Syst. Biol. **9**, 226–233 (2015)

27. Lones, M.A., Smith, S.L., Alty, J.E., Lacy, S.E., Possin, K.L., Jamieson, D.S., Tyrrell, A.M.: Evolving Classifiers to recognize the movement characteristics of Parkinson's disease patients. IEEE Trans. Evol. Comput. **18**, 559–576 (2014)
28. Lones, M.A., Alty, J.E., Lacy, S.E., Jamieson, D.S., Possin, K.L., Schuff, N., Smith, S.L.: Evolving classifiers to inform clinical assessment of Parkinson's disease. In: IEEE Symposium on Computational Intelligence in Healthcare and e-health (CICARE) 2013, pp. 76–82. IEEE (2013)
29. Lones, M.A., Alty, J.E., Duggan-Carter, P., Turner, A.J., Jamieson, D., Smith, S.L.: Classification and characterisation of movement patterns during levodopa therapy for parkinson's disease. In: Proceedings of the 2014 Conference Companion on Genetic and Evolutionary Computation Companion, pp. 1321–1328. ACM (2014)
30. Shrivastava, J.P., Mangal, K., Woike, P., Marskole, P., Gaur, R.: Role of FNAC in diagnosing thyroid neoplasms-A retrospective study. IOSR J. Dent. Med. Sci. (IOSR-JDMS) **1**, 13–16
31. Kendall, C., Isabelle, M., Bazant-Hegemark, F., Hutchings, J., Orr, L., Babrah, J., Baker, R., Stone, N.: Vibrational spectroscopy: a clinical tool for cancer diagnostics. Analyst **134**, 1029–1045 (2009)
32. Lones, M., Smith, S.L., Harris, A.T., High, A.S., Fisher, S.E., Smith, D.A., Kirkham, J.: Discriminating normal and cancerous thyroid cell lines using implicit context representation cartesian genetic programming. In: IEEE Congress on Evolutionary Computation (CEC) 2010, pp. 1–6. IEEE (2010)

Part III
Chemistry and Development

Chemical Computing Through Simulated Evolution

Larry Bull, Rita Toth, Chris Stone, Ben De Lacy Costello and Andrew Adamatzky

Abstract Many forms of unconventional computing, i.e., massively parallel computers which exploit the non-linear material properties of their substrate, can be realised through simulated evolution. That is, the behaviour of non-linear media can be controlled automatically and the structural design of the media optimized through the nature-inspired machine learning approach. This chapter describes work using the Belousov-Zhabotinsky reaction as a non-linear chemical medium in which to realise computation. Firstly, aspects of the basic structure of an experimental chemical computer are evolved to implement two Boolean logic functions through a collision-based scheme. Secondly, a controller is evolved to dynamically affect the rich spatio-temporal chemical wave behaviour to implement three Boolean functions, in both simulation and experimentation.

1 Introduction

Following early work in evolvable hardware with programmable silicon devices [26], Miller and Downing [17] suggested the same general approach might be applied to aid the design and programming of unconventional computers. That is, evolutionary computing techniques may be able to harness the as yet only partially understood intricate dynamics of non-linear media to perform computations more effectively than with traditional architectures. Previous theoretical and experimental studies have shown that reaction-diffusion chemical systems are capable of information processing. Experimental prototypes of reaction-diffusion processors have been used to solve a wide range of computational problems, including image processing, path planning, robot navigation, computational geometry and counting (see

L. Bull (✉) · C. Stone · B. De Lacy Costello · A. Adamatzky
Unconventional Computing Group, University of the West of England,
Bristol BS16 1QY, UK
e-mail: larry.bull@uwe.ac.uk

R. Toth
High Performance Ceramics, EMPA, Dubendorf, Switzerland

© Springer International Publishing AG 2018
S. Stepney and A. Adamatzky (eds.), *Inspired by Nature*, Emergence,
Complexity and Computation 28, https://doi.org/10.1007/978-3-319-67997-6_13

[2] for an overview). As such, we have been exploring the use of simulated evolution to design such chemical systems which exploit collision-based computing (e.g., [1]).

The directly analogous approach to that typically adopted in evolvable hardware is to use the evolutionary search process to determine how to configure the basic setup of the computational medium before inputs are applied and the resultant outputs interpreted. For example, Harding and Miller [8] searched the space of control line voltages for a liquid crystal display for various tasks, and this general approach has been used in all known subsequent work (see [18] for an overview). This chapter begins by describing the use of the same approach with a spatially-distributed light-sensitive form of the Belousov-Zhabotinsky (BZ) [29] reaction in gel which supports travelling reaction-diffusion waves and patterns. In particular, the spatial location of inputs (waves of excitation) is defined by evolution.

In our other approach, we exploit the photoinhibitory property of the reaction dynamically. That is, the chemical activity (amount of excitation on the gel) can be controlled by an applied light intensity, namely it can be decreased by illuminating the gel with high light intensity and vice versa. In this way a BZ network can be created via light and here controlled using (heterogeneous) Cellular Automata (CA) [27] designed using simulated evolution. This architecture is adapted from the system described by Wang et al. [28] and experimental chemical computers have again been realised, after simulation, as is described below.

2   Chemical Computing

Excitable and oscillating chemical systems have been used to solve a number of computational tasks such as implementing logical circuits [24], image processing [14], shortest path problems [23] and memory [20]. In addition chemical diodes [3], coincidence detectors [7] and transformers where a periodic input signal of waves may be modulated by the barrier into a complex output signal depending on the gap width and frequency of the input [21] have all been demonstrated experimentally.

A number of experimental and theoretical constructs utilising networks of chemical reactions to implement computation have been described. These chemical systems act as simple models for networks of coupled oscillators such as neurons, circadian pacemakers and other biological systems [13]. Ross and co-workers (e.g., [10]) produced a theoretical construct suggesting the use of "chemical" reactor systems coupled by mass flow for implementing logic gates neural networks and finite-state machines. In further work Hjelmfelt and Ross [9] simulated a pattern recognition device constructed from large networks of mass-coupled chemical reactors containing a bistable iodate-arsenous acid reaction. They encoded arbitrary patterns of low and high iodide concentrations in the network of 36 coupled reactors. When the network is initialized with a pattern similar to the encoded one then errors in the initial pattern are corrected bringing about the regeneration of the stored pattern. However, if the pattern is not similar then the network evolves to a homogenous state signalling non-recognition.

In related experimental work Laplante et al. [15] used a network of eight bistable mass coupled chemical reactors (via 16 tubes) to implement pattern recognition operations. They demonstrated experimentally that stored patterns of high and low iodine concentrations could be recalled (stable output state) if similar patterns were used as input data to the programmed network. This highlights how a programmable parallel processor could be constructed from coupled chemical reactors. This described chemical system has many properties similar to parallel neural networks. In other work Lebender and Schneider [16] described methods of constructing logical gates using a series of flow rate coupled continuous flow stirred tank reactors (CSTR) containing a bistable nonlinear chemical reaction. The minimal bromate reaction involves the oxidation of cerium (III) (Ce^{3+}) ions by bromate in the presence of bromide and sulphuric acid. In the reaction the Ce^{4+} concentration state is considered as "0" "false" ("1" "true") if a given steady state is within 10% of the minimal (maximal) value. The reactors were flow rate coupled according to rules given by a feedforward neural network run using a PC. The experiment is started by feeding in two "true" states to the input reactors and then switching the flow rates to generate "true"-"false", "false"-"true" and "false"-"false". In this three coupled reactor system the AND (output "true" if inputs are both high Ce^{4+}, "true"), OR (output "true" if one of the inputs is "true"), NAND (output "true" if one of the inputs is "false") and NOR gates (output "true" if both of the inputs are "false") could be realised. However to construct XOR and XNOR gates two additional reactors (a hidden layer) were required. These composite gates are solved by interlinking AND and OR gates and their negations. In their work coupling was implemented by computer but they suggested that true chemical computing of some Boolean functions may be achieved by using the outflows of reactors as the inflows to other reactors, i.e., serial mass coupling.

As yet no large scale experimental network implementations have been undertaken mainly due to the complexity of analysing and controlling many reactors. That said, there have been many experimental studies carried out involving coupled oscillating and bistable systems (e.g., see [4, 5, 11, 25]). The reactions are coupled together either physically by diffusion or an electrical connection or chemically, by having two oscillators that share a common chemical species. The effects observed include multistability, synchronisation, in-phase and out of phase entrainment, amplitude or "oscillator death", the cessation of oscillation in two coupled oscillating systems, or the converse, "rhythmogenesis", in which coupling two systems at steady state causes them to start oscillating [6].

3   Experimental Setup

Sodium bromate, sodium bromide, malonic acid, sulphuric acid, tris(bipyridyl) ruthenium (II) chloride, 27% sodium silicate solution stabilized in 4.9 M sodium hydroxide were purchased from Aldrich (U.K.) and used as received unless stated otherwise. To create the gels a stock solution of sodium silicate was prepared by mixing 222 mL of the purchased sodium silicate solution with 57 mL of 2 M sulphuric acid and 187 mL of deionised water. $Ru(bpy)_3SO_4$ was recrystallised from the chloride salt with sulphuric acid. Gels were prepared by mixing 2.5 mL of the acidified silicate solution with 0.6 mL of 0.025 M $Ru(bpy)_3SO_4$ and 0.65 mL of 1.0 M sulphuric acid solution. Using capillary action, portions of this solution were quickly transferred into a custom-designed 25 cm long 0.3 mm deep Perspex mould covered with microscope slides. The solutions were left for 3 h to permit complete gellation. After gellation the adherence to the Perspex mould is negligible leaving a thin gel layer on the glass slide. After 3 h the slides were carefully removed from the mould and the gels on the slides were washed in deionised water at least five times to remove by-products. The gels were 26-by-26 mm, with a wet thickness of approximately 300 μm. The gels were stored under water and rinsed just before use.

The catalyst-free reaction mixture was freshly prepared in a 30 mL continuously-fed stirred tank reactor (CSTR), which involved the in situ synthesis of stoichiometric bromomalonic acid from malonic acid and bromine generated from the partial reduction of sodium bromate. This CSTR in turn continuously fed a thermostated open reactor with fresh catalyst-free BZ solution in order to maintain a nonequilibrium state. The final composition of the catalyst-free reaction solution in the reactor was: 0.42 M sodium bromate, 0.19 M malonic acid, 0.64 M sulphuric acid and 0.11 M bromide. The residence time was 30 min.

An InFocus Model LP820 Projector was used to illuminate the computer-controlled image. Images were captured using a Lumenera Infinity 2 USB 2.0 scientific digital camera. The open reactor was surrounded by a water jacket thermostated at 22 °C. Peristaltic pumps were used to pump the reaction solution into the reactor and remove the effluent. A diagrammatic representation of the experimental setup is shown in Fig. 1.

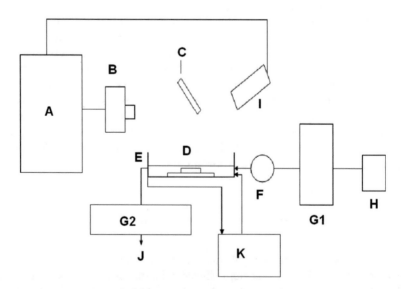

Fig. 1 A block diagram of the experimental setup where A: computer, B: projector, C: mirror, D: microscope slide with the catalyst-laden gel, E: thermostatted Petri dish, F: CSTR, G1 and G2: pumps, H: stock solutions, I: camera, J: effluent flow, K: thermostatted water bath

4 Configuration Design

A spatially distributed excitable field was created on the surface of the gel by the projection of a 7-by-7 cell grid pattern generated using a computer. The image comprised of cells with three possible intensity levels of 0.35 (level 0), 1.0 (level 1) and 3.5 mW cm^{-2} (level 2), representing excitable, subexcitable, and non-excitable domains respectively. The grid pattern was projected onto the catalyst-laden gel through a 455 nm narrow bandpass interference filter and 100/100 mm focal length lens pair and mirror assembly. The size of the projected grid was approximately 14 mm square. Every 10 s, the grid pattern was replaced with a uniform grey level of 3.5 mW cm^{-2} for 10 ms during which time an image of the BZ fragments on the gel was captured. The purpose of removing the grid pattern during this period was to allow activity on the gel to be more visible to the camera and assist in subsequent image processing of chemical activity.

Captured images were processed to identify chemical wave activity. This was done by differencing successive images on a pixel by pixel basis to create a black and white thresholded image. Each pixel in the black and white image was set to white, corresponding to chemical activity; if the intensity of the red or blue channels differed in successive images by more than 5 out of 256 pixels (1.95%). Pixels at locations not meeting this criterion were set to black. The images were cropped to the grid location and the grid superimposed on the thresholded images to aid analysis of the results.

Around the bottom, left and right sides of the grid is a channel through which chemical activity in the form of excitation fragments can travel. Chemical fragments are generated at the centre of this channel by means of a small area of darkness and these travel along the channel and up the sides of the grid approximately symmetrically (Fig. 2a). Three sides of the projected grid are used to provide inputs to the system. A one-cell boundary around the edge of the grid is set by default to level 2. This light level acts as a barrier to prevent spurious entry of chemical fragments into the centre of the grid where computation is to occur. The boundary at the left and right sides of the grid is modulated to allow entry of fragments. Under program control the left and right sides of the grid are treated as being the two binary inputs to the system. There are thus four possible states—00, 01, 10 and 11—for which chemical activity must be supplied to the grid in order to test its operation implementing Boolean logic. For a '0' input, light at the appropriate boundary remains at level 2 prohibiting fragments from passing from the initiation channel into the grid. For an input of value '1' light level at the left and/or right side boundary of the grid is controlled by the genome used by the simple evolutionary algorithm. The genome is a binary string of length 10 bits, with 5 bits coding for the 5 active boundary cells along the left side of the grid and 5 for the right side. Depending upon the genome, the light level associated with each cell in the active boundary is given the level 0 or 2. A light level of 0 allows fragment activity to pass into the 5-by-5 cell centre area of the grid, which is illuminated with a uniform light level of 1. In this way the evolutionary algorithm is able to influence the spatial and temporal

Fig. 2 a Initiation pattern, **b** example input configuration representing input logic states and **c** collisions of fragments occurring during configuration (**b**). The black area (level 0) in (**a**) and (**b**) represents the excitable medium (where chemical waves can fully develop) whilst the white area (level 2) is non excitable (where the formation of waves is inhibited). The grey area (level 1) in the centre of the grid is subexcitable where wave fragments just manage to propagate

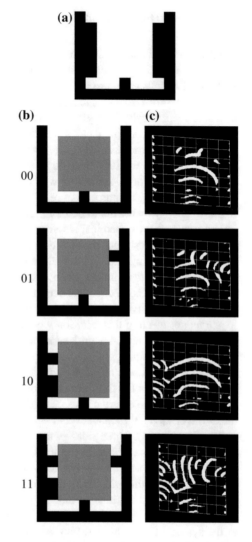

dynamics of chemical activity in the centre of the grid with the aim of inducing 'useful' activity in the form of computation. In addition to the two inputs controlled by the EA, a constant truth input is provided to the grid by means of a single cell in the centre of the bottom boundary that has a fixed light level of 0.

The evolutionary algorithm used is a simple genetics-based hill climber: starting from an initial random solution, an offspring is evaluated and adopted as the parent if at least as fit as the original solution. This choice is motivated by the impracticality of testing multiple individuals in a population with a real (slow) chemical system such as used here. Bitwise mutation is the only variation operator, with a single gene being mutated after each complete cycle of four logical input presentations.

The output of computations occurring in the centre of the grid is obtained by monitoring the 25 cells in the centre of the grid using the image processing techniques described above. Each output cell is tested 300 s after the system has been presented with one of the four possible Boolean input configurations and is assigned the Boolean output value of 1 if the total chemical activity in the cell is at a level of 20% or more and a value of 0 otherwise. The centre of the grid thus implements (potentially) 25 logic functions in parallel. To determine the fitness of each cell position with respect to a target logic function, each of the four possible input configurations is presented in turn and a fitness of 1 is assigned to the cell if it presents the correct output for that configuration. A complete logic function is discovered when a single cell presents all four correct outputs in a single learning episode. For this work we investigated the ability of the system to discover two-input AND and NAND gates. These gates are complementary in function and both were investigated to avoid the possibility of results being biased by particular properties of the chemical system used.

Figure 3a shows the discovery of correctly functioning AND gates in at least one of the cells in the central sub-excitable area of the grid. We were able to run a maximum of 30 input configurations (which required about 6 h) in the experiments because after that time the excitability of the system changed due to the

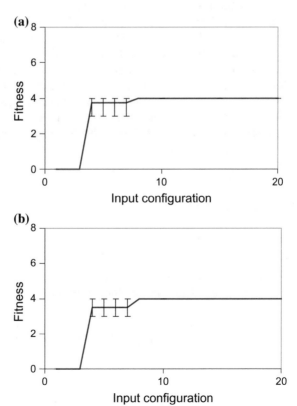

Fig. 3 Number of input configurations necessary to discover the first (**a**) AND and (**b**) NAND gate. Gate is found when fitness reaches four. Fitness is an average of 3 runs. Error bars show minimum and maximum fitness

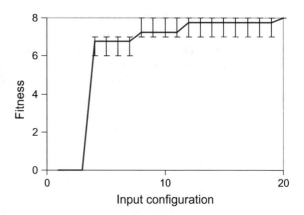

Fig. 4 Number of input configurations necessary to discover the first two cells implementing concurrently functioning AND and NAND gates. Both gates are found when fitness reaches eight. Fitness is an average of 3 runs. Error bars show minimum and maximum fitness

"desensitization" effect of high intensity light. Figure 3b shows similar results for the design of single NAND gates.

From these results it is evidently somewhat trivial for the EA to design grid boundary conditions such that single instances of correctly functioning two-input logic gates can be produced by collision-based computing. We therefore wanted to attempt a more testing problem, namely the simultaneous design of multiple concurrently functioning logic gates. Figure 4 shows that it is possible to design concurrently functioning AND and NAND gates using the same technique with only a marginal increase in difficulty. A second general approach is explored next.

From these results it is evidently somewhat trivial for the EA to design grid boundary conditions such that single instances of correctly functioning two-input logic gates can be produced by collision-based computing. We therefore wanted to attempt a more testing problem, namely the simultaneous design of multiple concurrently functioning logic gates. Figure 4 shows that it is possible to design concurrently functioning AND and NAND gates using the same technique with only a marginal increase in difficulty. A second general approach is explored next.

5 Dynamic Control

5.1 Simulated Gels

Features of the chemical system are simulated using a two-variable Oregonator model modified to account for photochemistry:

$$\frac{\partial u}{\partial t} = \frac{1}{\varepsilon} \left(u - u^2 - (fv + \Phi) \frac{u - q}{u + q} \right) + D_u \nabla^2 u$$

$$\frac{\partial v}{\partial t} = u - v.$$

The variables u and v represent the instantaneous local concentrations of the bromous acid autocatalyst and the oxidized form of the catalyst, $HBrO_2$ and tris (bipyridyl) Ru (III), respectively, scaled to dimensionless quantities. The rate of the photo-induced bromide production is designated by Φ, which also denotes the excitability of the system in which low simulated light intensities facilitate excitation while high intensities result in the production of bromide that inhibits the process. The system was integrated using the Euler method with a five-node Laplacian operator, time step $\Delta t = 0.001$ and grid point spacing $\Delta x = 0.62$. The diffusion coefficient, D_u, of species u was unity, while that of species v was set to zero as the catalyst is immobilized in the gel. The kinetic parameters were set to $\varepsilon = 0.11$, $f = 1.1$ and $q = 0.0002$. The medium is oscillatory in the dark which made it possible to initiate waves in a cell by setting its simulated light intensity to zero. At different Φ values the medium is excitable, subexcitable or non-excitable.

5.2   *Evolutionary Control*

We have used Cellular Automata to control such chemical systems, i.e., finite automata are arranged in a two-dimensional lattice with aperiodic boundary conditions (an edge cell has five neighbours, a corner cell has three neighbours, all other cells have eight neighbours each). Use of a CA with such a two-dimensional topology is a natural choice given the spatio-temporal dynamics of the BZ reaction (Fig. 5). Each automaton updates its state depending on its own state and the states of its neighbours. States are updated in parallel and in discrete time. In standard CA all cells have the same state transition function (rule), whereas in this work the

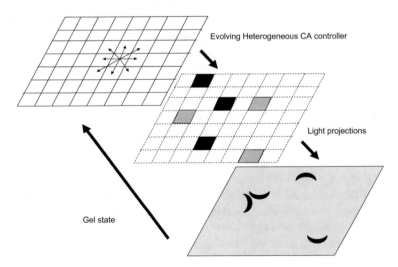

Fig. 5 Relationship between the CA controller, applied grid pattern and chemical system comprising one process control cycle

CA is heterogeneous, i.e., each cell/automaton has its own state transition function. The transition function of every cell is evolved by a simple evolutionary process. This approach is very similar to that presented by Sipper [22]. However, his reliance upon each cell having access to its own fitness means it is not applicable in the majority of chemical computing scenarios we envisage. Instead, fitness is based on emergent global phenomena in our approach (as in [19], for example). Thus, following Kauffman [12], and as above, we use a simple approach wherein each automaton of the two-dimensional CA controller is developed via a simple genetics-based hillclimber. After fitness has been assigned, some proportion of the CA's genes are randomly chosen and mutated. Mutation is the only variation operator used here to modify a given CA cell's transition rule to allow the exploration of alternative light levels for the grid state. For a CA cell with eight neighbours there are 2^9 possible grid state to light level transitions, each of which is a potential mutation site. After the defined number of such mutations have occurred, an evolutionary generation is complete and the simulation is reset and repeated. The system keeps track of which CA states are visited since mutation. On the next fitness evaluation (at the end of a further 25 control cycles) mutations in states that were not visited are discarded on the grounds that they have not contributed to the global fitness value and are thus untested. We also performed control experiments with a modified version to determine the performance of an equivalent random CA controller. This algorithm ignored the fitness of mutants and retained all mutations except those from unvisited states.

For a given experiment, a random set of CA rules is created for a two-dimensional array of size 10-by-10, i.e., 100 cells. The rule for each cell is represented as a gene in a genome, which at any one time takes one of the discrete light intensity values used in the experiment. As previously mentioned, the grid edges are not connected (i.e., the grid is planar and does not form a toroid) and the neighborhood size of each cell is of radius 1; cells consider neighborhoods of varying size depending upon their spatial position, varying from three in the corners, to five for the other edge cells, and eight everywhere else. In the model each of the 100 cells consists of 400 (20-by-20) simulation points for the reaction. The reaction is thus simulated numerically by a lattice of size 200-by-200 points, which is divided into the 10-by-10 grid.

To begin examining the potential for the evolution of controllers for such temporally dynamic structures in the continuous, non-linear 2D media described we have designed a simple scheme to create a number of two-input Boolean logic gates. Excitation is fed in at the bottom of the grid into a branching pattern. To encode a logical '1' and '0' either both branches or just one branch of the two "trees" shown in Fig. 6 are allowed to fill with excitation, i.e., the grid is divided into two for the inputs. These waves were channelled into the grid and broken up into 12 fragments by choosing an appropriate light pattern as shown in Fig. 7. The black area represents the excitable medium whilst the white area is non-excitable. After initiation three light levels were used: one is sufficiently high to inhibit the reaction; one is at the sub-excitable threshold such that excitation just manages to propagate; and the other low enough to fully enable it. The modelled chemical system was run for 600

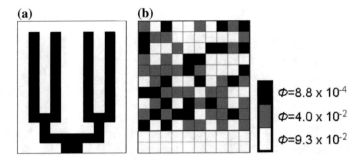

Fig. 6 Showing initiation pattern (**a**) and a typical example of a coevolved light pattern (**b**)

iterations of the simulator. This value was chosen to produce network dynamics similar to those obtained in experiment over 10 s of real time.

A colour image was produced by mapping the level of oxidized catalyst at each simulation point into an RGB value. Image processing of the colour image was necessary to determine chemical activity. This was done by differencing successive images on a pixel by pixel basis to create a black and white thresholded image. Each pixel in the black and white image was set to white (corresponding to excitation) if the intensity of the red or blue channels in successive colour images differed by more than 5 out of 256 pixels (1.95%). Pixels at locations not meeting this criterion were set to black. An outline of the grid was superimposed on the black and white images to aid visual analysis of the results.

The black and white images were then processed to produce a 100-bit description of the grid for the CA. In this description each bit corresponds to a cell and it is set to true if the average level of activity within the given cell is greater than a pre-determined threshold of 10%. Here, activity is computed for each cell as the fraction of white pixels in that cell. This binary description represents a high-level depiction of activity in the BZ network and is used as input to the CA. Once cycle of the CA is performed whereby each cell of the CA considers its own state and that of its neighbours (obtained from the binary state description) to determine the light level to be used for that grid cell in the next time step. Each grid cell may be illuminated with one of three possible light levels. The CA returns a 100-digit trinary action string, each digit of which indicates whether high ($\Phi = 0.093023$), sub-excitable threshold ($\Phi = 0.04$) or low ($\Phi = 0.000876$) intensity light should be projected onto the given cell. The progression of the simulated chemical system, image analysis of its state and operation of the CA to determine the set of new light levels comprises one control cycle of the process. A typical light pattern generated by the CA controller is shown in Fig. 6b. Another 600 iterations are then simulated with those light-levels projected, etc. until 25 control cycles have passed. The number of active cells in the grid, that is those with activity at or above the 10% threshold, is used to distinguish between a logical '0' and '1' as the output of the system. For example, in the case of XOR, the controller must learn to keep the number of active cells below the specified level for the 00 and 11 cases but increase the number for the 01 and 10 cases.

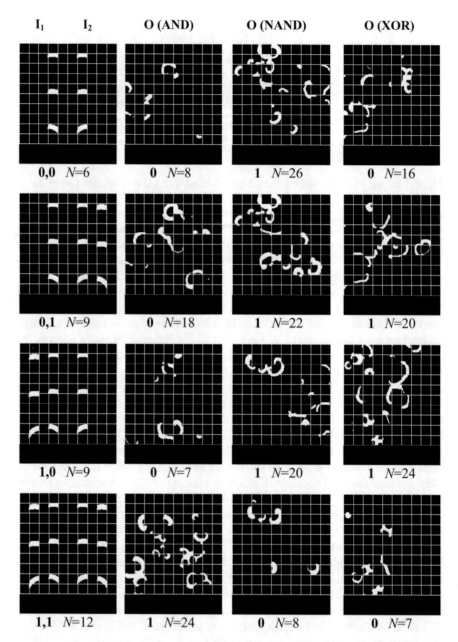

Fig. 7 Typical examples of solutions of AND, NAND and XOR logic gates in simulation, required active cells: 20, N: actual number of active cells. Input states I_1, I_2 for the logic gates are shown on the left and consist of two binary digits, spatially encoded using left and right "initiation trees" (Fig. 6). Input values of '0' are encoded using a single branch of the relevant tree resulting in 3 fragments, while binary '1' is encoded using both branches of the tree resulting in 6 fragments. Evolution found a solution in 56 (AND), 364 (NAND) and 1656 (XOR) generations

5.3   Results 1: Simulation

Figure 7 shows typical examples of each of the three logic gates learned using the simulated chemical system. Each of the four possible input combinations is presented in turn—00 to 11—and for each input combination the system is allowed to develop for 25 control cycles. Fitness for each input pattern is evaluated after the 25 control cycles with each correct output scoring 1, resulting in a maximum possible fitness of 4. Figure 8 shows the fitness averaged over ten runs for AND and NAND tasks with mutation rate 4000, and similar results for XOR are shown for mutation rate 6000. Favorable comparisons to an equivalent random controller are also shown in each case.

From these results, it is evidently somewhat trivial for the EA to design grid boundary conditions such that single instances of correctly functioning two-input logic gates can be produced by collision-based computing. We therefore wanted to attempt a more testing problem, namely the simultaneous design of multiple concurrently functioning logic gates. Figure 4 shows that it is possible to design concurrently functioning AND and NAND gates using the same technique with only a marginal increase in difficulty. A second general approach is explored next.

5.4   Results 2: Experimentation

Given the success of our approach using the simulated chemical reaction, we have attempted to implement the same tasks using a real chemical system constructed using a slight variation of the previous methodology described above. The spatially distributed excitable field on the surface of the gel was achieved by the projection of a 10-by-10 cell checkerboard grid pattern. The checkerboard image comprised of cells with three possible intensity levels of 0.35, 1.6 and 3.5 mW cm^{-2}, representing excitable, the subexcitable threshold, and non-excitable domains respectively. The size of the projected grid was approximately 20 mm square. Again, every 10 s, the checkerboard pattern was replaced with a uniform grey level of 3.5 mW cm^{-2} for 10 ms during which time an image of the BZ fragments on the gel was captured. Captured images were cropped to the grid location and processed to identify activity in the same manner as for the model (see above).

Figures 9 and 10 show how similar performance is possible on the real chemical system for each of the three logic functions. In order to produce working XOR and NAND gates from these experiments, it was necessary to use a value of 15 for the required number of active cells due to the relative difficulty of these tasks. All other parameters were the same as those used for numerical simulation.

Because of the limited lifetime of the medium these runs were seeded with CAs evolved during the simulated runs presented in Fig. 8. Runs using random initial controllers were also explored on the real chemical system (dashed lines on Fig. 10), but no successful runs were found over the 40 generations. This is not

Fig. 8 Showing the performance of evolving CA controllers for the three logic gate tasks considered. Dashed lines show the equivalent performance of random search

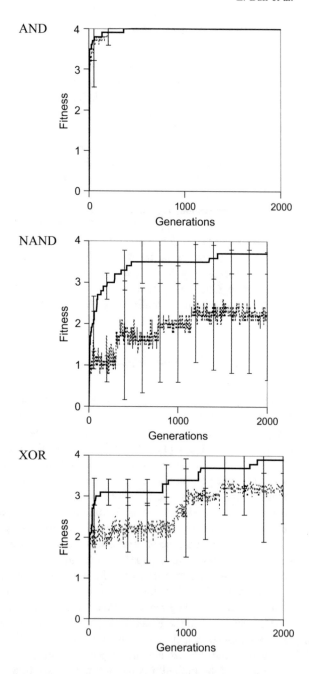

surprising considering that the average generations needed to find a good solution was higher than 40 in the simulations because of the relative increase in difficulty. Nevertheless the two systems are very similar since the runs with seeded CAs

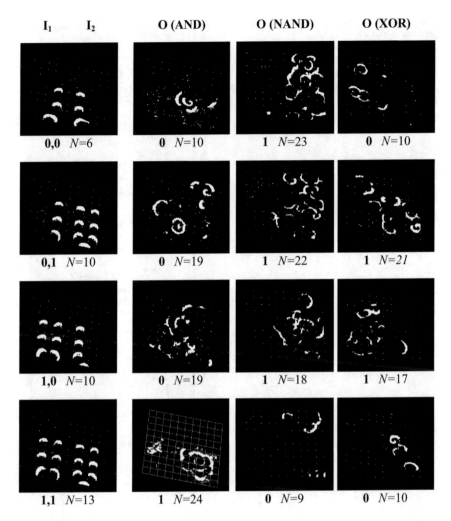

Fig. 9 Typical examples of solutions of AND, NAND and XOR logic gates in chemical experiment, required number of active cells: 15 (20 for AND), N: actual number of active cells. Input states I_1, I_2 for the logic gates are shown on the left and consist of two binary digits, spatially encoded using left and right "initiation trees" (Fig. 7). Input values of '0' are encoded using a single branch of the relevant tree resulting in 3 or 4 fragments, while binary '1' is encoded using both branches of the tree resulting in 6 or 7 fragments. The simulated evolution—seeded with a CA evolved during the simulated runs—found a solution in 16 generations in each case

evolved during the simulations found solutions in a very short time, namely in 16 or 20 generations. If a solution had been found in four generations it would have meant that the initial states of the simulated and real chemical system are perfectly identical. However, since there is a noticeable difference between the initial states, the solution found by the evolutionary algorithm in simulation was very close to the solution needed for the real chemistry, but a few generations of evolution were

Fig. 10 Showing the performance of evolving CA controllers for the three logic gate tasks considered. Dashed lines show the equivalent performance of random search

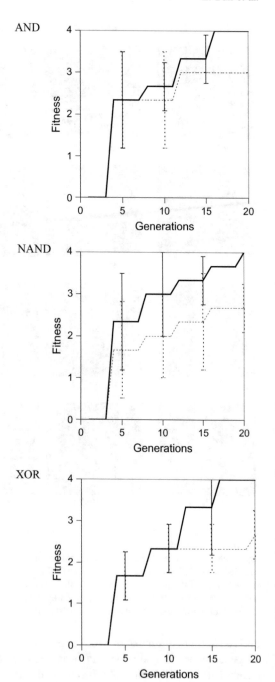

needed to adapt to the difference between the two systems. These results show that the approach is capable of adapting to small changes in its environment and finding a solution very quickly when presented with domain-specific knowledge obtained from modeling.

6 Conclusions

Excitable and oscillating chemical systems have previously been used to solve a number of simple computational tasks. However, the experimental design of such systems has typically been non-trivial. In this chapter we have presented results from a general methodology by which to achieve the complex task of designing such systems—through the use of simulated evolution. We have shown using both simulated and real systems that it is possible in this way to control dynamically the behavior of the BZ reaction, and to design the basic configuration of a chemical computer (after [17]). Which of these approaches is best able to exploit the properties of non-linear media for computation—or whether their use in combination is possible—remains open to future exploration.

References

1. Adamatzky, A. (ed.): Collision-based Computing. Springer, London (2002)
2. Adamatzky, A., De Lacy Costello, B., Asai, T.: Reaction-Diffusion Computers. Elsevier (2005)
3. Agladze, K., Aliev, R.R., Yamaguhi, T., Yoshikawa, K.: Chemical diode. J. Phys. Chem. **100**, 13895–13897 (1996)
4. Bar-Eli, K., Reuveni, S.: Stable stationary-states of coupled chemical oscillators: experimental evidence. J. Phys. Chem. **89**, 1329–1330 (1985)
5. Crowley, M.F., Field, R.J.: Electrically coupled Belousov-Zhabotinskii oscillators 1: experiments and simulations. J. Phys. Chem. **90**, 1907–1915 (1986)
6. Dolnik, M., Epstein, I.R.: Coupled chaotic oscillators. Phys. Rev. E **54**, 3361–3368 (1996)
7. Gorecki, J., Yoshikawa, K., Igarashi, Y.: On chemical reactors that can count. J. Phys. Chem. A **107**, 1664–1669 (2003)
8. Harding, S.L., Miller, J.F.: Evolution in materio: initial experiments with liquid crystal. In: Evolvable Hardware, NASA/DOD Conference on Computer Society, pp. 289–299. IEEE (2004)
9. Hjelmfelt, A., Ross, J.: Mass-coupled chemical systems with computational properties. J. Phys. Chem. **97**, 7988–7992 (1993)
10. Hjelmfelt, A., Weinberger, E.D., Ross, J.: Chemical implementation of neural networks and turing machines. PNAS **88**, 10983–10987 (1991)
11. Holz, R., Schneider, F.W.: Control of dynamic states with time-delay between 2 mutually flow-rate coupled reactors. J. Phys. Chem. **97**, 12239 (1993)
12. Kauffman, S.: The Origins of Order. Oxford (1993)
13. Kawato, M., Suzuki, R.: Two coupled neural oscillators as a model of the circadian pacemaker. J. Theor. Biol. **86**, 547–575 (1980)

14. Kuhnert, L., Agladze, K.I., Krinsky, V.I.: Image processing using light sensitive chemical waves. Nature **337**, 244–247 (1989)
15. Laplante, J.P., Pemberton, M., Hjelmfelt, A., Ross, J.: Experiments on pattern recognition by chemical kinetics. J. Phys. Chem. **99**, 10063–10065 (1995)
16. Lebender, D., Schneider, F.W.: Logical gates using a nonlinear chemical reaction. J. Phys. Chem. **98**, 7533–7537 (1994)
17. Miller, J.F., Downing, K.: Evolution in materio: looking beyond the silicon box. In: Proceedings of the NASA/DOD Conference on Evolvable Hardware, Computer Society, pp. 167–176. IEEE (2002)
18. Miller, J., Harding, S., Tufte, G.: Evolution in-materio: evolving computation in materials. Evol. Intel. **7**(1), 49–67 (2014)
19. Mitchell, M., Hraber, P., Crutchfield, J.: Revisiting the edge of chaos: evolving cellular automata to perform computations. Complex Syst. **7**, 83–130 (1993)
20. Motoike, I.N., Yoshikawa, K., Iguchi, Y., Nakata, S.: Real time memory on an excitable field. Phys. Rev. E **63**, 1–4 (2001)
21. Sielewiesiuk, J., Gorecki, J.: Passive barrier as a transformer of chemical frequency. J. Phys. Chem. A **106**, 4068–4076 (2002)
22. Sipper, M.: Evolution of Parallel Cellular Machines. Springer (1997)
23. Steinbock, O., Toth, A., Showalter, K.: Navigating complex labyrinths: optimal paths from chemical waves. Science **267**, 868–871 (1995)
24. Steinbock, O., Kettunen, P., Showalter, K.: Chemical wave logic gates. J. Phys. Chem. **100**, 18970–18975 (1996)
25. Stuchl, I., Marek, M.: Dissipative structures in coupled cells: experiments. J. Phys. Chem. **77**, 2956–2963 (1982)
26. Thompson, A.: An evolved circuit, intrinsic in silicon, entwined with physics. In: Higuchi, T., Iwata, M., Weixin, L. (eds.) Proceedings of 1st International Conferences on Evolvable Systems (ICES'96), LNCS. pp. 1259:390–405. Springer (1997)
27. Von Neumann, J.: The Theory of Self-Reproducing Automata. University of Illinois (1966)
28. Wang, J., Kádár, S., Jung, P., Showalter, K.: Noise driven avalanche behavior in subexcitable media. Phys. Rev. Lett. **82**, 855–858 (1999)
29. Zaikin, A.N., Zhabotinsky, A.M.: Concentration wave propagation in two-dimensional liquid-phase self-oscillating system. Nature **225**, 535–537 (1970)

Sub-Symbolic Artificial Chemistries

Penelope Faulkner, Mihail Krastev, Angelika Sebald
and Susan Stepney

Abstract We wish to use Artificial Chemistries to build and investigate open-ended systems. As such, we wish to minimise the number of explicit rules and properties needed. We describe here the concept of *sub-symbolic Artificial Chemistries* (ssAChems), where reaction properties are emergent properties of the internal structure and dynamics of the component particles. We define the components of a ssAChem, and illustrate it with two examples: RBN-world, where the particles are Random Boolean Networks, the emergent properties come from the dynamics on an attractor cycle, and composition is through rewiring the components to form a larger RBN; and SMAC, where the particles are Hermitian matrices, the emergent properties are eigenvalues and eigenvectors, and composition is through the non-associative Jordan product. We conclude with some ssAChem design guidelines.

1 Introduction

Artificial Chemistries (AChems) are examined for many reasons. In the context of Artificial Life (ALife) they can form an underpinning technology. In these cases, the systems are often carefully crafted, with atoms and reaction rules hand tuned to produce the desired properties, such as replication and evolution.

Another use of AChems is to study open-ended systems in general [6]. In such cases it is important to ensure that the open-ended properties are not specifically

P. Faulkner · A. Sebald
Department of Chemistry, University of York, York, UK
e-mail: pf550@york.ac.uk

A. Sebald
e-mail: angelika.sebald@york.ac.uk

M. Krastev · S. Stepney (✉)
Department of Computer Science, University of York, York, UK
e-mail: susan.stepney@york.ac.uk

M. Krastev
e-mail: mk599@york.ac.uk

© Springer International Publishing AG 2018
S. Stepney and A. Adamatzky (eds.), *Inspired by Nature*, Emergence,
Complexity and Computation 28, https://doi.org/10.1007/978-3-319-67997-6_14

designed in, but rather emerge from the properties underlying system. Such behaviour is needed to allow the possibility of multiple levels of emergence.

In 2009, Faulconbridge, Stepney, Miller and Caves presented the first work on a *subsymbolic artificial chemistry* (ssAChem), called bRBN-world [11], with subsequent work described in [9, 10]. The core idea of ssAChems is that reaction properties should be emergent properties of the internal structures of the relevant atoms and molecules, analogous to the way that these are emergent properties of the electronic structures of physical atoms and molecules. The behavioural specifics emerge from these structures, rather than being solely defined by external rules.

We have continued work on ssAChems, enriching and diversifying the original concept. In this chapter we brings together some of the recent work. We start with some background material on AChems in general (Sect. 2). We then give a formal definition of ssAChems (Sect. 3). We use this definition to introduce a new ssAChem, SMAC, based on matrix algebra (Sect. 4). We also use the definition to summarise bRBN-world, and add a new feature: an environmental temperature model (Sect. 5). We finish with some guidelines to follow when designing a new ssAChem (Sect. 6).

2 Why AChems?

2.1 What Chemistry Can Give ALife/Complexity Science

When we take a good look at the contemporary activities carried out under the broad umbrella of 'chemistry' we are confronted with a vast range of fascinating developments and output. An increasing proportion of these research and development activities no longer fits the traditional boxes of the various science disciplines, but instead has moved to cover areas in between classical disciplines, ranging from shared approaches with physics, engineering, materials science, biology, pharmacology, archaeology, forensic sciences all the way to medical sciences.

This contemporary state of affairs can act as an encouragement for our endeavour to design subsymbolic artificial chemistries (we come back to various definitions in a moment): whatever theoretically underpins this wide range of natural activities must surely cater well for the emerging properties of a rich system.

Traditionally, the discipline has been subdivided into the classical domains of organic, inorganic and physical chemistry. This subdivision is still maintained in many areas of undergraduate teaching and in organisational structures of many chemistry departments. However, its mainly output/substance focus cannot help us with finding inspiration for the rational design of artificial chemistries.

If we want to take inspiration from chemistry for the design of algorithms with emergent properties we need to take a step back, and take a dispassionate and perhaps slightly unconventional look at the very core of what forms the basis of the conceptual and descriptive framework that helps us to deal with chemistry. Stripping matters back to the essentials yields the following picture.

Nature provides a set of possibilities: the elements in the periodic table. Their properties in turn provide variability, and their number forms the basis of endless options for combinatorics constrained by reaction properties. In practice much of the rich behaviour of chemical systems relies on a fairly small subset of chemical elements. Biochemistry produces its own larger 'units' from this small subset of chemical atoms: the four unit genetic code, and the 20 unit amino acid code. Primary linear combinatoric systems (DNA and proteins) then exploit spatial properties to gain functional secondary and tertiary structures. In each case, the properties of the units emerges from their internal structure. Further exotic units can be added to these small alphabets in certain circumstances. The properties of these further units emerge in a similar manner.

Faced with having to understand, handle, predict or manipulate a massive set of possibilities, the discipline of chemistry has developed empirical rules to come to grips with this richness in nature. The basic set of rules in use is limited and fairly straightforward. The energetics of the systems considered are described by thermodynamics; kinetics describe rates of change (and often reaction mechanisms) of the systems as macroscopic entities. The laws of quantum mechanics describe microscopic properties such as electronic structures of atoms and molecules. Statistics, mainly in the form of statistical thermodynamics, provides the necessary link between macroscopic and microscopic properties.

This small but effective set of rules in operation deals with essentially the entire vast range of 'chemistry'. Note that (i) none of these rules applied to 'chemistry' make any reference to chemical properties, and (ii) all of the many and varied properties of chemical systems emerge from the combination of a set of possibilities with a set of rules. This 'chemistry rule book' seems to be quite capable of dealing with the rich natural system.

Let us translate this highly minimalistic view of the working of the foundations of chemistry to our task of recreating similarly rich emergent properties in silico. To do this we need recognise that chemistry uses descriptive rules that are a scientific model built from observation of what the system does; such a model may be wrong or partial. AChems, on the other hand, implement underlying rules that govern what the simulation system does; such rules cannot be wrong, cannot be broken. Thus we must chose our simulation rules judiciously, not to rigidly enforce one level of behaviour, but to provide sufficient richness of possibilities that higher level rule-based behaviours can emerge.

A stripped down view of the underpinning rules of chemistry lends itself to being applied in the rational design of algorithms with emergent properties: we need (i) to pick a sufficiently large and versatile set of possibilities ('atoms') and (ii) combine this set with a small but powerful set of rules to govern the possibilities and steer the system toward a sweet spot of just enough complexity. Nature does this fabulously well, so there is no immediately obvious reason why we should not be able to recreate such behaviours in silico, as long as we are able to do so in a computationally feasible manner.

2.2　Quick Definition of AChems

An artificial chemistry (AChem) is a computational system that is analogous to certain aspects of the dynamics of atomic and molecular level interactions of real chemistry. AChems are not typically used as *simulations* of real chemistry, but rather exploit combinatorics, dynamics, and other properties in a computational setting. Often, AChems are used to explore aspects of Artificial Life, and aspects of the origin of life. Here we are interested in exploring how open-ended systems can develop as a consequence of emergent properties.

Dittrich et al. [7] formally define an AChem as a triple $(\mathbb{S}, \mathbb{R}, \mathbb{A})$, where $\mathbb{S}$ is the set of possible molecules, $\mathbb{R}$ is the set of rules for interacting the molecules, and $\mathbb{A}$ is an algorithm describing the dynamics of the environment, the reaction vessel, and how the rules are applied to the molecules. $\mathbb{S}$ and $\mathbb{R}$ are more fundamental to the AChem, and may be thought of as the underlying 'physics' of the system; $\mathbb{A}$ may then be varied to see how the specified molecules and rules behave in different environments or contexts, for example, spatial *versus* aspatial, closed *versus* chemostat.

$\mathbb{S}$ and $\mathbb{R}$ can be defined explicitly (by listing all possible molecules, and all possible interactions), or implicitly, using a procedural algorithm or declarative expression. For an open-ended system, both $\mathbb{S}$ and $\mathbb{R}$ need to be defined implicitly, as novel molecules and their interactions can continually arise.

2.3　Historical Context

AChems were originally created as an addition to the study of artificial life in the hopes that they would illuminate the transition from inanimate to animate matter through complex chemistry. Three properties are thought to be the basis for open-ended evolution and complexification of life [2, 8, 24]

1. *self-replication*: a property that directly or indirectly propagates the creation of copies of itself [14, 15, 23]
2. *metabolism*: a lifeform's ability to change and maintain itself in its life time, by processing energy to reproduce, repair damage, create or destroy structures [16, 23]
3. *mutability*: the capability for minor changes to occur during replication [8], needed for life capable of evolutionary behaviours

From these requirements of life the need for several high level emergent properties of chemistries have been inferred [11]: autocatalytic sets [17] (a precursor of self-reproduction); hypercycles [8]; and heteropolymers (giving an information-bearing molecule made up from a set of repeating subunits, for example RNA and DNA [22]).

2.4   Desirable Properties of an AChem

Complex *emergent* properties such as self-replication and metabolism should not be designed *into* an AChem. Instead, we can define a set of low level requirements of a chemistry that can be tested at the level of small 'molecules' comprising only a few atoms [9, 11, 25]. These properties are not sufficient for these and other higher level properties to emerge, but are necessary to allow them to exist.

2.4.1   Macroscopic Properties

- *unbounded molecular size*: an open-ended system must have some aspect of its state space unbounded in size; a bounded system could (at least potentially) exhaust its state space.
- *conservation of mass*: some form of conservation law (energy, mass, particle number) is needed to stop the system simply being a white hole; we conserve atoms, in a context where our reaction vessel may have inflows and outflows.

2.4.2   Microscopic Properties

- *synthesis*: the forming of a bond between two molecules, producing a new larger molecule: $A + B \rightarrow D$. The products of synthesis (and decomposition, below) should be the same kind of objects as the reactants, so that these can then react without the need to introduce new kinds of rules.
- *self-synthesis*: the forming of a bond between two identical molecules: $A + A \rightarrow D$. The overall goal is start from a relatively small set of atoms, so we must allow some atoms and molecules to bind with copies of themselves. However, self-synthesis should not be universal, as we do not want every randomly selected set of atoms to form a molecule. If all molecules were possible, the system would be the basic combinatorics of all possible combinations of the atoms. We want the system to have the same complexity as real chemistry, so we need implicit restrictions to the set of possible molecules.
- *decomposition*: the removal of a bond in a molecule (which may affect other bonds in the molecule): $D \rightarrow A + B$. Decomposition allows for the potential formation of structures that cannot be formed directly by synthesis, for example by allowing the use of scaffolding. It also allows cycles in reaction networks, for example, metabolic cycles as a combination of anabolic (building up) and catabolic (breaking down) reactions.

2.4.3 Environmental and Contextual Properties

Molecules and reactions should be able to be influenced by environmental or contextual conditions, such as space and temperature analogues. This allows the same 'physics' of the molecules to exhibit different behaviours in different contexts, and so for that behaviour to potentially be controlled, regulated, or influenced.

Additionally, molecules and reactions should be able to influence their environment, permitting feedback cycles so that the system can influence its own dynamics. For example, endothermic and exothermic reactions can affect the environmental temperature, which in turn can affect reaction rates.

2.5 Rationale for Sub-symbolic AChems

In real-world chemistry, whether an atom or molecule bonds with another is a result of a complex set of factors based on the states of the reactants, such as the number of electrons and energy states. These are emergent properties that cannot be inferred from the chemical formula (symbolic form) of a molecule.

In order to build a system with such properties we use a new approach, of defining atoms with internal structure. Rather than assigning an atom a symbol and then defining its bonding behaviour purely in the rule system, we define an atom with its own structure and emergent properties. We then define bonding rules in terms of these emergent properties.

We call AChems that display this property of controlling binding based on emergent properties of the molecule *sub-symbolic AChems* (ssAChems) [9, 11]. These ssAChems move beyond treating atoms as structureless symbols as in traditional AChems.

We use the same $(\mathbb{S}, \mathbb{R}, \mathbb{A})$ formalism to define our ssAChems below, but our systems have more of their semantics and behaviour provided in the molecules in $\mathbb{S}$ (since it emerges from their internal structure), and less in the rules $\mathbb{R}$; behaviour can emerge in a uniform manner across molecules. This makes the definitions in ssAChems more implicit than in symbolic AChems.

2.5.1 Terminology

To help prevent confusion between the properties of real chemical molecules and our AChem objects, and to prevent the abuse of chemistry terminology, we use the following distinct terminology:

- The objects of interest are **particles**; these are either atomic particles (**atoms**) or composite particles (**composites**).
- Particles can be joined together, or composed, with **links**; links can be broken to decompose composite particles; no operations within the system can decompose an atom.

2.5.2 Requirements and Design Principles

We place several requirements on the kinds of things that can be used as the basis of an ssAChem.

- *structure and/or dynamics*: the particles should have an underlying structure and/or dynamics, from which linking properties can emerge
- *multiple emergent properties*: in a system with a rich set of emergent properties, different properties can be selected to serve a variety of functions
- *single type of particle*: atomic particles can be linked to form composite particles that are the same type of thing as the atoms, with the same kind of emergent properties, so that these can in turn link to other atoms and composites, without the need to introduce new rules
- *everything emergent*: any design decision (for example, linking probability) is based on an emergent property, not on a property of the underlying representation (for example, a property should not rely on using the first item in a list, but instead the item in the list with, say, the maximum or minimum value of an emergent property)
- *computational tractability*: we wish to build computational systems, and so we need the calculation of emergent properties, and of the composition rules, to be tractable.

3   Definition of an ssAChem

In this section we define a generic ssAChem, in terms of its ($\mathbb{S}$, $\mathbb{R}$, $\mathbb{A}$). This provides a detailed framework that can be instantiated with specific ssAChems. This framework is then instantiated with two example ssAChems, one based on Hermitian matrices (Sect. 4), and one based on Random Boolean Networks (Sect. 5).

3.1   The Set of Possible Particles $\mathbb{S}$

The particles are defined in terms of the underlying **structure** of particles, their **behavioural model**, and the **emergent properties** of that model. Particles in an ssAChem have their own emergent behaviours that are exploited by the rules. Some of the 'physics' of the system has been moved from $\mathbb{R}$ into $\mathbb{S}$.

3.1.1   Structure

S defines the underlying **structure** of the members of the particle set.
   Example structures are:

- Binary trees: $S := A \mid S \times S$, so S is the set of possible particles comprising the atoms A and all pairwise linked particles $S \times S$. This structure is used in the matrix chemistry of Sect. 4.
- General n-ary trees: $S := A \mid S^+$, so S is the set of possible particles comprising the atoms A and all chains of linked particles S^+ (we also require the chain length to be ≥ 2). This structure is used in the RBN chemistry of Sect. 5.
- General graphs: $S := (E, V)$, where the vertices are the constituent atoms A, and the edges are the relevant links. This structure is a model for real world molecules.

Other structures can be defined. We require structures to have a 'memory' of the underlying constituent atoms. That is, any structure defined must conserve atomic constituents:

$$A + B \rightleftharpoons C \Rightarrow bag(A) \oplus bag(B) = bag(C) \tag{1}$$

where $bag(P)$ denotes the bag (multiset) of atoms in particle P, and $\oplus$ denotes bag addition.

3.1.2 Behavioural Model

B is the **behavioural model** of the particles, which provides the basis of the chemical properties of interest of the particles.

Each particle structure $s \in S$ has a behavioural model instance $b \in B$, derived via linking rules from the behavioural model instances of its constituent atoms.

Some models are static mathematical constructs (for example, matrices, Sect. 4). Others may be discrete dynamical systems (for example, RBNs, Sect. 5), which can have an associated current **state**, Σ.

3.1.3 Emergent Properties

Behavioural models have one or more **emergent properties**: $e : B \times \Sigma \rightarrow X$. The ssAChem-specific type X is typically the real numbers, but it may be, say, a vector, or any other type of interest.

These emergent properties can be used for a variety of purposes: here we use them mainly for linking probabilities.

It is desirable for these emergent properties to be efficiently computable.

3.1.4 Atoms and Composites

Each of the **atoms**, $A \subset S$, is assigned a unique base atomic model $b_a \in B$. Where relevant, each atom instance is initialised to a particular or random initial state, $\sigma_0 \in \Sigma$.

Each atom is assigned a unique symbol as its name, enabling us to write down the structure of specific particles. This name is typically an arbitrary letter, or a structured name reflecting some of its underlying emergent properties. The name may be decorated with a tag to indicate the current state of an atom instance.

Each composite is defined by its structure, behavioural model, and current state: $C = S \times B \times \Sigma$.

Composites may have an internal composition mirroring their structure S; that is, subcomposites corresponding to substructures may also have behavioural models and state. These submodels and substates are related to the overall model and state in a way defined by the linking algorithms (see later).

One possibility we have yet to explore is associating multiple diverse models with a given structure, with $C = S \times (B \times \Sigma)^N$. For example, one model might express micro/particle level behaviour, and another macro/ensemble level behaviour. Or one might express linking properties, another functional properties, allowing composites to have some derived behaviour. Or the models might be combined in some way to jointly express a single property.

3.2 The Rules, $\mathbb{R}$

The rules define the result of linking two particles, and of decomposing a composite particle. Each of these is defined through a precondition (whether the rule applies), and an operation (the result of applying the rule).

3.2.1 Ancillary Information

Ideally, all the properties of a particle emerge from its behavioural model. However, in some cases extra information is needed to make deterministic the definition of how two particles link, or how a composite particle decomposes. This ancillary information is provided to operations by the parameter $\lambda \in L$.

The ancillary information $\lambda \in L$ might provide some internal position within the structure of C where the link is to be established or decomposed; it might provide details of which linking site is to be used. The specific value of λ is provided by the algorithm.

For example, in bRBN-world (Sect. 5), each particle has two linking sites, and the ancillary information of which site is to be used in the linking attempt is provided by the algorithm (a random choice if both are available). This choice is made emergently in the later Spiky-RBN ssAChem [20].

3.2.2 Linking Criterion

The **linking criterion**, $K : \mathbb{P}((C \times L) \times (C \times L))$, defines whether a pair of particles can in principle link, based on their emergent properties, plus any ancillary information $\lambda \in L$. K defines the domain of applicability, or precondition, of the linking operation $\frown$.

Whether particles that can link in principle then link in practice is given by the *linking algorithm*, which depends on further factors.

3.2.3 Linking Operation

The **linking operation**, $\frown : (C \times L)^2 \to C$, takes two particles, plus ancillary information, and gives their linked composition. The linking operation can be applied only if the linking criterion holds.

It is desirable that the linking operation to be either non-commutative ($a\frown b \neq b\frown a$ in general), or non-associative ($a\frown(b\frown c) \neq (a\frown b)\frown c$ in general), or both [12]. An operation that is both commutative and associative cannot capture *isomers*, composites comprising the same collection of atoms but with different structures and properties. This is because with associativity we can omit the brackets, so $a\frown(b\frown c) = a\frown b\frown c$, and then with commutativity we can swap adjacent atoms: $= a\frown c\frown b = c\frown a\frown b$. By this means we can get all permutations of a, b, c; the linked system is merely an *unstructured* bag of atoms, not a structured composite, and so loses much of the possible combinatoric power.

3.2.4 Decomposition Criterion

The **decomposition criterion**, $K_d : \mathbb{P}(C \times L)$, defines whether a composite particle can in principle decompose, based on its emergent properties, plus any ancillary information $\lambda \in L$. K_d defines the domain of applicability, or precondition, of the decomposition operation.

Decomposition acts to break links between subcomposite components: it does not cleave atomic particles. Hence no atomic particle is in K_d.

Whether composite particles that can decompose in principle then decompose in practice is given by the *decomposition algorithm*, which depends on further factors.

3.2.5 Decomposition Operation

The **decomposition operation**, $D : C \times L \to C^2$, takes a composite particle, plus ancillary information, and gives its decomposed products resulting from breaking a single link. The decomposition operation can be applied only if the decomposition criterion holds.

There is no requirement for the products to be able to link to form the original composite. Thus decomposition may allow the indirect formation of composites that cannot form by linking alone.

3.3 The Algorithm, $\mathbb{A}$

The overall reactor algorithm describes the *dynamics* of the reaction system, including the effect of the environment. Environmental effects may include such things as spatial structure (well-mixed, grid, etc.), inflows and outflows of particles, and (analogues of) energetics. The environmental state can affect the probability of an attempted linking or decomposition succeeding.

3.3.1 Environment

The **environment**, $\mathcal{E}$, captures those properties of the system within which reactions happen that are not captured by particle properties alone.

For example, the environment might be aspatial, or a spatial system, allowing varying concentrations and movement of particles; it might have sources and sinks of particles (as in a chemostat); it might have a temperature analogue.

3.3.2 Linking Probability

The **linking probability**, $Pr_b : \mathcal{E} \rightarrow (C \times L)^2 \rightarrow C \rightarrow \mathfrak{R}$, defines the probability that, in a given environment, an attempted linking operation $(c_1, \lambda_1) \frown (c_2, \lambda_2) = c'$ will be successful.

If the linking criterion is not satisfied, that is, if $((c_1, \lambda_1), (c_2, \lambda_2)) \notin K$, then the linking operation is not applicable, and we say that $Pr_b(e)((c_1, \lambda_1), (c_2, \lambda_2))(c') = 0$. The linking algorithm can distinguish the case of not linkable in principle (not in K) from linkable but with zero probability in this case (in K, but nevertheless $Pr_b = 0$).

3.3.3 Linking Algorithm

The **linking algorithm** defines how to use the linking rule and linking probability to perform a linking attempt. The behaviours of a given particle structure and rules can be explored in a variety of algorithms.

3.3.4   Decomposition Probability

The **decomposition probability**, $Pr_d : \mathcal{E} \to C \times L \to C^2 \to \mathfrak{R}$, defines the probability that, in a given environment, an attempted decomposition operation, $(c, \lambda) \to (c_1, c_2)$, will be successful.

3.3.5   Decomposition Algorithm

The **decomposition algorithm** defines how to use the decomposition rule and decomposition probability to perform a decomposition attempt.

3.3.6   Reactor Algorithm

The **reactor algorithm** defines how to use the linking and decomposition algorithms, and the environment state, to describe the overall behaviour of the reacting system. It defines which particles are in the initial system, which are chosen for linking and decomposition attempts, and if and how particles are added to or removed from the system during execution.

The algorithm for choosing particles can cover all the possibilities: exhaustive pair-wise search, random collisions via a Gillespie-style algorithm [13] for a well-mixed reaction vessel, proximity in a spatial grid of diffusing particles, and more. There might be a fixed number of atoms (conservation of mass), or a chemostat-style inflow/outflow, or other non-physical simulation possibilities. Each of these choices can used the same 'physics' of the system (the same $\mathbb{S}$ and $\mathbb{R}$), merely changing the reaction vessel setup, $\mathbb{A}$, to investigate different behaviours and properties of the AChem.

3.4   Summary of the ssAChem Framework

1. the set of possible particles $\mathbb{S}$, and their underlying properties, defined by

   a. S, the underlying **structure** of the particles
   b. B, the **behavioural model** of the particles
   c. e, the **emergent properties** of B

2. the rules $\mathbb{R}$, defining how particles link and decompose, given by

   a. the **linking rule**, comprising:
      i. the **linking criterion**, which defines whether a pair of particles can in principle link, based on their emergent properties
      ii. the **linking operation**, which takes two particles that satisfy the linking criterion, and gives their linked composition

b. the **decomposition rule**, comprising:

    i. the **decomposition criterion**, which defines whether a composite particles can in principle decompose, based on its emergent properties

    ii. the **decomposition operation**, which takes a composite particle that satisfies the decomposition criterion, and gives its decomposition products

3. the algorithm $\mathbb{A}$, using the following components to define the overall behaviour of the reacting system:

    a. the **environment**, defining properties of the system within which reactions occur (e.g., space, temperature)

    b. the **linking probability** that, in a given environment, an attempted linking operation will succeed

    c. the **linking algorithm**, which uses the linking rule under the linking probability to perform a linking attempt

    d. The **decomposition probability**, that, in a given environment, an attempted decomposition operation will succeed

    e. The **decomposition algorithm**, which uses the decomposition rule under the decomposition probability to perform a decomposition attempt

    f. The overall **reactor algorithm**, which uses the linking and decomposition algorithms, and defines how particles are chosen for reaction attempts.

4 SMAC: Sub-symbolic Matrix Artificial Chemistry

In our first example we illustrate the basics of the ssAChem concepts in a relatively simple system that has no dynamic state, but a lot of rich mathematical structure.

We consider a mathematical system with rich algebraic structure that can be employed to define a ssAChem. The basic particle is the *matrix*. Given the richness of emergent properties of matrices and the fact that most computers are optimised for matrix arithmetic, the matrix seems to be a natural base for a sub-symbolic AChem.

There are various so-called matrix AChems in the literature. For example, the Matrix-multiplication chemistry [3–5] works over binary strings and folds them in to matrices to perform linking through matrix multiplication. However, this makes no use of the many emergent properties or algebraic structure of mathematical matrices. In general, existing matrix AChems make little or no use of the mathematical properties of matrices, and would be better named 'array-based' AChems.

4.1 SMAC's Set of Possible Particles, $\mathbb{S}$

4.1.1 SMAC Structure

The structure of SMAC particles is defined as $S ::= A \mid S{\smallsmile}S$, that is, the structure of the set of possible particles comprises the atoms A and all pairwise linked particles $S{\smallsmile}S$. So SMAC particles have an underlying binary tree structure.

4.1.2 SMAC Behavioural Model

The SMAC behavioural model B is the set of d dimensional Hermitian matrices (here $d = 3$). Particles are static; they have no current state Σ.

Matrix Notation

We define a vector $\mathbf{v}$ of dimension d in terms of its components v_i, $1 \leq i \leq d$; it has magnitude $|\mathbf{v}| = v$. We define a matrix $\mathbf{M}$ of dimension d in terms of its components M_{ij}, $1 \leq i, j \leq d$. Vector dot product $\mathbf{v}.\mathbf{w}$, matrix addition $\mathbf{M} + \mathbf{N}$, matrix-vector multiplication $\mathbf{Mv}$, and matrix multiplication $\mathbf{MN}$ have their usual definitions.

A Hermitian matrix is one equal to its conjugate transpose: $\mathbf{M} = \mathbf{M}^\dagger \iff M_{ij} = \bar{M}_{ji}$ where $M_{ij} \in \mathbb{C}$.

4.1.3 SMAC Emergent Properties

Matrices have several emergent properties, that is, properties of the matrix as a whole, rather than of its individual components. These include:

- Eigenvalues λ_i and eigenvectors $\mathbf{v}_i$, solutions of $\mathbf{Mv} = \lambda\mathbf{v}$. A d-dimension matrix in general has d eigenvalue-eigenvector pairs. Each of the eigenvectors describes a direction, while the corresponding eigenvalue provides a magnitude to that direction. This provides an inherent geometry to the matrix. In general eigenvalues can be complex numbers (they are solutions of the d-order characteristic polynomial of the matrix). Here we restrict ourselves to Hermitian matrices, which have real eigenvalues.
- Trace: $Tr(\mathbf{M}) = \sum_{i=1}^d M_{ii}$. The sum of the eigenvalues equals the trace: $\sum_{i=1}^d \lambda_i = Tr(\mathbf{M})$, hence the trace of a Hermitian matrix is real.
- Rank: the dimension spanned by the matrix's rows (or columns).
- Determinant: $|\mathbf{M}|$
- Similar properties of submatrices of $\mathbf{M}$

Here we restrict our investigations to $d = 3$, both for tractability (the complexity of eigenvalue calculations is $> O(d^2)$), and for a convenient geometric interpretation of the eigenvectors.

The SMAC emergent properties are the absolute values of the three eigenvalues of the behavioural model matrix, truncated to the nearest integer towards zero:

$$e_i := \lfloor |\lambda_i| \rfloor \tag{2}$$

The trace is not necessarily equal to the sum of these truncated values.

We also use the length-normalised eigenvectors $\hat{v}_i$ in calculating linking probabilities.

4.1.4 SMAC Atoms and Composites

We create an atomic set of 3×3 Hermitian matrices. We restrict the entries in our atomic set to have real and imaginary parts of 0 or ± 1, and further restrict the leading diagonal elements to be real (because Hermitian).

We disallow the 'singleton matrices' (those with a single non-zero entry) and the traceless matrices (which includes the zero matrix), since these have trivial linking properties (later).

We generate our set of atomic behavioural models as:

$$
\begin{aligned}
B_a = \{ \mathbf{A} \mid & A_{jj} \in \{0, \pm 1\}; \\
& A_{jk} \in \{0, \pm 1, \pm i, \pm 1 \pm i\}, j < k; \\
& A_{jk} = \bar{A}_{kj}, k < j; \\
& \#(A_{jk} \neq 0) > 1; \\
& Tr(\mathbf{A}) \neq 0 \}
\end{aligned}
\tag{3}
$$

This gives a set of $3^9 - 6 - 7 \times 3^6 = 14574$ 'atoms', which is rather too large to be investigated fully. So we partition this set into 66 equivalence classes of atoms, where atoms in the same class have the same integer values of their eigenvalues rounded to zero, and same traces. We take a single arbitrarily chosen[1] entry from each class. This provides us with a 'periodic table' of 66 elements, classified by eigenvalue set and by trace (Fig. 1).

The textual representation of large composites is not very readable. For example, one of the composites generated by our system (see later) is:

$$
\begin{aligned}
M1000 = & ((((FFd \frown LLf) \frown Vc) \frown Lb) \frown ((AAd \frown (Nc \frown Re)) \\
& \frown ((Re \frown (FFd \frown LLf)) \frown ((Nc \frown GGf) \frown (JJf \frown (Aa \frown Dc))))))
\end{aligned}
\tag{4}
$$

---

[1]To follow our design criteria, we should choose a non-arbitrary element. Future work includes developing a choice criterion for this.

Unique Absolute Eigenvalues

Legend: trace / Element / e'vals

ε	3,2	3	2			1		2,1		3,1
−3	Ca −3,0,0	La −2,0,0			Sa −1,−1,−1	Ga −2,−2,1	Ia −2,−1,1 · Ja −2,−1,1	Ba −3,−1,1	Aa −3,−1,0	Da −3,0,1 · Fc −3,1,1
−2	Cb −3,0,0	Lb −2,0,0	Hb −2,−2,2		Ub −1,−1,1 · Tb −1,−1,0	Gb −2,−2,1	Vb −1,−1,2 · Jb −2,−1,1	Kb −2,−1,2	Mb −2,0,1	Db −3,0,1
−1	Ec −3,0,2	Lc −2,0,0	Nc −2,0,2 · Yc −1,0,1		Uc −1,−1,1 · Xc −1,0,0	Zc −1,0,2	Vc −1,−1,2 · Jc −2,−1,1	Kc −2,−1,2	Mc −2,0,1	Dc −3,0,1 · Fc −3,1,1
1	Od −2,0,3	GGd 0,0,2	Nd −2,0,2 · Yd −1,0,1		BBd −1,1,1 · FFd 0,0,1	Pd −2,1,1	Qd −1,0,2 · Zd −1,0,2	CCd −1,1,2	Md −2,0,1	AAd −1,0,3 · Wd −1,−1,3
2	HHe 0,0,3	GGe 0,0,2	Re −2,2,2		BBe −1,1,1 · IIe 0,1,1	Pe −2,1,1	Qe −2,1,2 · Ze −1,0,2	CCe −1,1,2	EEe −1,2,2 · AAe −1,0,3	
3	HHf 0,0,3	GGf 0,0,2			LLf 1,1,1	JJf 0,1,2		CCf −1,1,2	EEf −1,2,2 · AAf −1,0,3	DDf −1,1,3 · KKf 0,1,3

Fig. 1 Periodic table: the arrangement of the 66 SMAC elements by trace and eigenvalue set. We name each element using one or two upper case letters (A–Z, AA–LL) to indicate which of the 37 distinct sets of rounded eigenvalues, and a lower case letter (a–f) to indicate which of the 6 distinct trace values is associated with that element

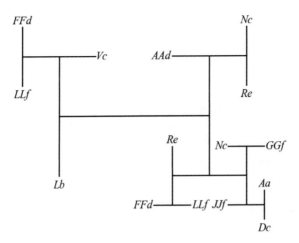

Fig. 2 Graph representation of $M1000$. This demonstrates a property of SMAC that differs from real chemistry: links can form between links, as well as between atoms

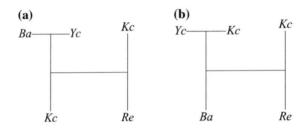

Fig. 3 SMAC graphs showing the atomic structure of two composites containing the same atoms **a** $(((Ba{\frown}Yc){\frown}Kc){\frown}(Kc{\frown}Re))$ **b** $(((Yc{\frown}Kc){\frown}Ba){\frown}(Kc{\frown}Re))$

This is difficult to unpick, so we introduce a graph representation of composites. The graph of the above composite is shown in Fig. 2.

Figure 3 shows two isomers (comprising the same atoms, but linked in different structures) in the graphical representation.

Since the behavioural model has no state, a composite is completely defined by its binary tree structure and its matrix; $C = S \times B$. The matrix at the root of the structure is the only one involved in linking; sub-structures have their own associated component matrices which need to be remembered, however, as they may become the root matrix as a result of decomposition. See Fig. 4.

Fig. 4 The underlying structure of the SMAC compound $((A\frown B)\frown(C\frown D))$. The structure S is given by the binary tree with A, B, C, D as the atomic leaves. The behavioural model at each sub-tree is given by the relevant matrix product defined in Sect. 4.2.2. The matrix at the root of the structure is the only one involved in the linking algorithm. Sub-structure matrices may become the root of decomposed composites

4.2 SMAC Rules

4.2.1 SMAC Linking Criterion

The SMAC emergent properties used in linking are the set of d (truncated) eigenvalue and eigenvector pairs. A SMAC particle can be considered to have d potential binding sites, labelled $1, \ldots, d$. For SMAC, ancillary information is provided by the linking criterion, which determines which (if any) specific site of these d is linkable in each particle: $L = 1, \ldots, d$. Here $d = 3$.

Consider two particles A and B, with corresponding emergent properties of sets of truncated eigenvectors $\{b_i^A\}$ and $\{b_j^B\}$. For the particles to link, we require them to have

- a non-zero eigenvalue in common: $\{b_i^A\} \cap \{b_j^B\} \notin \{\emptyset, \{0\}\}$
- a non-zero trace of their Jordan product, $Tr(A \circ B) \neq 0$ (in order to apply trace scaling; see Sect. 4.2.2).

4.2.2 SMAC Linking Operation

For SMAC, no ancillary information is needed for the linking operation, so we have $\frown : C^2 \to C$.

We impose the following algebraic requirements on our linking operation as applied to our behavioural model of Hermitian matrices:

- It should preserve the property of being Hermitian: $\mathbf{M} = \mathbf{M}^\dagger$ and $\mathbf{N} = \mathbf{N}^\dagger \implies \mathbf{M} \frown \mathbf{N} = (\mathbf{M} \frown \mathbf{N})^\dagger$.
- It should be commutative: $\mathbf{M} \frown \mathbf{N} = \mathbf{N} \frown \mathbf{M}$. This is inspired by real chemistry; for example, carbon monoxide could be written as $C \frown O$ or $O \frown C$, but is still the same molecule. Note that matrix addition is commutative and matrix multiplication is not commutative.
- It should therefore be non-associative: $\mathbf{M} \frown (\mathbf{N} \frown \mathbf{P}) \neq (\mathbf{M} \frown \mathbf{N}) \frown \mathbf{P}$ in general (see Sect. 3.2.3). Note that both matrix addition and matrix multiplication are associative.

To meet these requirements, we use a *Jordan algebra* [21] based on Hermitian matrices, and define the linking operation using the Jordan product:

$$\mathbf{M} \circ \mathbf{N} = \tfrac{1}{2}(\mathbf{MN} + \mathbf{NM}) \tag{5}$$

This is clearly commutative (since matrix addition is commutative). A little algebra demonstrates that it is non-associative, and that the product preserves the Hermitian property.

The Jordan product itself is not suitable for a linking operation that uses eigenvalue matching as its linking property. The eigenvaules of Jordan product matrices tend to grow exponentially fast with the number of Jordan products used. This would result in composites predominantly linking only with composites containing similar numbers of atoms: we would not, for example, be able to see a single atom linking to a large composite.

So our linking operation incorporates trace-scaling, which exhibits matching eigenvalues over a wide range of particle sizes.

$$A \frown B := \frac{|Tr(A)| + |Tr(B)|}{|Tr(A \circ B)|} A \circ B \tag{6}$$

Consequently, we require composites to have a non-zero trace.

4.2.3 SMAC Decomposition Criterion

Currently, we have not defined a SMAC decomposition criterion. We are investigating the products of decomposition given that a link breaks, but not yet the criterion for allowing that link breaking.

4.2.4 SMAC Decomposition Operation

Although the SMAC structure is a binary tree, we wish to have a decomposition operation that allows 'internal' links to break. This potentially allows the formation of structures that cannot form directly.

The simplest case is breaking an outermost link. $A \frown B \rightarrow A + B$. Both A and B may be complex structures.

Breaking a link one deep in the tree is also simple to define: $A \frown (B \frown C) \rightarrow A + B + C$.

When we break a link two deep, we need to 'fix up' the resulting mal-formed binary tree: $A \frown (B \frown (C \frown D)) \rightarrow A \frown B + C + D$. This forms composite $A \frown B$ indirectly, potentially allowing new composites that could not form directly, or have only a low probability of forming directly.

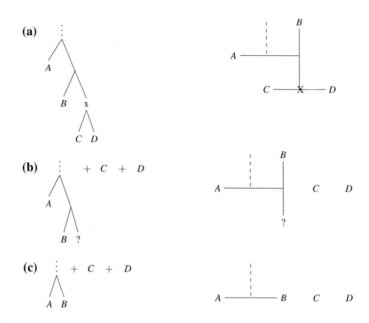

Fig. 5 Breaking the link between C and D in the composite subcomponent $A{\sim}(B{\sim}(C{\sim}D))$. **a** the link to be broken (left shows the binary tree representation; right shows the SMAC notation); **b** the link broken, and the resulting components: B has no partner; **c** B migrates up the tree, resulting in $A{\sim}B$; in the SMAC notation, we refer to this as 'link straightening'

This break and fix-up pattern also holds for more deeply nested links: $A{\sim}(B{\sim}(C{\sim}(D{\sim}E))) \rightarrow A{\sim}(B{\sim}C) + D + E$. The underlying operation is illustrated in Fig. 5.

4.3 SMAC Algorithm

4.3.1 SMAC Environment

The version of SMAC described here has no environmental input.

4.3.2 SMAC Linking Probability

Consider two particles A and B, with corresponding sets of truncated eigenvalues $\{b_k^A\}$ and $\{b_k^B\}$, and length-normalised eigenvectors $\{\hat{\mathbf{v}}_k^A\}$ and $\{\hat{\mathbf{v}}_k^B\}$.

If the linking criterion holds, then we have a matching eigenvalue $b_i^A = b_j^B = b$. Let the eigenvectors associated with this matching eigenvalue be $\hat{\mathbf{v}}_i^A$ and $\hat{\mathbf{v}}_j^B$.

We construct a linking probability from their dot product, such that anti-parallel eigenvectors have the highest linking probability of 1, and parallel eigenvectors have the lowest linking probability of 0:

$$Pr(A \frown B) = \frac{1}{2} \left(1 - \hat{\mathbf{v}}_i^A . \hat{\mathbf{v}}_i^B\right) \tag{7}$$

If there is more than one pair of matching eigenvalues, we choose the pair whose eigenvectors are most nearly anti-parallel ($\min\{\hat{\mathbf{v}}_i^A . \hat{\mathbf{v}}_i^B\}$), and so produce the highest linking probability.

This choice of parallel eigenvectors resulting in zero linking probability ensures that not every atom can undergo self synthesis. In order to get self-synthesis we must have a repeated eigenvalue.

4.3.3 SMAC Linking Algorithm

For particles to be able to link, they must have non-zero eigenvalues in common, and the trace of the resulting composite must be non-zero. If these criteria are met, then linking can occur on this attempt with probability as given by Eq. 7.

4.3.4 SMAC Decomposition Probability

Currently, we have not specified a decomposition probability.

Our decomposition operation can produce composites that cannot form directly by synthesis, because certain links do not have matching eigenvalues. This implies the decomposition probability needs to be different from the linking probability, to ensure such composites do not immediately disintegrate.

4.3.5 SMAC Decomposition Algorithm

Currently, we have not specified a decomposition algorithm.

4.3.6 SMAC Reactor Algorithm

Here we are in the early stages of investigation, and use a simple reactor algorithm involving linking only, and no decomposition. We generate 10,000 composites using Algorithm 1. We then analyse the resulting composites (see Sect. 4.5).

Algorithm 1 SMAC 10,000 composite synthesis

 1: mols := list of the 66 atoms
 2: tries := 0
 3: **repeat**
 4:     $A, B :\in$ mols
 5:     $C := A \frown B$
 6:     **if** C **then**
 7:         append C to mols list
 8:     **end if**
 9:     ++tries
10: **until** # mols $= 10,066$

4.4   Summary of the SMAC ssAChem

4.4.1   Definition

- structure S

  - structure: binary tree: $S ::= A \mid S \frown S$
  - behavioural model: 3×3 Hermitian matrices, $Tr \neq 0$; no state
  - emergent properties: eigenvalues and eigenvectors; trace
  - atoms: 66 specific matrices

- linking rules R

  - linking criterion: matching (truncated, absolute-value) eigenvalues and $Tr(A \circ B) \neq 0$
  - linking operation: $A \frown B = \frac{|Tr(A)| + |Tr(B)|}{|Tr(A \circ B)|}(A \circ B); A \circ B = \frac{1}{2}(AB + BA)$
  - decomposition criterion: not defined
  - decomposition operation: 'link straightening', see Fig. 5

- algorithm A

  - environment: none, for now
  - linking probability: $\frac{1}{2}\left(1 - \hat{\mathbf{v}}_i^A . \hat{\mathbf{v}}_i^B\right)$
  - linking algorithm: the linking operation is applied with the linking probability
  - decomposition probability: not defined
  - decomposition algorithm: not defined
  - reactor algorithm: see Algorithm 1; used for our initial investigations

Fig. 6 The isomers $(A^m \frown X) \frown A^n$ and $A^m \frown (X \frown A^n)$, where X is an atom or a composite, have identical properties in this ssAChem

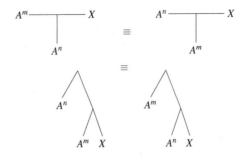

4.4.2 Structure v Model

The commutative but non-associative linking operation used in SMAC allows isomers exist (Fig. 3). However, the linking operation is associative in some special cases. For example, it is clear from Eq. (5) that $A \circ A = AA = A^2$, and that therefore

$$\text{Power associative:} \qquad A^{\circ m} A^{\circ n} = A^{\circ(m+n)} = A^{m+n} \quad \forall m, n \geq 0 \qquad (8)$$

where $A^{\circ n} = A \circ A \circ \dots \circ A(n \text{ times}) = A^n$. Hence all composites containing n copies of just a single atom type A have identical properties in this ssAChem.

It can readily be checked that the following associativity condition also holds:

$$\text{Jordan identity:} \qquad (A^m \circ X) \circ A^n = A^m \circ (X \circ A^n) \quad \forall m, n \geq 0 \qquad (9)$$

Hence the two isomers (different structures) in Fig. 6 have identical properties (model values) in this ssAChem.

This demonstrates the difference between structure and properties. Two particles with the same properties can nevertheless be different particles, in that they have a different internal structure. In particular, they may decompose differently.

4.5 SMAC Results

We have performed a variety of experiments to investigate SMAC's low-level chemical properties of synthesis and decomposition, to evaluate its promise as the basis for a sub-symbolic AChem. We summarise the results here.

4.5.1 Synthesis and Self-synthesis

The linking probability is zero for parallel eigenvectors, so in order to get self-synthesis we must have a repeated eigenvalue. In such cases there will be (at least two) eigenvectors orthogonal to one another (because Hermitian), and hence $\cos \theta = 0, p = 0.5$. Hence self-synthesis probabilities are either be 0 (no repeated eigenvalues: 10 atom classes) or 0.5 (repeated eigenvalues, using the orthogonal eigenvectors: 28 atom classes). Hence we do have atoms capable of self-synthesis, and atoms not capable of self-synthesis.

The probability distribution for general atomic synthesis across all possible atom pairs shows a broad range of probabilities from zero to one.

To investigate the synthesis properties of composites we use the reactor Algorithm 1 to generate 10,000 composites. This process generates a wide range of composite sizes, including example composite $M1000$ (Eq. 4, Fig. 2), and large composites comprising over 200 atoms.

The composite generator took 119,757 tries at forming links to generate 10,000 composites. This implies an 8.35% probability of successful linking in this setup. The probability distribution for this form of composite synthesis also shows a broad range of linking probabilities from zero to one. Additionally, there does not seem to be a strong effect of trace size (and hence composite size) on these linking probabilities. This indicates that large composites can still link effectively.

We observe that large and small composites react with each other, as do similarly sized composites. For any particular size of composite s the set of reactions creating composites of that size have reactants ranging in size evenly from 1 to $s - 1$. Thus there is no bias to linking either similarly or differently sized composites.

4.5.2 Decomposition

We have examined the larger composites produced in the synthesis experiment, and have found several cases where the decomposition operation could form composites that cannot be created directly (because of no matching eigenvalues), or only with a very low probability (because of near-parallel eigenvectors).

A particular composite formation by decomposition in shown in Fig. 7; the product composite cannot be formed by synthesis alone.

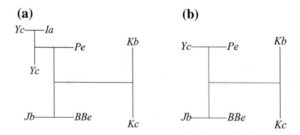

Fig. 7 **a** Initial composite before (Yc–Ia) link decomposed; **b** The two atoms Yc and Ia, plus the composite shown, result from link breaking and link straightening. The new composite cannot form directly because (Yc–Pe) and (Jb–BBe) do not have matching eigenvalues

4.6 SMAC *History* Versus *Presentation*

The definition of SMAC here following the ssAChem framework starts with matrices as particles, then moves to Jordan products for linking rules. The invention of SMAC followed a different route [12]: the requirement for non-associativity led to the investigation of Jordan algebras, and then the Hermitian matrices as a suitably rich model of these algebras. The overall aim of the SMAC work is to use mathematical structures (starting with the algebraic structure of the particles and their links) as a rich source of inspiration and results for exploitation by ssAChems.

5   The bRBN-World ssAChem

Our second example to illustrate the ssAChem concepts exploits a different approach from the algebraic matrix structures in SMAC. Here, we use Random Boolean Networks as our particles and network manipulations as our linking operation: these systems have rich dynamical structure but less overt mathematical foundations.

This work is based mainly on the bRBN models discussed in [9–11], augmented with a first investigation of a temperature model. Current work is exploring more emergent linking sites, to produce 'spiky RBNs' [20].

5.1   bRBN-World Particles

5.1.1   bRBN-World Structure

We define this structure inductively as $S ::= A \mid S^+$, with $n > 1$. That is, the set of possible particles comprises the atoms A and all (finite) lists of linked particles $S \frown \ldots \frown S$ ($n > 1$ terms). This results in bRBN-world particles having an underlying n-ary tree structure.

5.1.2   bRBN-World Behavioural Model

The behavioural model B is the set of $K = 2$ 'linking' Random Boolean Networks (bRBNs).

Random Boolean Network Definition

A Random Boolean Network (RBN) is a discrete dynamical system with the following structure. It has N nodes. Each node i has:

- a binary valued state, which at time t is $s_{i,t} \in \mathbb{B}$
- K inputs, assigned randomly from K of the N nodes (including possibly itself), defining its neighbourhood, a K-tuple of node labels with no duplicates
- a randomly assigned state transition rule, a boolean function from its neighbourhood state to its own next state: $\phi_i \in \mathbb{B}^K \to \mathbb{B}$

The state of node i's neighbourhood at time t is $\chi_{i,t} \in \mathbb{B}^K$, a K-tuple of the binary states of the neighbourhood nodes. At each timestep, the state of all the nodes updates in parallel: $s_{i,t+1} = \phi_i(\chi_{i,t})$.

As with any finite discrete dynamical system, the state of an RBN eventually falls on an attractor: a repeating cycle of states. Kauffman [18, 19] discovered that $K = 2$ RBNs rapidly converge to relatively few relatively short attractors: they are complex, but not chaotic.

bRBNs: RBNs Modified for Linking

$K = 2$ RBNs are a computationally tractable system with rich microdynamics (of the entire microstate) and complex macrodynamics (the attractor space). We take such $K = 2$ RBNs as the basis for an ssAChem.

bRBN-world [9–11] uses a modification of basic RBNs, called bRBNs, as the behavioural model. A bRBN atom is constructed from a plain RBN as follows: add b linking nodes to the RBN (for this work, $b = 2$). For each linking node, select an RBN node at random, and change one of its K inputs to come from the linking node. Linking nodes do not have inputs, so have no need for an associated random boolean function; instead they have a fixed boolean state (0 or 'cleared' if unlinked, 1 or 'set' if linked; see later for linking details). See Fig. 8.

The linking algorithm can access the microstate of the RBN (to calculate the emergent linking property), each of the b linking nodes and their state, and the 'wiring' between nodes (to perform the link).

A given particle can exist in different states (the microstate of the underlying RBN at time t), dependent on its initial condition.

5.1.3 bRBN-World Emergent Properties

An RBN is a discrete dynamical system which exhibits many rich emergent properties. We can look at properties at different levels of resolution. There are properties that merely count the number of timesteps needed to traverse between two states:

- transient length e_t: the number of timesteps to move from some initial state (for example, all zeros or all ones) to an attractor
- attractor cycle length e_c

Then there are detailed properties of the microstates during these macro-transitions, such as:

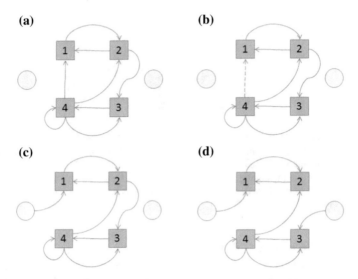

Fig. 8 Constructing a bRBN: **a** a $K = 2, N = 4$ RBN, plus $b = 2$ unattached linking nodes (circles); **b** connecting the first bnode: randomly select rnode 1; randomly select south input (dashed); **c** connecting the first bnode: replace south input with input from b1; **d** connecting the second bnode: randomly select rnode 3; randomly select north input; replace with input from b2

- flashing: how many nodes change state during the macro-transition
- flashes: total number of state changes during the macro-transition
- total: the sum of the state values during the macro-transition
- proportion: the proportion of nodes that are 'on', averaged over the macro-transition

We have investigated each of these properties for attractor cycles to determine their suitability as emergent properties in bRBN-world [9, 10]. The suitable choices, which also depend on choice of linking criterion, are discussed in Sect. 5.5.1.

5.1.4   bRBN-World Particles

The atoms are chosen from the set of size N RBNs. We discuss the choice of N in Sect. 5.5.1.

Initially [10, 11] we chose small atomic RBNs at random. Later investigations [9] use an evolutionary algorithm to search for an 'interesting' set of atomic bRBNs; see Sect. 5.5.2.

We use arbitrary alphabetical symbols to name the atoms. We use parenthesised strings of atoms to display the composites. We distinguish the different basin of attraction, where necessary, with a superscript digit. For example (adapted from [9, Fig. 8.5.3]):

$$((((C^1 \frown C^1)^2 \frown (C^1 \frown C^1)^1)^3 \frown (C^1 \frown C^6)^4)^1 \frown B^3)^1 \tag{10}$$

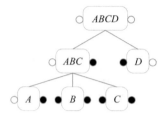

Fig. 9 The underlying structure of the RBN-world compound $((A\frown B\frown C)\frown D)$. The structure S is given by the n-ary tree with A, B, C, D as the atomic leaves. The behavioural model at each subtree is given by the relevant linked RBNs (Sect. 5.2.2 and Fig. 10). The bRBNs at each level of the structure are involved in the linking algorithm. A white circle denotes a linking site not yet linked, with value set to 0; a black circle denotes a linking site that has been linked (in a parent node), with value here set to 1

An RBN-world particle is defined by its n-ary tree structure, its bRBN, and its current state; $C = S \times B \times \Sigma$. At the root of each substructure is a single composite bRBN; the intermediate nodes contains smaller composite bRBNs; the leaves contain the atomic bRBNs (Fig. 9). The properties of the bRBNs at each level of the structure are involved in the linking algorithm. The fact that properties of lower level structures are used in the linking algorithm stops the linking operation from being associative: $((A\frown B\frown C)\frown D) \neq ((A\frown B)\frown(C\frown D)) \neq (A\frown B\frown C\frown D)$ in general.

5.2 bRBN-World Rules

5.2.1 bRBN-World Linking Criterion

An RBN is a discrete dynamical system which exhibits a rich possible set of emergent properties that we can use for defining linking criteria (Sect. 5.1.3).

These numerical properties can be compared in a variety of ways. We have investigated [9, 10]:

- equal: the same values (to within a small tolerance)
- similar: like equal, but with a larger tolerance bound
- different: not similar
- sum one: the properties sum to one (to within a small tolerance)
- sum zero: the properties sum to zero (to within a small tolerance), hence are opposites

The suitable choices, which also depend on choice of emergent property, are discussed in Sect. 5.5.1.

5.2.2   bRBN-World Linking Operation

Two individual bRBNs are linked into a larger composite as shown in Fig. 10. Note that although the result is a bRBN, its RBN component does not have the structure of a 'typical' RBN: the communication is channeled through specific links, and the result is not a small world network. In particular, large composites comprising long chains of RBNs will have different dynamics from a typical RBN.

Two structured compounds are linked, at a given linking site in each, as illustrated in Figs. 11 and 12. The link is formed, and the structures are 'zipped' together moving up the trees, to form a single tree.

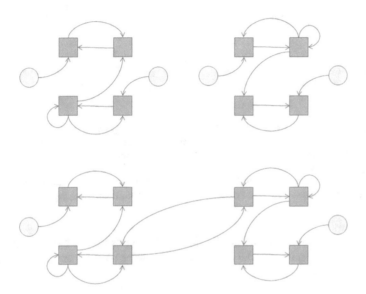

Fig. 10 Linking two bRBNs. The unlinked bRBNs are stored in the lower nodes of the structure tree; the linked composite bRBN is stored in the higher level of the tree

Fig. 11 Two composites, $(A(B(CDE)))$ and $(((FG)H)J)$, to be linked at the linking sites indicated by the arrow

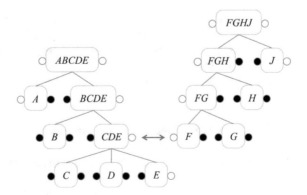

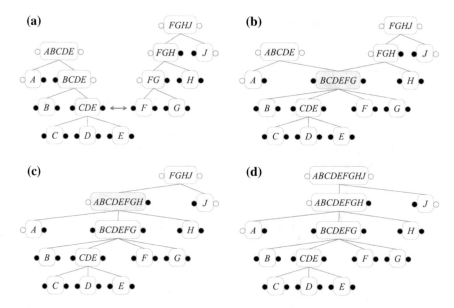

Fig. 12 The linking operation illustrated: linking $(A(B(CDE)))$ and $(((FG)H)J)$ to produce the linked compound $((A(B(CDE)FG)H)J)$: **a** set linking sites here and below to 'linked' black dots; **b** merge parent nodes; create bRBN that is the join of all the new node's children (shown in grey); move up to the newly merged parent; **c** while the node has two parents, merge the parent nodes; create the newly merged node's bRBN from its children; move up to the newly merged parent; **d** while the node has one parent, create the node's bRBN from its children; move up to the parent

Ancillary information used for linking is: S, the position in the structure where the link is to be formed; b, which linking node is to be used; σ, the current state (for the linking property).

5.2.3 bRBN-World Decomposition Criterion

The decomposition criterion on a formed link is the same as the linking criterion for forming the link. So, for example in Fig. 10, the criterion for decomposing the top level link uses the two bRBNs ABC and D. Although these had to fulfil the linking criterion when the link was formed, they might no longer fulfil it, because now their linked input nodes are set to 1 rather than 0, and so their dynamics have been changed.

5.2.4 bRBN-World Decomposition Operation

The decomposition operation is the reverse of the linking operation (Sect. 5.3.3). The designated link is broken, the higher level nodes are 'unzipped' to form two separate trees, and the binding site nodes of the lower levels set to 0.

5.3   bRBN-World Algorithm

5.3.1   Environment

In previously published work on bRBN-world [9–11], there are no environmental inputs. In particular, all reactions occur in an aspatial environment.

We report here some preliminary results related to reaction in a constant temperature heat bath.

5.3.2   bRBN-World Linking Probability

In the previously published work on bRBN-world [9–11], $Pr = 1$. That is, if the linking criterion holds, and the link is attempted, then it succeeds with no further criterion to satisfy.

We report here some preliminary results related to reaction probabilities given by a temperature analogue, inspired by reaction kinetics. We use transient lengths as the emergent properties for calculating a temperature-dependent probability. Let X_0, X_1 be the length of the transient from the all-zeros and all-ones state of particle X. Then consider the linking operation attempting to form composite C from particles A, B. We define two energies; the linking energy ΔE_b and the decomposition energy ΔE_d (Fig. 13):

$$\Delta E_b = (A_1 + B_1 + C_1) - (A_0 + B_0) \tag{11}$$

$$\Delta E_d = (A_1 + B_1 + C_1) - C_0 \tag{12}$$

The binding and decomposition probabilities use these energy-analogues in a Boltzmann-like factor, similar to that used in simulated annealing:

$$Pr = \begin{cases} 1 & ; \ \Delta E < 0 \\ \exp(-\Delta E/T) & ; \ 0 \le \Delta E \end{cases} \tag{13}$$

Fig. 13 'Energy' levels of reactants and linked particles, used to calculate probabilities

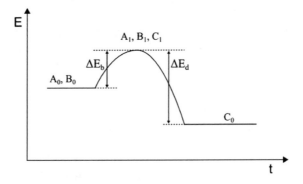

The previously published work can be thought of as the infinite temperature limit, with $Pr = 1$.

5.3.3 bRBN-World Linking Algorithm

Choose one of the linking nodes in each particle, at random.

Get the linking location for these nodes: starting at atomic level, progressively move up the composite tree structure until the linking property holds; use the first linking location found, or FAIL if no linking location is found.

Apply the linking operation (Sect. 5.2.2) at the linking location with probability given by Eq. 13.

5.3.4 bRBN-World Decomposition Probability

In the previously published work on bRBN-world [9–11], $Pr = 1$. That is, if the linking criterion does not hold, and decomposition is attempted, then it succeeds with no further criterion to satisfy.

We report some preliminary results related to reaction probabilities given by a temperature analogue, inspired by reaction kinetics. This allows otherwise 'unstable' links that do not meed the linking criterion to nevertheless persist. See Sect. 5.3.2 for the linking and decomposition probabilities.

5.3.5 bRBN-World Decomposition Algorithm

Decomposition proceeds as follows. For a given composite, all the links of its linked component bRBNs are examined. Any link the no longer fulfils the linking criterion (the linking property may have changed on linking) is broken, with probability given by Eq. 13. Any products are similarly decomposed.

5.3.6 bRBN-World Reactor Algorithm

The full reactor algorithm is as follows. Two distinct composites are selected; a linking is attempted (Sect. 5.3.3). Decomposition is then attempted (Sect. 5.3.5), either on the initial reactants, or on the successfully linked product. The overall result is the outcome of these two attempts.

A newly-formed composite may decompose because the original linking no longer satisfies the linking criterion (due to changing the state of the relevant linking nodes at the linking location), or because the linking criterion no longer holds elsewhere in the compound. If no link takes place, one or both initial composite particles might themselves decompose, if the linking attempt somehow falsified a linking criterion somewhere in the stateful composite.

5.4 Summary of bRBN-World ssAChem

- set $\mathbb{S}$

  - structure: n-ary tree: $S ::= A \,|\, S^+, n > 1$
  - model: bRBN with $b = 2$ linking sites; cycle evolution experiments use $N = 10$, $K = 2$
  - emergent properties: various transient and attractor cycle properties (Sect. 5.1.3); cycle evolution experiments use 'proportion'
  - atoms: specific small bRBNs

- rules $\mathbb{R}$

  - linking criterion: matching emergent properties of the bRBNs (Sect. 5.2.1); cycle evolution experiments use 'sum one'
  - linking operation: form a larger bRBN at the top of the structure tree, zipping together component RBNs lower down the tree
  - decomposition criterion: linking criterion fails to hold for linked structures
  - decomposition operation: break the link, and unzip the trees

- algorithm $\mathbb{A}$:

  - environment: aspatial heat bath
  - linking probability: Boltzmann factor from transient emergent properties
  - linking algorithm: link those that pass the linking criterion, with linking probability
  - decomposition probability: Boltzmann factor from transient emergent properties
  - decomposition algorithm: break internal links that fail the linking criterion, with decomposition probability
  - reactor algorithm: link, then decompose.

5.5 bRBN-World Results

5.5.1 Choice of Parameters, Linking Property and Criterion

There are many parameters and properties to choose from: do any result in a good ssAChem?

In our initial work on bRBN-world [11], we arbitrarily chose attractor cycle length as our emergent property, to establish that bRBNs exhibited sufficiently rich behaviours to be the basis for an ssAChem.

In subsequent investigations [9, 10] we determined a better choice, through exploration of the parameter space. That exploration demonstrates that atomic network size N and connectivity K has little influence, and that the best choice of emergent property (Sect. 5.1.3) and linking criterion (Sect. 5.2.1) is either 'proportion' as property and 'sum one' as criterion, or 'total' as property and 'sum zero' as criterion. We use 'proportion/sum one' for further work on bRBNs.

5.5.2 Evolutionary Search for Atomic bRBNs

The space of possible atomic RBNs is vast. The preliminary bRBN-world experiments [10, 11] sample bRBNs at random. To find richer behaviours from small atomic sets, [9] employs a search over the bRBN atomic space, using a genetic algorithm.

The richness of the exhibited behaviour of a candidate atomic set is measure by a fitness function. This function examines the reaction network generated by the atoms for 'loops'; cyclic reactions such as $A \rightarrow A(BCD) \rightarrow A(BC) \rightarrow AB \rightarrow A$. What counts as a 'loop' has to be carefully defined to exclude trivial loops such as $A^1 \rightarrow A^2 \rightarrow A^1$ (where the superscript indicates an atom in a different attractor state), and meaningless loops such as $A \rightarrow BC \rightarrow A$ (where there is no common element around the loop). See [9, Chap. 8] for details.

A well-mixed reactor vessel is populated with 1000 atoms of each of five types. A reaction network is formed by running the vessel for a sufficient time, using a Gillispie-style algorithm. The fitness function is applied to the resulting network.

The genetic algorithm uses a population size of 100 vessels, and is run for 300 generations. Mutation changes the details of the five types of atoms used. See [9, Chap. 8] for further details.

This evolutionary approach successfully discovered atomic sets capable of producing complex reaction networks. For example, one reaction network analysed in detail exhibited 1,286 reactions and 645 different particles, and a longest reaction loop comprising 8 reactions.

This demonstrates that the bRBN-world ssAChem can continue to support more complex behaviours.

5.5.3 Addition of a Temperature Analogue

Following on from the work presented in [9–11], we have performed some preliminary experiments with the temperature analogue described in Sect. 5.3.2.

At very low temperatures ('absolute zero', $T = 0$) essentially no reactions occur, as the probability is always zero (except when $\Delta E < 0$, which happens only rarely).

At very high temperatures ($T = 15$) the system essentially reduces to the previously published bRBN-world behaviours: all reactions that merely satisfy the linking criterion occur, as do all the possible decompositions (no 'unstable' links can survive).

At intermediate temperatures ($2 < T < 10$) we have a reduced reaction rate, as expected. We see a mix of behaviours, with different composite species thriving in different temperature ranges. Additionally, we see new composites that cannot occur at higher temperatures, because they would decompose.

We have also performed preliminary experiments where the temperature of the system changes between three temperatures (low $T = 2$, medium $T = 5$, high $T = 10$) for five cycles. These systems show a higher degree of reactivity than a similar system held at the medium temperature.

These preliminary experiments demonstrate that a temperature analogue has an interesting effect on the behaviour of the system, by allowing otherwise unstable composites to persist, and engage in further reactions. Having demonstrated the promise of the approach, we need to perform larger scale experiments to quantify the effects.

6   ssAChem Design Guidelines

We have provided a framework within which ssAChems can be defined. SMAC and bRBN-world provide two different instantiations of this framework. SMAC focusses on the underlying mathematical structure provided by Jordan algebras and their realisation in the Hermitian matrix model. RBN-world focusses on the underlying dynamical system that provides the atomic properties. These results demonstrate that the ssAChem concept is generically useful. It is not only the original dynamic bRBN-world implementation that demonstrates interesting chemistry: other systems with less dynamics but deeper mathematical structure also demonstrate interesting chemistry.

The process of designing a new ssAChem can proceed as follows:

1. start from the ssAChem framework provided here
2. instantiate it with a particular *model*: an atomic set, the emergent properties, the linking rules
3. develop a computational implementation
4. perform initial experiments, to demonstrate that the chosen model has rich behaviour, and to find good parameter values
5. perform full experiments, to explore full system behaviours

Acknowledgements Faulkner is funded by an York Chemistry Department Teaching PhD studentship. Krastev is funded by a York Computer Science Department EPSRC DTA PhD studentship. We thank Leo Caves for some insightful comments on this work. We thank Michael Krotosky, Andrew Balin, and Rudi Mears for their work in exploring some earlier versions of ideas presented here.

References

1. ALife XV, Cancun, Mexico. MIT Press (2016)
2. Anet, F.A.L.: The place of metabolism in the origin of life. Curr. Opin. Chem. Biol. **8**(6), 654–659 (2004)
3. Banzhaf, W.: Self-replicating sequences of binary numbers—Foundations I: General. Biol. Cybern. **69**(4), 269–274 (1993)
4. Banzhaf, W.: Self-replicating sequences of binary numbers—Foundations II: Strings of length $N = 4$. Biol. Cybern. **69**(4), 275–281 (1993)
5. Banzhaf, W.: Self-organization in a system of binary strings. In: Proceedings of Artificial Life IV, pp. 109–118 (1994)
6. Banzhaf, W., Baumgaertner, B., Beslon, G., Doursat, R., Foster, J.A., McMullin, B., de Melo, V.V., Miconi, T., Spector, L., Stepney, S., White, R.: Requirements for evolvability in complex systems. Theory Biosci. **135**(3), 131–161 (2016)
7. Dittrich, P., Ziegler, J., Banzhaf, W.: Artificial chemistries–a review. Artif. Life **7**(3), 225–275 (2001)
8. Eigen, M., Schuster, P.: A principle of natural self-organization. Naturwissenschaften **64**(11), 541–565 (1977)
9. Faulconbridge, A.: RBN-world: sub-symbolic artificial chemistry for artificial life. Ph.D. thesis, University of York, UK (2011)
10. Faulconbridge, A., Stepney, S., Miller, J.F., Caves, L.: RBN-world: The hunt for a rich AChem. In: ALife XII, Odense, Denmark, pp. 261–268. MIT Press (2010)
11. Faulconbridge, A., Stepney, S., Miller, J.F., Caves, L.S.D.: RBN-World: a sub-symbolic artificial chemistry. In: ECAL 2009, Budapest, Hungary. LNCS, vol. 5777, pp. 377–384. Springer (2011)
12. Faulkner, P., Sebald, A., Stepney, S.: Jordan algebra AChems: exploiting mathematical richness for open ended design. In: ALife XV, Cancun, Mexico [1], pp. 582–589 (2016)
13. Gillespie, D.T.: Exact stochastic simulation of coupled chemical reactions. J. Phys. Chem. **81**(25), 2340–2361 (1977)
14. Hutton, T.J.: Evolvable self-replicating molecules in an artificial chemistry. Artif. Life **8**(4), 341–356 (2002)
15. Hutton, T.J.: Evolvable self-reproducing cells in a two-dimensional artificial chemistry. Artif. Life **13**(1), 11–30 (2007)
16. Kauffman, S.A.: Metabolic stability and epigenesis in randomly constructed genetic nets. J. Theoret. Biol. **22**(3), 437–467 (1969)
17. Kauffman, S.A.: Autocatalytic sets of proteins. J. Theoret. Biol. **119**(1), 1–24 (1986)
18. Kauffman, S.A.: Requirements for evolvability in complex systems. Physica D **42**, 135–152 (1990)
19. Kauffman, S.A.: The Origins of Order. Oxford University Press (1993)
20. Krastev, M., Sebald, A., Stepney, S.: Emergent bonding properties in the Spiky RBN AChem. In: ALife XV, Cancun, Mexico [1], pp. 600–607 (2016)
21. McCrimmon, K.: Jordan algebras and their applications. Bull. Am. Math. Soc. **84**(4), 612–627 (1978)
22. Ogawa, A.K., Yiqin, W., McMinn, D.L., Liu, J., Schultz, P.G., Romesberg, F.E.: Efforts toward the expansion of the genetic alphabet: information storage and replication with unnatural hydrophobic base pairs. J. Am. Chem. Soc. **122**(14), 3274–3287 (2000)
23. Ono, N., Ikegami, T.: Model of self-replicating cell capable of self-maintenance. In: Advances in artificial life, pp. 399–406. Springer (1999)
24. Pross, A.: Causation and the origin of life: metabolism or replication first? Orig. Life Evol. Biosph. **34**(3), 307–321 (2004)
25. Suzuki, H., Ono, N., Yuta, K.: Several necessary conditions for the evolution of complex forms of life in an artificial environment. Artif. Life **9**(2), 153–174 (2003)

Discovering Boolean Gates in Slime Mould

Simon Harding, Jan Koutník, Júrgen Schmidhuber
and Andrew Adamatzky

Abstract Slime mould of *Physarum polycephalum* is a large cell exhibiting rich spatial non-linear electrical characteristics. We exploit the electrical properties of the slime mould to implement logic gates using a flexible hardware platform designed for investigating the electrical properties of a substrate (*Mecobo*). We apply arbitrary electrical signals to 'configure' the slime mould, i.e. change shape of its body and, measure the slime mould's electrical response. We show that it is possible to find configurations that allow the Physarum to act as any 2-input Boolean gate. The occurrence frequency of the gates discovered in the slime was analysed and compared to complexity hierarchies of logical gates obtained in other unconventional materials. The search for gates was performed by both sweeping across configurations in the real material as well as training a neural network-based model and searching the gates therein using gradient descent.

1 Introduction

Slime mould *Physarum polycephalum* is a large single cell [23] capable for distributed sensing, concurrent information processing, parallel computation and decentralized actuation [1, 7]. The ease of culturing and experimenting with Physarum makes this slime mould an ideal substrate for real-world implementations of unconventional sensing and computing devices [1]. A range of hybrid electronic devices were implemented as experimental working prototypes. They include Physarum self-routing and self-repairing wires [3], electronic oscillators [6], chemical sensor [33], tactical sensor [4], low pass filter [34], colour sensor [5], memristor [13, 24], robot controllers [12, 30], opto-electronics logical gates [20], electrical oscillation frequency logical gates [32], FPGA co-processor [21], Shottky diode [10], transistor [24]. There prototypes show that Physarum is amongst most prospective candidates

S. Harding · A. Adamatzky (✉)
Unconventional Computing Centre, University of the West of England, Bristol, UK
e-mail: andrew.adamatzky@uwe.ac.uk

J. Koutník · J. Schmidhuber
IDSIA, USI&SUPSI, Manno-Lugano, CH, Switzerland

for future hybrid devices, where living substrates physically share space, interface with and co-function with conventional silicon circuits.

So far there are four types of Boolean gates implemented in Physarum.

- Chemotactic behaviour based morphological gate [28]; logical False is an absence of tube in a particular point of substrate and logical True is a presence of the tube.
- Ballistic morphological gate [2] based on inertial propagation of the slime mould; logical False is an absence of tube in a particular point of substrate and logical True is a presence of the tube.
- Hybrid slime mould and opto-electronics gate [20] where growing Physarum acts as a conductor and the Physarum's behaviour is controlled by light.
- Frequency of oscillation based gate i [32], where frequencies of Physarum's electrical potential oscillations are interpreted as Boolean values and input data as chemical stimuli.

These logical circuits are slow and unreliable, and are difficult to reconfigure. Thus we aimed to develop an experimental setup to obtain fast, repeatable and reusable slime mould circuits using the 'computing in materio' paradigm.

The Mecobo platform has been designed and built within an EU-funded research project called NASCENCE [9]. The purpose of the hardware and software is to facilitate 'evolution in materio' (EIM)—a process by which the physical properties of a material are exploited to solve computational problems without requiring a detailed understanding of such properties [22].

EIM was inspired by the work of Adrian Thompson who investigated whether it was possible for unconstrained evolution to evolve working electronic circuits using a silicon chip called a Field Programmable Gate Array (FPGA). He evolved a digital circuit that could discriminate between 1 or 10 kHz signal [26]. When the evolved circuit was analysed, Thompson discovered that artificial evolution had exploited physical properties of the chip. Despite considerable analysis and investigation Thompson and Layzell were unable to pinpoint what exactly was going on in the evolved circuits [25]. Harding and Miller attempted to replicate these findings using a liquid crystal display [14]. They found that computer-controlled evolution could utilize the physical properties of liquid crystal—by applying configuration signals, and measuring the response from the material—to help solving a number of computational problems [16].

In the work by Harding and Miller, a flexible hardware platform known as an 'Evolvable Motherboard' was developed. The NASCENCE project has developed a new version of this hardware based on low-cost FPGAs that more flexible than the original platform used with liquid crystal.

In this paper, we describe the Mecobo platform (Sect. 1) and how it is interfaced to Physarum (Sect. 1) to perform an exhaustive search over a sub-set of possible configurations. In Sect. 2 the results from this search are then searched to confirm whether Physarum is capable of acting as Boolean gates. We find that all possible 2-input gates can be configured this way. Section 2.2.1 further explores the complexity of the computation by modelling the material using a neural network model.

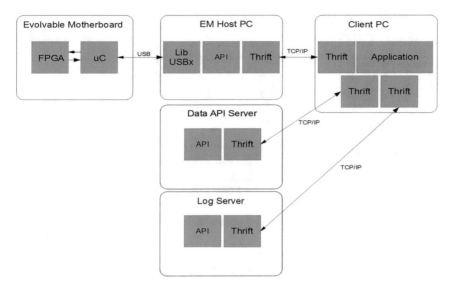

Fig. 1 Overview of the Mecobo hardware software architecture

The Mecobo Platform

Mecobo is designed to interface a host computer to a candidate computational substrate, in this instance the Physarum. The hardware can act as a signal generator, and route these signals to arbitrary 'pins'—which are typically electrodes connected to the candidate material. The hardware can also record the electrical response from the substrate. Again, this recording (or recordings) can be linked to any 'pin'. Both applied signals (used for configuration) and measurements can be analog or digital.

The core of the hardware is a microprocessor and an FPGA that run a 'scheduler', illustrated in Fig. 2. A series of actions, such as "output 1 Hz square wave on pin 5, measure on pin 3" are placed into a queue, and the queue executed. Complex series of actions can be scheduled onto the Mecobo. The Mecobo connects, via USB, to a host PC. The host PC runs software implementing the server side of the Mecobo Application Programming Interface (API). The API work flow mirrors the hardware's scheduler, and allows for applications to interface to the hardware without understanding all of the underlying technical details. Implemented using THRIFT, the API is also language and operating system agnostic, with the additional benefit that it can executed remotely over a network. The API software also includes functionality for data processing and logging of collected data for later analysis.

Figure 1 shows a simplified architectural overview of the Mecobo hardware and software. A full description, of both hardware and software can be found in [19] Fig. 3. The hardware, firmware and software for the Mecobo are open source and can be accessed at the project website: http://www.nascence.eu/.

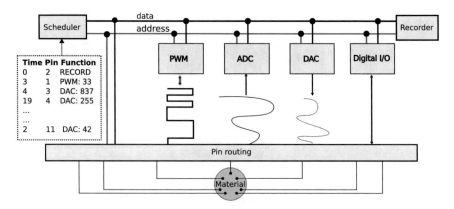

Fig. 2 Overview of the low-level Mecobo hardware architecture

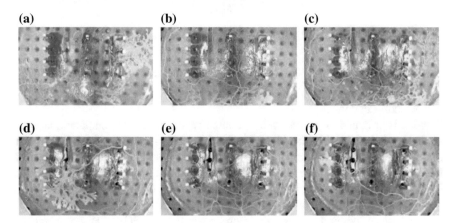

Fig. 3 Time-lapse of Physarum growing on agar with electrodes. Images are approximately 6 h apart

Interfacing Physarum to Mecobo

A crude containing dish with electrodes was constructed from matrix board and gold-plated header pins, as shown in Fig. 4a. Onto these pins, we pushed down a section of 2% agar (with or without Physarum) so that the pins punctured the agar, and were visible above the surface. This allowed the Physarum to come into direct contact with the electrodes. For experiments where the substrate was intended to be only Physarum, a small amount of molten agar was painted over the contacts and some of the matrix board. This minimized the amount of agar, in order to reduce its influence. Small amounts of damp paper towel were placed around the container to maintain humidity.

A video available at https://www.youtube.com/watch?v=rymwltzyK88 shows the Physarum growing and moving around the electrodes.

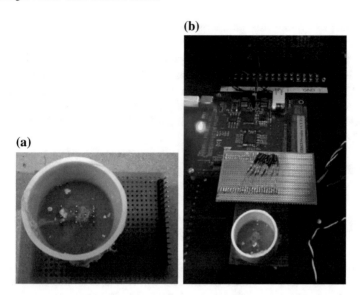

Fig. 4 **a** Electrodes with agar and Physarum. The white spots are oat flakes, which are used to feed the Physarum. **b** Agar dish connected to Mecobo

As shown in Fig. 4b the electrodes were then connected to the digital outputs of the Mecobo. Each electrode was connected to the Mecobo interface using a 4.7 kΩ resistor to limit the amount of applied current. The digital outputs of the Mecobo provide either 0 v (low) or 3.3 v (high).

2   Data Collection

The possible configuration space for the Mecobo platform is vast, and it would be infeasible to attempt to apply all possible configurations to the Physarum. In the Nascence project, evolutionary algorithms are used as a practical method to search through the configuration space and find configurations that perform a particular task. Just as with the initial experiments in the Nascence project [19], we start with an exhaustive search over a small configuration space.

The exhaustive search procedure consisted of applying all possible binary combinations of various frequency pairs to 9 pins (which is a practical amount for time purposes). One pin was used as an output from the material, with the other 8 pins acting as inputs.

For each binary combination, each pair of frequencies was tried with one frequency representing a 'low' and the other representing a 'high' input. The frequency pairs were combinations of square waves of either (250 Hz, 500 Hz, 1 kHz, or 2.5 kHz).

The order of applied configurations was shuffled to help prevent similar configurations being applied sequentially.

The Mecobo measured the digital response from the Physarum and/or agar for 32 ms. The digital threshold is 0.75 v for high, with voltages below this being classed as low. The sampling frequency was twice the highest input frequency applied.

In total, 49,152 states are applied, in a process that takes approximately 2 h to complete. The majority of the time is spent configuring the Mecobo and transferring sampled data back to the host PC.

Exhaustive Search for Boolean Gates

2.1   Method

Boolean logic gates are the building block for many circuits, sensors and computational devices. Using an exhaustive search, we were able to data mine within the collected data to look for configurations that acted as Boolean gates.

The data was searched for all two-input Boolean functions. Two pins were selected as the inputs to the gates (A and B), and one pin was selected as the gate output. The output response of the material (i.e. *true* or *false*) was determined from the output frequency of the material. A Fast Fourier Transform (FFT) was performed, and the frequency of the highest magnitude response was determined. The output was *true* if this frequency was nearer to the *true* input frequency, and *false* if nearer in value to the *false* input frequency. The remaining pins were then treated as configuration pins for the Physarum/agar. Each possible combination of configuration, input pair and output were then compared to see what logical operation it performed.

If for a given configuration (using the same frequency pairs), the output for all combinations of inputs A and B matched the expected behaviour of a gate, then it was judged that the Physarum and/or agar could implement that gate. As the configurations are temporally spaced, and each configuration applied multiple times (as A and B could be swapped over), a discovered gate would likely be 'stable' over the time taken to run the experiment. As the Physarum grows and moves, or as the agar dehydrates and shrinks, the physical substrate will vary. Therefore, we would not expect the system to be stable over long periods of time. Repeated measurements that are temporally spaced also reduces the possibility of measurement noise strongly influencing the results.

2.2   Results

As detailed in Table 1, agar on its own was unable to produce the universal gates NAND or NOR. It was also unable to produce the non-linear gates XOR or NXOR. Substrates that involve Physarum can be seen to produce more types of gates. It is

Table 1 Number of gates mined from the frequency responses of the Physarum. The *Agar* column contains a number of such configurations (input pins having values *true*, T, or *false*, F) found in the bare agar substrate and the *Physarum* column contains number of the configuration found in the dish containing the Physarum. We can see that the Physarum actually performs the logic functions whereas the sole agar is not capable of that

No.	Inputs xy				Agar	Physarum	Physarum + Agar	Gate
	FF	FT	TF	TT				
1	F	F	F	F	76 104	121 144	15864	Constant False
2	T	F	F	F		86	8	x NOR y
3	F	T	F	F		201	13	NOT x AND y
4	T	T	F	F		60	2	NOT x
5	F	F	T	F		201	13	x AND NOT y
6	T	F	T	F		60	2	NOT y
7	F	T	T	F		66	4	x XOR y
8	T	T	T	F		112	10	x NAND y
9	F	F	F	T	43 546	13 268	564	x AND y
10	T	F	F	T		52	0	x XNOR y
11	F	T	F	T	15 249	18 259	3 707	y
12	T	T	F	T		260	9	NOT x AND NOT y OR y
13	F	F	T	T	15 249	18 259	3 707	x
14	T	F	T	T		260	9	x OR NOT y
15	F	T	T	T	43 564	13 128	536	x OR y
16	T	T	T	T	74 996	113 266	13 448	Constant True

particularly interesting to note that far fewer AND or OR gates were found when using Physarum+Agar, compared to only agar.

Using Physarum, both with the agar and with minimal agar, the search was able to find many types of logic gates.

When the amount of agar was minimised, the fewest gates were found. This result was expected as the Physarum did not appear to connect between many of the electrodes, and therefore would only be able to participate in a smaller number of the configurations.

Table 2 Number of XOR gates found for given Input pin configurations

Input Pin B	Input Pin A							
	0	2	3	4	5	6	7	8
0		24	24	16	40	24		8
2	24			16	8			
3	24				8		16	
4	16	16			24	8	16	8
5	40	8	8	24		8		
6	24			14	8		8	
7			16	16		8		
8	8			8				

2.2.1 Detailed Analysis XOR

Considering only the XOR solutions that occur in the Physarum+Agar substrate, we can investigate the collected data in some detail. Table 2 shows the frequency of times that particular pins are used as inputs (the symmetry is a result of the fact that in this instance A and B can be swapped and still produce a valid gate).

Table 2 shows that some electrode pins are used much more frequently than others as inputs, and that only two electrodes ever get used successfully as outputs. This strongly suggests that the physical structure of the Physarum is important, and that a uniform mass would not be effective.

The experiment ran for 7322 s (122 min). During this time we observed the Physarum did move around on the agar. We generated a histogram that shows the number of times an applied configuration was used as part of XOR against time. The chart in Fig. 5 indicates that most of the results were found earlier on in the experiment. In addition to the movement of the Physarum, this may be caused by the agar drying out and becoming less conductive. It therefore seems likely that there might be more possible gate configurations, but that the search was too slow to discover them before the characteristics of the Physarum+agar changed.

Mould Modelling Using Neural Networks

This section uses neural network (NN) models [17], trained by gradient descent [11, 18, 31] to approximate the mapping from input voltages to the output voltage by training them on many randomly chosen examples measured during the exhaustive search. By solving that task a NN becomes a differentiable model of the—potentially complex—structures inside the material. Having the NN model, one can assess the complexity of the material based on trainability of the NN model.

A NN consist of a sequence of layers, where each layer computes an affine projection of its inputs followed by the application of a point-wise non-linear function ψ:

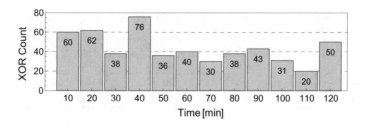

Fig. 5 XOR density. The chart shows how many XOR functions were found in the Physarum material over the course of 120 min long experiment. We can see that most of the gates was found in the beginning

$$\mathbf{h} = \psi(\mathbf{W}\mathbf{x} + \mathbf{b}) \tag{1}$$

where $\theta = \{\mathbf{W}, \mathbf{b}\}$ are the parameters of the layer. By stacking these layers one can build non-linear functions of varying expressiveness that are differentiable. In theory they can approximate any function to arbitrary precision given enough hidden units.

2.2.2   Data Preprocessing

The random search collected data was preprocessed in order to form a supervised training set. The input frequencies were ordered and translated to integer numbers in range from 1 to 5 forming an input vector $\mathbf{x} \in \{1, 2, 3, 4, 5\}^9$, where the input frequency was present. The value of 0 was used in the case of a grounded input pin or when the pin was used as the output pin. The target output vector had 9 entries as well ($\mathbf{t} \in [0, 1]^9$) but only the one that corresponds to the output pin was used at a time.

A *middle* section of the output buffer (1/4 to 3/4 section of the signal length) was preprocessed and transformed to a desired output target in one following three ways in the three experiments:

- **Ratio**: The proportion of ones in the signal.
- **Peak Frequency**: The peak of the frequency spectrum normalised such that 0.0 corresponds to a constant signal and 1.0 corresponds to the sequence 010101 …. The peak was obtained by computing the Fourier spectrum of the middle section of the signal, removing the DC component (first frequency component), computing the absolute values of the complex spectrum and using the first half (the spectrum is symmetric) to find the position of the maximum peak index.
- **Compressibility**: Relative length of the output buffer compressed using LZW to the maximum length of the compressed buffer. High values point to irregular outputs.

The networks have been trained to predict the output vector given the inputs, where we only trained for the one active output while ignoring the other 8 predictions

Table 3 Best hyperparameter settings for each of the three tasks along with the resulting Mean Squared Error (MSE)

Task	# Layers	# Units	Act. Fn.	Learning rate	MSE
Ratio	8	200	ReLU	0.055	6.058×10^{-4}
Frequency	5	200	ReLU	0.074	2.68×10^{-2}
Compressibility	5	200	ReLU	0.100	7.09×10^{-4}

of the network. Note that this essentially corresponds to training 9 different neural networks with one output each, that share the weights of all but their last layer.

The network weights were trained using Stochastic Gradient Descend (SGD) with a minibatch-size of 100 to minimise the Mean Squared Error on the active output. We kept aside 10% of the data as a validation set. Training was stopped after 100 epochs (full cycles through the training data) or once the error on the validation set didn't decrease for 5 consecutive epochs.

Neural networks have certain hyperparameters like the learning rate for SGD and the network-architecture that need to be set. To optimize these choices we performed a big random search of 1650 runs sampling the hyperparameters as follows:

- learning rate η log-uniform from 10^{-3} to 10^{-1}
- number of hidden layers uniform from $\{1, 2, 3, 4, 5, 6, 7, 8\}$
- number of hidden units in each layer from the set $\{50, 100, 200, 500\}$
- The activation function of all hidden units from the set $\{$tanh, ReLU,[1] logistic sigmoid$\}$

The best networks for the three tasks can be found in Table 3.

2.2.3 Modelling Results

On all three tasks the networks prediction error on the validation set decreased significantly during training. Figure 6 depicts the loss, the target and the prediction of the network for each example of the validation for all three tasks sorted by their loss. It can be seen that in all three cases the targets (*true* outputs) have a characteristic structure consisting of horizontal lines. This is due to the fact that certain target values are very likely. For each such value there are easy cases that the networks predict with high accuracy (the left side of the bar), and there seem to be difficult cases (right part of the bars) for which the predictions bifurcate (see prediction plots at the bottom of Fig. 6). This essentially means that the output distribution has sharp peaks which get smoothed out a bit by the networks. Overall, the networks predictive performance is very good (apart from the high values in the frequencies task which all fall into the high-loss region). The worst performance is on the frequencies task, while the best performance is achieved on the first task (predicting the ratios).

---

[1] ReLU: Rectified Linear Unit.

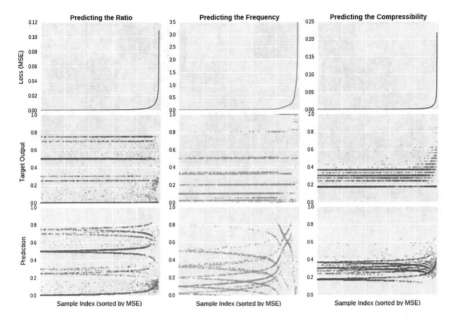

Fig. 6 Loss, true output, and prediction for all validation samples on each of the three tasks. The examples are sorted by their associated training loss

2.2.4   Searching for Logic Gates in the NN Model

The best network for the first task (predicting the ratio) was searched for the following logic functions: AND, OR, NAND, NOR, XOR, and XNOR. We encode the values for *True* and *False* for the inputs as 4 and 1 respectively (corresponding to high and low input frequency). For the output values we define 0 as *False* and 0.5 as *True*, because these two values are easy to distinguish and are the most common output values.

First, a set of examples with different combinations of input values but which share the same (random) values for the configuration pins along with the desired output values.

Gradient descent is then used to minimise the MSE by adjusting the values for the configuration pins, but we keep their values the same for all examples. So formally, given our neural network model $\hat{f}$ of the nano-material, we define an error over our N input/output pairs $(I_1^{(i)}, I_2^{(i)}, O^{(i)})$:

$$E = \sum_{i=1}^{N} \frac{1}{2}(\hat{f}(I_1^{(i)}, I_2^{(i)}, \theta) - O^{(i)})^2 \qquad (2)$$

The gradient is then calculated using backpropagation of the error:

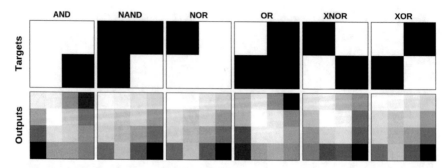

Fig. 7 The functions found by our search procedure all with pins 1 and 2 for inputs and pin 4 as output. The results for AND and NOR are a good fit, while the search for OR and XOR clearly failed to produce usable functions

$$\frac{\partial E}{\partial \theta} = (\hat{f}(I_1^{(i)}, I_2^{(i)}, \theta) - O^{(i)})\frac{\partial \hat{f}}{\partial \theta} \tag{3}$$

where $\theta = \{c_1, \ldots, c_6\}$.

At the end of this gradient descent process the configuration pin values are not in the allowed set $\{1, 2, 3, 4\}$ anymore. To make them valid inputs we round and clip them. This most likely increases the error, but hopefully still remains in a region where the network computes the desired function. It is unlikely that this procedure will always work, so in addition to this local search we also perform a global search.

One problem with the method described above is that it only performs a local search, which means that the solution that it converges to, might correspond to a bad local minimum. Furthermore the discretization we need to perform at the end of the local search might lead to a configuration which doesn't approximate the desired function very well.

To mitigate these problems we first sample 1 000 random starting points (settings of the configuration leads) and perform just 10 iterations of our local search on them. Only the starting point that lead to the lowest error is then optimised further for 500 epochs to obtain the final solution. In this way we reduce the risk of getting stuck in a poor local minimum. There is another free parameter which we haven't optimised yet, which is the assignments of input and output pins. We therefore repeat the above procedure for all 224 possible allocations of input and output pins. The found gates and their responses are summarised in Fig. 7.

3 Discussion

We do not know what is exact physical mechanism of the gate operations implemented by the slime mould. We can speculate that the following factors might contribute to the particular responses observed:

- The slime mould is conductive. Every single protoplasmic tube can be considered as a wire [3].
- The resistance of the slime mould wire depends on biochemical and physiological state of slime, thickness of tubes, and the Physarum's reaction to environmental conditions. The resistance also oscillates, due to contractile activity and shuffling of cytoplasm in the protoplasmic tubes [6].
- Physarum exhibits memristive properties when tested in current-voltage cycles [13]. Thus some pathways in Physarum could become non-conductive under AC stimulation.
- Morphology of Physarum can be modified by application of AC [29]. That is the slime mould can physically move between electrodes in response to stimulation by input electrical signals.
- In certain conditions Physarum allows the conduction of AC frequencies with attenuation profile similar to a low pass filter [34, 35].

The frequency of gates' discovery in Physarum varies substantially between the gates, with XOR and XNOR being the most rarely observable gates. Is it typical for all non-linear systems implementing gates or just Physarum? Let gates g_1 and g_2 be discovered with occurrence frequencies $f(g_1)$ and $f(g_2)$, we say gate g_1 is easy to develop or evolve than gate g_2: $g_1 \rhd g_2$ if $f(g_1) > f(g_2)$. The hierarchies of gates obtained using evolutionary techniques in liquid crystals [15], light-sensitive modification of Belousov-Zhabotinsky system [27] are compared with Physarum generated gates and morphological complexity of configurations of one-dimensional cellular automata governed by the gates [8]:

- Gates in liquid crystals [15]: {OR, NOR} $\rhd$ AND $\rhd$ NOT $\rhd$ NAND $\rhd$ XOR
- Gates in Belousov-Zhabotinsy medium [27]: AND $\rhd$ NAND $\rhd$ XOR
- Gates in cellular automata [8]: OR $\rhd$ NOR $\rhd$ AND $\rhd$ NAND $\rhd$ XOR
- Gates in Physarum: AND $\rhd$ OR $\rhd$ NAND $\rhd$ NOR $\rhd$ XOR $\rhd$ XNOR

We see that in all systems quoted the gate XOR is the most difficult to find, develop or evolve. Why is it so? A search for an answer could be one of the topics for further studies.

The search of gate configurations in the differentiable material model renders to be an efficient method of finding the correct configurations, that generates robust solutions. It remains to be confirmed whether these configurations work at the real mould again and close the loop between the modelling and the real material.

Acknowledgements The research leading to these results has received funding from the EC FP7 under grant agreements 317662 (NASCENCE project) and 316366 (PHYCHIP project).
The authors would like to acknowledge the assistance of Odd Rune Lykkebø for his technical assistance with Mecobo. Simon and Andy prepared the mould and performed the exhaustive search experiments on the hardware platform, Jan, Klaus and Jürgen contributed with the mould neural network modelling.

References

1. Andrew, A.: Physarum Machines: Computers from Slime Mould, vol. 74. World Scientific (2010)
2. Andrew, A.: Slime Mould Logical Gates: Exploring Ballistic Approach. arXiv:1005.2301, (2010)
3. Adamatzky, A.: Physarum wires: self-growing self-repairing smart wires made from slime mould. Biomed. Eng. Lett. **3**(4), 232–241 (2013)
4. Adamatzky, A.: Slime mould tactile sensor. Sens. Actuat. B Chem. **188**, 38–44 (2013)
5. Adamatzky, A.: Towards slime mould colour sensor: recognition of colours by Physarum polycephalum. Organ. Elect. **14**(12), 3355–3361 (2013)
6. Adamatzky, A.: Slime mould electronic oscillators. Microelect. Eng. **124**, 58–65 (2014)
7. Andrew, A.: Advances in Physarum Machines: Sensing and Computing with Slime Mould. Springer (2016)
8. Adamatzky, A., Bull, L.: Are complex systems hard to evolve? Complexity **14**(6), 15–20 (2009)
9. Broersma, H., Gomez, F., Miller, J.F., Petty, M., Tufte, G.: NASCENCE project: nanoscale engineering for novel computation using evolution. Int. J. Unconvent. Comput. **8**(4), 313–317 (2012)
10. Angelica,, C., Dimonte, A., Berzina, T., Erokhin, V.: Non-linear bioelectronic element: Schottky effect and electrochemistry. Int. J. Unconvent. Comput. **10**(5–6), 375–379 (2014)
11. Dreyfus, S.E.: The computational solution of optimal control problems with time lag. IEEE Trans. Automat. Control **18**(4), 383–385 (1973)
12. Ella, G., Andrew, A.: Translating slime mould responses: a novel way to present data to the public. In: Adamatzky, A.: (ed.), Advances in Physarum Machines. Springer (2016)
13. Gale, E., Adamatzky, A., De Lacy Costello, B.: Slime mould memristors. BioNanoScience **5**(1), 1–8 (2013)
14. Simon, H., Julian, F. Miller. Evolution in materio: A tone discriminator in liquid crystal. In: Proceedings of the Congress on Evolutionary Computation 2004 (CEC'2004), vol. 2, pp. 1800–1807, (2004)
15. Harding, S.L., Miller, J.F.: Evolution in materio: evolving logic gates in liquid crystal. In Proc. Eur. Conf. Artif. Life (ECAL 2005), Workshop on Unconventional Computing: From Cellular Automata to Wetware, pp. 133–149. Beckington, UK, (2005)
16. Harding, S.L., Miller, J.F.: Evolution in materio: evolving logic gates in liquid crystal. Int. J. Unconvent. Comput. **3**(4), 243–257 (2007)
17. Haykin, S.: Neural Networks: A Comprehensive Foundation, 2nd edn. Prentice Hall PTR, Upper Saddle River, NJ, USA (1998)
18. Linnainmaa, S.: The representation of the cumulative rounding error of an algorithm as a Taylor expansion of the local rounding errors. Master's thesis, University of Helsinki (1970)
19. Lykkebo, O.R., Harding, S., Tufte, G., Miller, J.F.: Mecobo: a hardware and software platform for in materio evolution. In: Ibarra, O., Kari, L., Kopecki, S. (eds.) Unconventional Computation and Natural Computation, LNCS, pp. 267–279. Springer International Publishing (2014)
20. Mayne, R., Adamatzky, A.: Slime mould foraging behaviour as optically coupled logical operations. Int. J. Gen. Syst. **44**(3), 305–313 (2015)
21. Mayne, R., Tsompanas, M.-A., Sirakoulis, G., Adamatzky, A.: Towards a slime mould-FPGA interface. Biomedical. Eng. Lett. **5**(1), 51–57 (2015)
22. Miller, Julian F., Harding, Simon L., Tufte, Gunnar: Evolution-in-materio: evolving computation in materials. Evolution. Intelligen. **7**, 49–67 (2014)
23. Stephenson, S.L., Stempen, H., Ian, H.: Myxomycetes: A Handbook of Slime Molds. Timber Press Portland, Oregon, (1994)
24. Giuseppe, T., Pasquale, D'.A., Cifarelli, A., Dimonte, A., Romeo, A., Tatiana, B., Erokhin, V., Iannotta, S.: A hybrid living/organic electrochemical transistor based on the physarum polycephalum cell endowed with both sensing and memristive properties. Chem. Sci. **6**(5), 2859–2868 (2015)

25. Thompson, A., Layzell, P.: Analysis of unconventional evolved electronics. Commun. ACM **42**(4), 71–79 (1999)
26. Thompson, A.: Hardware Evolution–Automatic Design of Electronic Circuits in Reconfigurable Hardware by Artificial Evolution. Springer, (1998)
27. Toth, R., Stone, C., Adamatzky, A., De Lacy Costello, B., Larry, B.: Dynamic control and information processing in the Belousov-Zhabotinsky reaction using a coevolutionary algorithm. J. Chem. Phys. **129**(18), 184708 (2008)
28. Tsuda, S., Aono, M., Gunji, Y-P.: Robust and emergent physarum logical-computing. Biosystems **73**(1), 45–55 (2004)
29. Tsuda, S., Jones, J., Adamatzky, A., Mills, J.: Routing physarum with electrical flow/current. arXiv:1204.1752, (2012)
30. Tsuda, S., Zauner, K.P., Gunji, Y.-P.: Robot control: from silicon circuitry to cells. In: Biologically Inspired Approaches to Advanced Information technology, pp. 20–32. Springer, (2006)
31. Werbos, P.J.: Beyond Regression: New Tools for Prediction and Analysis in the Behavioral Sciences. PhD thesis, Harvard University, (1974)
32. Whiting, J.G.H., De Lacy Costello, B., Adamatzky, A.: Slime mould logic gates based on frequency changes of electrical potential oscillation. Biosystems **124**, 21–25 (2014)
33. Whiting, J.G.H., De Lacy Costello, B., Adamatzky, A.: Towards slime mould chemical sensor: Mapping chemical inputs onto electrical potential dynamics of physarum polycephalum. Sens. Actuat. B Chem. **191**, 844–853 (2014)
34. Whiting, J.G.H., De Lacy Costello, B., Adamatzky, A.: Transfer function of protoplasmic tubes of Physarum polycephalum. Biosystems **128**, 48–51 (2015)
35. Whiting, J.G.H., Mayne, R., Moody, N., De Lacy Costello, B., Adamatzky, A.: Practical circuits with physarum wires. arXiv:1511.07915 (2015)

Artificial Development

Tüze Kuyucu, Martin A. Trefzer and Andy M. Tyrrell

Abstract Development as it occurs in biological organisms is defined as *the process of gene activity that directs a sequence of cellular events in an organism which brings about the profound changes that occur to the organism*. Hence, the many chemical and physical processes which translate the vast genetic information gathered over the evolutionary history of an organism, and put it to use to create a fully formed, viable adult organism from a single cell, is subsumed under the term "development". This also includes properties of development that go way beyond the formation of organisms such as, for instance, mechanisms that maintain the stability and functionality of an organism throughout its lifetime, and properties that make development an adaptive process capable of shaping an organism to match—within certain bounds—the conditions and requirements of a given environment. Considering these capabilities from a computer science or engineering angle quickly leads on to ideas of taking inspiration from biological examples and translating their capabilities, generative construction, resilience and the ability to adapt, to man-made systems. The aim is thereby to create systems that mimic biology sufficiently so that these desired properties are emergent, but not as excessively as to make the construction or operation of a system infeasible as a result of complexity or implementation overheads. Software or hardware processes aiming to achieve this are referred to as *artificial developmental models*. This chapter therefore focuses on motivating the use of artificial development, provides an overview of existing models and a recipe for creating them, and discusses two example applications of image processing and robot control.

T. Kuyucu (✉)
Disruptive Technologies, Trondheim, Norway
e-mail: tuze.kuyucu@disruptive-technologies.com

M.A. Trefzer · A.M. Tyrrell
Department of Electronics, University of York, York, UK
e-mail: martin.trefzer@york.ac.uk

A.M. Tyrrell
e-mail: andy.tyrrell@york.ac.uk

S. Stepney and A. Adamatzky (eds.), *Inspired by Nature*, Emergence,
Complexity and Computation 28, https://doi.org/10.1007/978-3-319-67997-6_16

1   Introduction

"The development of multicellular organisms from a single cell—the fertilized egg—is a brilliant triumph of evolution" notes Wolpert in his book, Principles of Development [71]. The structure of a single cell may arguably be the most complex part of any organism, but despite the evolutionary adaptation a unicellular organism is still vastly limited in the tasks it can achieve and is vulnerable to environmental threats. A multicellular organism is capable of multi-tasking using division of labour amongst the cells, and it is able to protect itself from environmental threats better than a unicellular organism would, since the loss of a cell or few cells does not necessarily harm the organism. Although multicellularity could have arisen through cell division failure or chance mutation in the evolutionary history of organisms, multicellularity is a key process harnessed by nature to create complex and intelligent biological organisms capable of executing sophisticated behaviours and surviving harsh and changing environmental conditions [6]. Multicellular organisms are a product of a process called development that builds these organisms from a single cell. Wolpert defines development as "The process of gene activity that directs a sequence of cellular events in an organism which brings about the profound changes that occur to the organism [71]". Development is how all the genetic information gathered over the evolutionary history of an organism is put to use to create an adult organism from a single cell. However, development is also a mechanism that maintains the stability and functionality of an organism throughout its lifetime, and not merely a genotype to phenotype mapping mechanism of biology for creating complex multicellular organisms. Thanks to multicellular development, an organism is capable of surviving damage and loss of its physical parts, which otherwise would be lethal to the organism.

1.1   Models of Development in Evolutionary Computation

Multicellular biological organisms have been a topic of interest in the computer science and engineering fields as inspiration of models of intelligent systems. Their ability to be robust, adaptive, and scalable while they develop, make them interesting in computational intelligence as these properties are difficult to design using traditional approaches to computation or engineering. Biological organisms can grow from a single cell into a multicellular organism using the same genotype for all cells. These cells can then specialize to form different parts of an organism. Although the process of development in biology is clearly defined, its definition in Evolutionary Computation (EC) varies significantly [8, 13, 16, 18, 27, 38, 49, 59, 67]. While some artificial algorithms try to closely model the biological development, others are simply inspired by a mechanism of biological development. In the latter cases, artificial development is defined by its main inspiration and the task it is used for. Roggen's diffusion-based developmental model [56], the self modifying cartesian genetic programming by Harding et al. [27], and Stanley's pattern producing networks [59], are examples of systems that use simple inspirations from select biological developmental mechanisms. Whereas the developmental models such as those presented in [16, 19, 64], are examples of systems that model the biological development more closely. Although most commonly referred to as artificial development, its name can also take many other forms; computational embryology, artificial embryology, artificial embryogeny, artificial ontogeny, computational development [8]. In the remainder of this article *artificial development* is used to refer to a computational system that models the biological developmental process for the uses of understanding biology and/or aiding EC applications.

1.2 Benefits of Artificial Developmental Systems

Multicellular development is a key element of nature in achieving complex, robust, adaptive and intelligent organisms, Fig. 1. Biological development makes use of the changing environment that is internal as well as external to the organism to bring about complex properties that EC strives to solve: scalability, fault tolerance, and adaptivity. Furthermore, such embedded environmental awareness in the biological systems lead to complex higher level behaviours such as self-organization, despite their seemingly architectured design [14].

A multicellular design approach in EC benefiting from the decentralized organizational mechanism achieved in biological organisms could also bring about similar benefits to EC. Scalability, fault tolerance and adaptivity are the three possible benefits discussed in further detail in this subsection.

1.2.1 Scalability

Over the years of EC research, the complexity of the evolved designs has not increased greatly. The inability of evolution to design systems at the desired level of complexity in a reasonable amount of time is a major problem. The ability to achieve higher complexity systems without a major increase in genotype size and within an acceptable time frame is referred to as *scalability*. Traditional system design methods in engineering and computer science build complex systems via the repeated use of functions or modules. Hence it has been acknowledged by many that introducing a mechanism that can achieve modular behaviour while evolving designs could relieve the scalability problem [1, 24, 31, 37, 52, 63, 68, 69]. An example of the use of modules in EC is the Automatically Defined Functions (ADF) [37]; ADF introduces reuse of parts of the genetic code during evolution. This adds the concept of modularity to Genetic Programming (GP), [36], aiming to speed up evolution, and increase the achievable complexity. Koza showed that ADFs increase the evolutionary speed of GP. A similar modularity was introduced by Walker and Miller [69], for Cartesian Genetic Programming (CGP) [50], to speed up the evolution of more complex

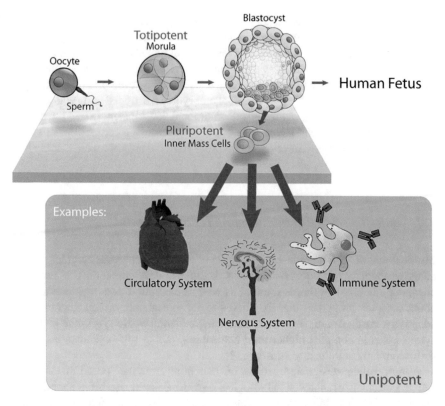

Fig. 1 The process of multicellular development brings a relatively simple cell to a system of many-cells that scale to enormously complex organisms, which are adaptive to their environment, and can also survive hostilities both at a cellular level and system (organism) level (taken from Wikipedia 6/2/14)

problems with CGP. It was shown that evolution of problems with modular CGP was much faster (20x in some cases), and scaled better for complex problems. The modularity in GP and CGP is done in a systematic way, where a modularity mechanism works in parallel with evolution to create modules from pieces of the evolved system which then can be reused by evolution. However, such explicitly defined mechanism that incorporates modularity into evolution can not solve the scalability problem. This is due to the presence of direct genotype[1]-phenotype[2] mapping. A direct genotype-phenotype encoding causes the genotype to grow in proportion to the phenotype. This creates a large search space as the target system gets more complex. Hence a direct mapping from genotype to phenotype is less effective in designing complex systems [1, 3, 20, 25, 51, 56, 68].

---

[1]Genetic information in a cell that is used to obtain a certain phenotype.

[2]The physical form and characteristics of an organism; the target system in EC.

In nature, biological organisms achieve phenotypes specified by genes that are orders of magnitude smaller. An example of this is the human genome, which comprises approximately 30,000 genes, yet a human brain alone has roughly 10^{11} neurons, [7, 9]. It is noteworthy that the number of distinct cell types in human body is around 200, and this number is as low as 13–15 for Hydra, which is a predatory animal with regenerative ability, [33]. The extremely complex structural and behavioural architecture of biological organisms is not through intelligent design, but the heavy reuse of cells and genes. Biology achieves a highly scalable mapping via multicellularity and gene reuse; each cell has the same copy of genotype, and each gene in a given genotype may have different effects depending on when and where they are expressed. Also, the same set of genes are used over and over again in building phenotypic structures of similar characteristics, e.g. limbs in animals. Taking inspiration from biology, the idea of multicellularity in achieving complex systems in EC has been implemented by many researchers. Although the ability of artificial development to be scalable has been demonstrated only in simple experiments, [3, 16, 20, 26, 41, 56], successful use of development in the design of systems at desired complexities that tackle real world problems is yet to be achieved.

1.2.2 Fault Tolerance

Biological organisms are robust creatures that can achieve a very high level of fault tolerance. The regenerative ability of plants is an excellent example of fault tolerance and recovery in biological organisms. Plants cells are classified as totipotent (the ability of a cell to grow and generate all the specialized parts of an organism and revert back to this stem-cell capability even after specialisation), hence, under the right conditions any plant cell would theoretically be able to grow into a fully developed adult plant [44]. Another example of a simple organism that features amazing regenerative properties is the Hydra, which has the ability to regenerate even when cut in half, producing two hydrae [4]. One of the main reasons for the type of repair and regeneration that happens in biological organisms is because of the lack of a central control mechanism. The ability of multiple cells to coordinate and organize themselves using various communication mechanisms provide an emergent adaptivity and fault tolerance to the whole organism.

Fault tolerant systems in electronics and computer science are highly important for remote, safety critical and hazardous applications. Almost all of the widely used techniques (N-modular redundancy [NMR] being the most popular) in achieving fault tolerance in electronics require a central control mechanism or a "golden" memory which is assumed to be failure-proof, [43]. A de-centralized multicellular architecture can provide the system designed with redundancy, allowing the destruction of a number of cells before failure. In a multicellular design all cells are essentially identical to one another, hence a cell has the potential to change specialization and replace a damaged cell in order to recover from faults. This multiple redundant behaviour in development can be used to create a system free of a single point of failure if each cellular structure is represented by an independent piece of hardware; hence removing the weakest link present in traditional redundancy designs.

In addition to cell redundancy, biological organisms also have functional redundancy in their genetic code. In biology this functional redundancy arising from different genetic codes is referred to as *degeneracy*. Edelman and Gally [15], define degeneracy as; "the ability of elements that are structurally different to perform the same function or yield the same output", and they also note that degeneracy "is a well known characteristic of the genetic code and immune systems." Edelman and Gally emphasize that degeneracy is a key mechanism for the robustness of complex mechanisms and that it is almost directly related with complexity. In biological organisms degeneracy is present at almost every functional level; from genes to high level behaviours like body movements and social behaviours, [15].

A developmental model can provide degeneracy both at the genotypic and phenotypic levels [40]. Due to the indirect mapping of genes to the target phenotype, a developmental system can have multiple genes that perform the same function. Depending on their location in the organism each cell would have a different gene activity, but some of these cells would still have the same phenotypic functionality. Degeneracy in a developmental system can provide a powerful fault tolerance mechanism, as it provides robustness to genetic perturbations. A well explained example of gene redundancy in biology is the control of platelet activation by collagen [54].

Artificial development has been shown to provide a smoother degradation to perturbed genetic code, [2, 42]; when the genetic code of an artificial organism is altered before mapping the genome to the respective phenotype, the damaged genome will provide a phenotype that shows a more "graceful" degradation for a developmental system in comparison to direct mapping of the genome. In fact in some cases it was shown that a small number of gene knock-outs did not affect the overall result of gene expression, [40, 55]. It has also been demonstrated that a developmental system may be able to recover from transient changes in the phenotype, despite sometimes not being explicitly trained to do so, [17, 46, 49, 55, 56]. However in order to benefit from the fault recovery properties of development, the developmental mechanism needs to be continuously running even when a fully functional phenotype is reached. In other words, the developmental system needs to have reached an *attractor*[3] that represents the desired phenotype. The ability of a developmental system to keep its phenotype unchanged after it has reached the target phenotypic form (i.e. represent the target phenotype as an attractor) is termed stability. To achieve a target phenotype at an attractor state of a developmental system (i.e. finding a stable system) can be a harder task than achieving the target phenotype at a *transient* state. Unfortunately, solutions that occur at transient states are of no use for fault tolerance or adaptivity.

---

[3] An attractor is a single (point attractor) or a group of states (cyclic attractor) to which a dynamical system settles after a time.

1.2.3 Adaptivity

Adaptive behaviour of biological organisms is another attractive quality that is aimed to be captured in EC. Designing systems that change their structure to adapt to their environments is a very challenging task, especially when a lot of the environmental factors can vary unpredictably. Multicellular organisms achieve adaptivity smoothly, and they change their structures or behaviours to fit the given environment for maximum survival chances. The stability of a multi cellular organism directly relies on the internal stability of its cells, i.e. the Gene Regulatory Network (GRN), as explored by [32]. An example of adaptivity in biology is a changing plant structure depending on sunlight: if a plant "discovers" that there is an obstruction in the way that blocks the sun, the plant will grow in a way to maximise sunlight exposure.

It is intended that by modelling multicellular development an adaptive system will be achieved. However as mentioned in Sect. 1.2.2, in order to achieve an adaptive system the developmental system needs to be at a stable state when it achieves a functional phenotype. Once a stable state is achieved, the developmental process can run continuously in the background and adapt to environmental changes. A developmental system can have several attractors, [67]; in an ideal case a dramatic change in the environment making the current configuration ineffective will cause the developmental system move to another attractor that would suit the current environmental conditions better.

Again, degeneracy in a developmental system can also allow the system to be more adaptive, because of the existence of multiple implementations of the same function. For example a change in an environmental condition may affect the activation of some genes in a developmental system, however the functionality of the organism would still be protected due to the existence of other genes that serve the same purpose as the affected genes. Systems with high degree of degeneracy have been observed to be very adaptable in biology as well, and favoured by natural selection, [15].

2 Artificial Developmental Systems

In the literature, artificial developmental systems have sometimes been classified into two categories as: *Grammatical* or *Cell Chemistry* developmental models, [8, 17, 18, 60]. However, there are developmental models that do not fit either of these classifications, e.g. in [23], the developmental system is designed in a way to function without the use of cell chemistry but the developmental model is far from a grammatical implementation. The reason for classifying developmental models into *Grammatical* and *Cell Chemistry* is because the grammatical models of development follow a high level abstraction of biology, whereas the models that involve cell chemistry follow a low-level abstraction of biological development, and these two models of development cover most of the artificial developmental models present in the literature. In order to make a similar but slightly clearer distinction amongst the present developmental models, we categorize the developmental systems as either *Macro*- or *Micro*-model developmental systems.

2.1 Macro-Model Developmental Systems

A macro-model developmental system models the biological development at a high abstraction level, considering the overall behaviour of a biological organism or a developmental mechanism. A macro-model system's implementation is largely different to its biological inspiration, since the aim is to model the characteristic behaviour of the target developmental system/mechanism. Simply put a macro-model developmental system does not model individual cells in a multi-cellular organism, but provides a developmental behaviour in the system by the inclusion of time and ability to self modify over time. A widely known example of a macro-model developmental system is the Lindenmayer Systems (L-Systems) [45]. L-Systems, a parallel rewriting system, was introduced for modelling the growth processes of plant development [45]. L-Systems model plant development using a set of rules via a grammar implementation, thus aiming to imitate biological development of plants using recursive functions. L-Systems have been applied to circuit design problems [25, 35], neural networks [5], and 3D morphology design [29, 58]. Another example of a grammatical developmental system is Cellular Encoding (CE) [21]. CE was designed to be used in the design of neural networks. Using CE, a neural net would learn recurrence and solve large parity problems such as a 51-bit parity problem [21].

An example of a non-grammatical macro-model developmental system is self-modifying Cartesian Genetic Programming (CGP), which models a CGP system that could alter its own structure over time after the evolution phase is complete [27]. One of the big advantages of using a macro-model developmental system is that they are generally computationally more efficient when compared to a micro-model developmental system. However, exceptions such as Cellular Automata (CA) [70], which is a computationally efficient to implement micro-model developmental system.

2.2 Micro-Model Developmental Systems

As a lower level model of the biological development, a micro-model developmental system uses a bottom-up approach to modelling development. This category of developmental systems can also be seen as the more biologically plausible implementations, which imitate biological development at a cellular level, see Fig. 2. Hence a micro-model developmental system involves the modelling of individual cells and their interactions, which together make up a whole organism. Each cell in a micro-model developmental system has the same genotype and inter-cellular

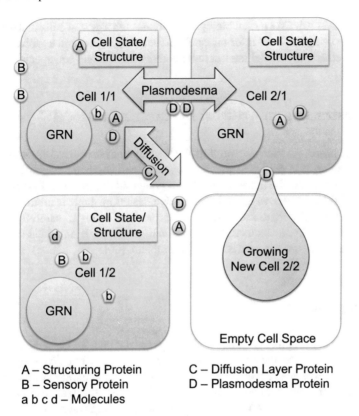

A – Structuring Protein   C – Diffusion Layer Protein
B – Sensory Protein   D – Plasmodesma Protein
a b c d – Molecules

Fig. 2 A micro-model developmental system considers the detailed process of multicellular development from the inner workings of a cell to its communication mechanisms with its neighbours

communication allows cells to specialize. All these cells together would form an organism which is the end product of development after each *developmental step*.[4]

Even though closer to biology, a micro-model developmental system does not necessarily model biological development accurately. In fact there is much work in this type of artificial development with diverse design constraints, those that model biology closely [16, 19, 30, 34, 39], those that aim to model biological development in a simplistic fashion [56, 62, 70], and models that are "in-between", [13, 20, 23, 48]. Mimicking biology closely should provide a developmental system with high evolvability, [3, 11], whereas a simplistic model would reduce the number of complicated processes that exist in biological development, reducing simulation times drastically. The first and one of the simplest examples to micro-model developmental

---

[4]The time it takes to carry out all the developmental processes—such as the processing of the genome and cell signalling for all cells– once, is referred to as a developmental step.

system is CA, [70]. CAs model biological systems with a grid of cells that determine their states using the local information from their neighbours and a global rule; this way, CAs effectively model inter-cellular communication and cell specialization.

3 Constructing an Artificial Developmental Model

Biological processes such as evolution and development are intricate systems. These partially understood processes provide engineers and computer scientists with new inspirations for new design techniques and computing paradigms. However, using models of these processes to achieve effective design methods is arduous. In order to exploit the full potential of these approaches, one needs to carefully construct their system. Each mechanism, however small, modelled is an added complexity, and another bias towards a certain type of problem, [40].

3.1 *Ingredients*

The process of designing an effective developmental model for EC reduces to implementing and sculpting the right biological mechanisms in an effective way. This aims to achieve an evolvable developmental system while maintaining a system that is not constrained or overwhelmed by undesirable biological processes. But how can we know which processes are useful and which are not? Implementations of artificial development in EC has been proposed since early 90's; e.g., [12, 19]. But most of the developmental models designed still rely on educated guesses, and various assumptions on the suitability of the biological developmental processes for EC applications. The need to investigate the behaviour and effective ways of implementing artificial development is already acknowledged by various researchers, [13, 22, 23, 60, 61]. Due to the lack of a solid understanding of (ADS)s, the practical use of them is still in infancy and many recent works still investigate and reproduce the known benefits and properties of ADSs in order to achieve a better understanding, [10, 28, 47, 53, 57].

Before the components of a developmental system are explored, the approach to modelling the developmental system needs to evaluated. Is macro-modelling appropriate for the problem at hand? Determining the right components of a successful developmental system starts with acquiring detailed knowledge of the problem that developmental system aims to solve. Properties such as scalability and adaptivity can often be conflicting, and mechanisms that favour one may often weaken the other [40]. Furthermore, the type symmetry (or asymmetry) that exists in the system also requires different dynamics from the developmental system and thus favours only some of the mechanisms available to a developmental system. In fact a pre-study of the problem being solved with the candidate system and the available mechanisms would be necessary for complex problems [40].

4 Case Studies

A detailed investigation into some of the poorly understood mechanisms and design constraints can be found in [42]. In this study, two dimensional patterns with different properties are used as experimental problems. The study shows that the type of problem being solved is highly correlated to the use of particular mechanisms. Conclusions then can be drawn from the resulting behaviour of a particular mechanism. The study explores the success of evolution in training an ADS, and concludes that the success of evolution is more likely when the evolution itself is given a larger control over the overall behaviour of the ADS. [42] provide a study in some of the common mechanisms used in micro-model developmental systems, such as cell signalling, chemical interactions, gene expression mechanisms, etc., where using the right combination of mechanisms provided up to four-fold improvement in success rates.

The study draws important conclusions on some of the fundamental mechanisms employed in ADSs. Understanding the capabilities of the mechanisms that accompany ADSs is an important step towards harnessing the full potential of ADSs for designing and optimizing engineering problems. With the advancing technology and a better understanding of what is needed from evolutionary computation, ADSs have the potential to be used in the evolutionary design of real life systems.

The conclusions drawn from this study led to the successful implementation of the practical applications, the following subsections detail a few examples.

4.1 *Robot Controller*

Navigation of a robotic system is a practical example with diverse challenges, which ADSs can be successfully applied. In [66], an ADS based controller is evolved, which is tested in simulation as well as on a real robot. The evolved control system is demonstrated to be robust and adaptive. The task is to achieve obstacle avoidance and reconnaissance; a developmental controller is evolved in evolution using the map shown in Fig. 3a.

In order to investigate whether the controller exhibits at least to a certain extent adaptive behaviour, it is tested in simulation on three different maps, shown in Fig. 3a, b, c. Despite these maps being considerably different, the robot successfully navigates around the obstacles and explores the map quite efficiently.

The controller evolved in Fig. 3a and tested in Fig. 3b, c, d, is also tested without alterations on a real robot, as shown in Fig. 3f. It is observed that the robot is trapped for an intial duration in the lower-right corner of the map shown in Fig. 3e, before it successfully resumes its primary wall-following behaviour, from then on without getting stuck in similar situations again, and navigates through the map. This demonstrates adaptation to the new environment when the controller is first deployed on a real robot. The described behaviour is achieved by evolving a developmental system

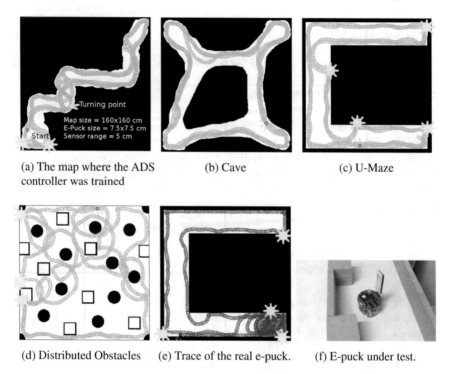

(a) The map where the ADS controller was trained

(b) Cave

(c) U-Maze

(d) Distributed Obstacles

(e) Trace of the real e-puck.

(f) E-puck under test.

Fig. 3 The figure shows a comparison of the behaviour of the same ADS based controller in different maps. The red and green lines show the robot's traces, and yellow marks show the collision points [66]

that controls the robots through the maze while being influenced by sensor readings. The ADS is allowed to run continuously in order to provide the adaptive behaviour demonstrated in the experiments.

4.2 Image Compression Using Artificial Development

One of the most prominent applications for studying ADSs is pattern generation and pattern formation, [13, 18, 45, 48]. The idea is simple; a two-dimensional array of identical entities (cells) achieve representation of a 2D pattern which is defined through the cell states or protein concentrations. This simple application can be extended to natural images, where instead of using the cell states as the only representation of images, the states at various points in the developmental time-line are used to represent parts of an image. The work by [65] introduces this application of ADSs to achieving competitive image compression rates.

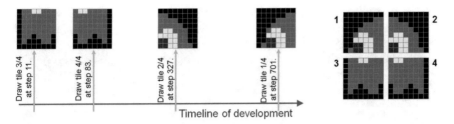

Fig. 4 Developing images as 8×8 tiles, where each tile is obtained at a certain developmental step [65]

An ADS is trained in a manner such that the resulting two dimensional array of cells form a pattern that matches a given set of target patterns as closely as possible at arbitrary developmental steps. For a defined starting point, the developed organisms are enumerated according to the order in which they occur while running the ADS. Hence, a sequence of complex states can be represented by the ADS and a sequence of integers. If the pattern at each developmental step is interpreted as a tile of a larger image, the image will be (re-)constructed during the developmental process. From a data compression point of view this is ideal since the states can become arbitrarily complex while the amount of compressed data remains constant for a fixed number of developmental steps. Figure 4 demonstrates the process of developing an image visually. The image is split into smaller tiles that fit the size of the developmental organism. The state of each cell at a certain developmental step represents a pixel of the whole image. The resulting tiles from each determined developmental step are then put together to form the complete image.

However, there are two major issues with this kind of data compression: first, optimisation of a dynamic system is time consuming and requires a large amount of computation. Second, it is not necessarily guaranteed that an arbitrary set of data can be exactly matched by any implementation of an ADS. While dynamic systems are probably not practical for lossless data compression, this work demonstrates the application of ADSs to lossy image data compression. In the latter case, it is not necessary to encode a perfect version of the input data, as long as sufficient image quality is achieved. In cases where smaller amounts of data are more important than maximum quality, the compression rate can be increased at the cost of losing information, i.e. losing image detail.

The resulting performance of the ADS as an image compression algorithm is impressive as it achieves better image compression than the most widely used image compression format: JPEG. The Genetically Compressed Images (GCI)—those that are the result of the evolved ADS—are shown side by side to JPEG compression of the same image in Fig. 5. With the shown image a 95% quality JPEG requires 150 kB of data storage, while the 6-pass GCI requires only 37 kB. The major drawback of the ADS based compression is the large amount of time needed for achieving the image compression (10 h on a 2.8 GHz Core2 desktop PC).

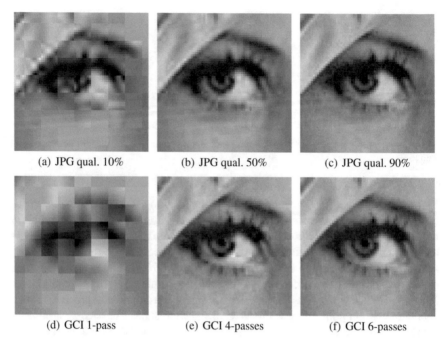

(a) JPG qual. 10%   (b) JPG qual. 50%   (c) JPG qual. 90%

(d) GCI 1-pass   (e) GCI 4-passes   (f) GCI 6-passes

Fig. 5 Comparison of JPG and GCI at low, medium and high quality settings [65]

The ADS's ability to achieve scalable designs is used in an innovative manner to achieve image compression rates not seen before.

5 Concluding Remarks

Artificial developmental systems present an attractive and advantageous medium to evolving systems that can solve difficult and complex problems by offering an evolvable genotypic landscape. There are vast amount of different approaches to modelling multicellular development, and none provides a silver bullet. To the contrary, each model with each configuration provides a different evolutionary landscape that makes its application suitable only for a selection of problems. Understanding the desired properties (and hence the application), and how to configure the ADS to meet those properties best is the key to a successful ADS. Prior to achieving the example applications presented in Sect. 4, it was necessary to go through a sequence of investigations in order to find the most suitable developmental mechanisms and appropriate parameters. It was also crucial to carefully study the application at hand in order to develop an artificial-development friendly genetic representation. It was only then when the full power of a generative genotype-to-phenotype mapping could be harnessed and achieve human-competitive results. Just like their biological example—and unlike any direct mapping ever will—artificial developmental mappings of genetic encoding to phenotype representing a solution are capable of learning and exploiting non-obvious and non-trivial similarities, regularities and patterns of a search space or solution. To conclude, therefore, it can be said that even though not as straight forward and one-size-fits all to design, generative mappings and artificial developmental processes hold the key to unlock efficient access to otherwise combinatorially infeasible large search spaces of complex, large-scale problems.

References

1. Banzhaf, W., Beslon, G., Christensen, S., Foster, J.A., Kepes, F., Lefort, V., Miller, J.F., Radman, M., Ramsden, J.J.: Guidelines: from artificial evolution to computational evolution: a research agenda. Nat. Rev. Genet. **7**, 729–735 (2006)
2. Bentley, P.: Investigations into graceful degradation of evolutionary developmental software. Nat. Comput. Int. J. **4**(4):417–437 (2005). doi:10.1007/s11047-005-3666-7
3. Bentley, P., Kumar, S.: Three ways to grow designs: a comparison of embryogenies for an evolutionary design problem. In: Proceedings of the Genetic and Evolutionary Computation Conference, pp. 35–43. Morgan Kaufmann, Orlando, Florida, USA. http://citeseer.ist.psu.edu/bentley99three.html (1999)
4. Bode, H.R.: Head regeneration in hydra. Dev. Dyn. **226**, 225–236 (2003)
5. Boers, E.J., Kuiper, H.: Biological Metaphors and the Design of Modular Artificial Neural Networks. Master's thesis, Leiden University (1992)
6. Bonner, J.T.: The origins of multicellularity. Integr. Biol. Iss. News Rev. **1**(1):27–36 (1998)
7. Braitenberg, V.: Brain size and number of neurons: an exercise in synthetic neuroanatomy. J. Comput. Neurosci. **10**(1):71–77 (2001)
8. Chavoya, A.: Foundations of Computational, Intelligence Volume 1, Studies in Computational Intelligence, vol. 201/2009, Springer Berlin / Heidelberg, Chap Artificial Development, pp. 185–215 (2009)
9. Claverie, J.M.: What if there are only 30,000 human genes? Science **291**(5507):1255–1257 (2001)
10. Cussat-Blanc, S., Harrington, K., Pollack, J.: Gene regulatory network evolution through augmenting topologies. IEEE Trans. Evol. Comput. **19**(6):823–837 (2015). doi:10.1109/TEVC.2015.2396199
11. Dawkins, R.: The evolution of evolvability. In: Kumar, S., Bentley, P. (eds.) On Growth Form and Computers, pp. 239–255. Elsevier Academic Press (2003)
12. Dellaert, F., Beer, R.D.: Toward an evolvable model of development for autonomous agent synthesis. In: Brooks, R.A., Maes, P. (eds.) Artificial Life IV, MIT Press Cambridge, pp. 246–257. http://citeseer.ist.psu.edu/dellaert94toward.html (1994)
13. Devert, A., Bredeche, N., Schoenauer, M.: Robust multi-cellular developmental design. In: GECCO '07: Proceedings of the 9th Annual Conference on Genetic and Evolutionary Computation, ACM, New York, NY, USA, pp. 982–989 (2007). doi:10.1145/1276958.1277156
14. Doursat, R., Sayama, H., Michel, O.: A review of morphogenetic engineering. Nat. Comput. **12**(4):517–535 (2013). doi:10.1007/s11047-013-9398-1
15. Edelman, G.M., Gally, J.A.: Degeneracy and complexity in biological systems. Proc. Nat. Acad. Sci. U.S.A. **98**(24):13,763–13,768 (2001). doi:10.1073/pnas.231499798
16. Eggenberger, P.: Evolving morphologies of simulated 3d organisms based on differential gene expression. In: Proceedings of 4th European Conference on Artificial Life, pp. 205–213 (1997)

17. Federici, D.: Using embryonic stages to increase the evolvability of development. In: Proceedings of the Genetic and Evolutionary Computation Conference (GECCO-2004), Springer (2004)
18. Flann, N., Jing, H., Bansal, M., Patel, V., Podgorski, G.: Biological development of cell patterns : characterizing the space of cell chemistry genetic regulatory networks. In: 8th European conference (ECAL), Springer, Berlin, Heidelberg (2005)
19. Fleischer, K., Barr, A.H.: A simulation testbed for the study of multicellular development: the multiple mechanisms of morphogenesis. In: Third Artificial Life Workshop Santa Fe, pp. 389–416. New Mexico, USA. http://citeseer.ist.psu.edu/fleischer93simulation.html (1993)
20. Gordon, T.G.W.: Exploiting Development to Enhance the Scalability of Hardware Evolution. Ph.D. thesis, University College London (2005)
21. Gruau, F.: Neural network synthesis using cellular encoding and the genetic algorithm. Ph.D. thesis, Ecole Normale Supirieure de Lyon, France. http://citeseerx.ist.psu.edu/viewdoc/summary? 10.1.1.29.5939 (1994)
22. Haddow, P., Hoye, J.: Investigating the effect of regulatory decisions in a development model. In: IEEE Congress on Evolutionary Computation CEC '09, pp. 293–300 (2009). doi:10.1109/CEC.2009.4982961
23. Haddow, P.C., Hoye, J.: Achieving a simple development model for 3d shapes: are chemicals necessary? In: Proceedings of the 9th Annual Conference on Genetic and Evolutionary Computation (GECCO), ACM, New York, NY, USA, pp. 1013–1020 (2007). doi:10.1145/1276958.1277160
24. Haddow, P.C., Tufte, G.: Bridging the genotype-phenotype mapping for digital fpgas. In: EH '01: Proceedings of the The 3rd NASA/DoD Workshop on Evolvable Hardware, IEEE Computer Society, Washington, DC, USA, p. 109 (2001)
25. Haddow, P.C., Tufte, G., van Remortel, P.: Shrinking the genotype: L-systems for EHW? In: ICES, pp. 128–139. http://citeseer.ist.psu.edu/554036.html (2001)
26. Harding, S., Miller, J.F., Banzhaf, W.: Self modifying cartesian genetic programming: Parity. In: 2009 IEEE Congress on Evolutionary Computation, pp. 285–292 (2009). doi:10.1109/CEC.2009.4982960
27. Harding, S.L., Miller, J.F., Banzhaf, W.: Self-modifying cartesian genetic programming. In: GECCO '07: Proceedings of the 9th Annual Conference on Genetic and Evolutionary Computation, ACM, New York, NY, USA, pp. 1021–1028 (2007). doi:10.1145/1276958.1277161
28. Harrington, K.I.: A circuit basis for morphogenesis. Theor. Comput. Sci. **633**:28–36 (2016). doi:10.1016/j.tcs.2015.07.002
29. Hornby, G.S., Pollack, J.B.: The Advantages of Generative Grammatical Encodings for Physical Design. IEEE Press, In. In Congress on Evolutionary Computation (2001)
30. Jakobi, N.: Harnessing morphogenesis. In: International Conference on Information Processing in Cells and Tissues, pp. 29–41. http://citeseer.ist.psu.edu/jakobi95harnessing.html (1995)
31. Kalganova, T.: Bidirectional incremental evolution in extrinsic evolvable hardware. In: Lohn, J., Stoica, A., Keymeulen, D. (eds.) The Second NASA/DoD Workshop on Evolvable Hardware, pp. 65–74. IEEE Computer Society, Palo Alto, California (2000)
32. Kauffman, S.: Metabolic stability and epigenesis in randomly constructed genetic nets. J. Theor. Biol. **22**(3):437–467 (1969). doi:10.1016/0022-5193(69)90015-0
33. Kauffman, S.: At Home in the Universe: The Search for the Laws of Self-Organization and Complexity. Oxford University Press (1996)
34. Kitano, H.: A simple model of neurogenesis and cell differentiation based on evolutionary large-scale chaos. Artif. Life **2**(1):79–99 (1995)
35. Kitano, H.: Building complex systems using developmental process: an engineering approach. In: ICES '98: Proceedings of the Second International Conference on Evolvable Systems, Springer, London, UK, pp. 218–229 (1998)
36. Koza, J.R.: Genetic Programming: On the Programming of Computers by Means of Natural Selection. MIT Press, Cambridge, MA, USA (1992)
37. Koza, J.R.: Genetic Programming II: Automatic Discovery of Reusable Programs. MIT Press, Cambridge, MA, USA (1994)

38. Kumar, S., Bentley, P. (eds.): On Growth. Elsevier Academic Press, Form and Computers (2003a)
39. Kumar, S., Bentley, P.J.: Biologically plausible evolutionary development. In: In Proceedings of ICES 03, the 5th International Conference on Evolvable Systems: From Biology to Hardware, pp. 57–68 (2003b)
40. Kuyucu, T.: Evolution of Circuits in Hardware and the Evolvability of Artificial Development. Ph.D. thesis, The University of York (2010)
41. Kuyucu, T., Trefzer, M., Miller, J.F., Tyrrell, A.M.: A scalable solution to n-bit parity via articial development. In: 5th International Conference on Ph.D. Research in Microelectronics & Electronics (2009)
42. Kuyucu, T., Trefzer, M.A., Miller, J.F., Tyrrell, A.M.: An investigation of the importance of mechanisms and parameters in a multicellular developmental system. IEEE Trans. Evolution. Comput. 15(3):313–345 (2011). doi:10.1109/TEVC.2011.2132724
43. Lala, P.K.: Self-checking and Fault-Tolerant Digital Design. Morgan Kaufmann Publishers Inc., San Francisco, CA, USA (2001)
44. Leyser, O., Day, S.: Mechanisms in Plant Development. Blackwell (2003)
45. Lindenmayer, A.: Mathematical models for cellular interactions in development. I. filaments with one-sided inputs. J. Theor. Biol. 280–299 (1968)
46. Liu, H., Miller, J.F., Tyrrell, A.M.: Intrinsic evolvable hardware implementation of a robust biological development model for digital systems. In: EH '05: Proceedings of the 2005 NASA/DoD Conference on Evolvable Hardware, IEEE Computer Society, Washington, DC, USA, pp. 87–92 (2005). doi:10.1109/EH.2005.32
47. Lopes, R.L., Costa, E.: Rencode: a regulatory network computational device. In: Genetic Programming: 14th European Conference, EuroGP, pp. 142–153 (2011)
48. Miller, J.F.: Evolving developmental programs for adaptation, morphogenesis, and self-repair. In: 7th European Conference on Artificial Life, Springer LNAI, pp. 256–265 (2003)
49. Miller, J.F.: Evolving a self-repairing, self-regulating, french flag organism. Genetic and Evolutionary Computation GECCO, pp. 129–139. Springer, Berlin / Heidelberg (2004)
50. Miller, J.F., Thomson, P.: Cartesian genetic programming. In: Genetic Programming, Proceedings of EuroGP'2000, Springer, pp. 121–132 (2000)
51. Miller, J.F., Thomson, P.: (2003) A developmental method for growing graphs and circuits. In: Evolvable Systems: From Biology to Hardware, 5th International Conference, pp. 93–104
52. Murakawa, M., Yoshizawa, S., Kajitani, I., Furuya, T., Iwata, M., Higuchi, T.: Hardware evolution at function level. In: PPSN IV: Proceedings of the 4th International Conference on Parallel Problem Solving from Nature. Springer, London, UK, pp. 62–71 (1996)
53. Öztürkeri, C., Johnson, C.G.: Self-repair ability of evolved self-assembling systems in cellular automata. Genet. Prog. Evol. Mach. 15(3):313–341 (2014). doi:10.1007/s10710-014-9216-2
54. Pearce, A.C., Senis, Y.A., Billadeau, D.D., Turner, M., Watson, S.P., Vigorito, E.: Vav1 and vav3 have critical but redundant roles in mediating platelet activation by collagen. Biol. Chem. 279(53):955–953, 962 (2004)
55. Reil, T.: Dynamics of gene expression in an artificial genome - implications for biological and artificial ontogeny. In: ECAL '99: Proceedings of the 5th European Conference on Advances in Artificial Life, Springer, London, UK, pp. 457–466 (1999)
56. Roggen, D.: Multi-cellular Reconfigurable Circuits: Evolution Morphogenesis and Learning. Ph.D. thesis, EPFL. http://citeseer.ist.psu.edu/roggen05multicellular.html (2005)
57. Schramm, L., Jin, Y., Sendhoff, B.: Evolution and analysis of genetic networks for stable cellular growth and regeneration. Artific. Life 18(4):425–444 (2012)
58. Sims, K.: Evolving 3d morphology and behavior by competition. Artif. Life 1(4):353–372 (1994)
59. Stanley, K.O.: Compositional pattern producing networks: a novel abstraction of development. Gen. Prog. Evolv. Mach. 8(2):131–162 (2007). doi:10.1007/s10710-007-9028-8
60. Stanley, K.O., Miikkulainen, R.: A taxonomy for artificial embryogeny. Artif. Life 9(2):93–130 (2003). doi:10.1162/106454603322221487

61. Steiner, T., Jin, Y., Sendhoff, B.: A cellular model for the evolutionary development of light-weight material with an inner structure. In: GECCO '08: Proceedings of the 10th annual conference on Genetic and evolutionary computation, ACM, New York, NY, USA, pp. 851–858 (2008). doi:10.1145/1389095.1389260

62. Tempesti, G., Mange, D., Petraglio, E., Stauffer, A., Thoma, Y.: Developmental processes in silicon: an engineering perspective. In: EH '03: Proceedings of the 2003 NASA/DoD Conference on Evolvable Hardware, IEEE Computer Society, Washington, DC, USA, pp. 255–264 (2003)

63. Torresen, J.: A divide-and-conquer approach to evolvable hardware. In: ICES '98: Proceedings of the Second International Conference on Evolvable Systems, Springer, London, UK, pp. 57–65 (1998)

64. Trefzer, M.A., Kuyucu, T., Miller, J.F., Tyrrell, A.M.: A model for intrinsic artificial development featuring structural feedback and emergent growth. In: IEEE Congress on Evolutionary Computation (2009)

65. Trefzer, M.A., Kuyucu, T., Miller, J.F., Tyrrel, A.M.: Image compression of natural images using artificial gene regulatory networks. In: GECCO'10 (2010a)

66. Trefzer, M.A., Kuyucu, T., Miller, J.F., Tyrrell, A.M.: Evolution and analysis of a robot controller based on a gene regulatory network. In: The 9th International Conference on Evolvable Systems: From Biology to Hardware (2010b)

67. Tufte, G.: The discrete dynamics of developmental systems. In: Evolutionary Computation, 2009. CEC '09. IEEE Congress on, pp. 2209–2216 (2009). doi:10.1109/CEC.2009.4983215

68. Vassilev, V.K., Miller, J.F.: Scalability problems of digital circuit evolution: Evolvability and efficient designs. In: EH '00: Proceedings of the 2nd NASA/DoD workshop on Evolvable Hardware, IEEE Computer Society, Washington, DC, USA, pp. 55–64 (2000)

69. Walker, J., Miller, J.: Evolution and acquisition of modules in cartesian genetic programming. EuroGp, Springer, Berlin, Heidelberg **3003**, 187–197 (2004)

70. Wolfram, S.: A New Kind of Science. Wolfram Media, Champaign, IL, USA (2002)

71. Wolpert,, L., Beddington, R., Jessell, T.M., Lawrence, P., Meyerowitz, E.M., Smith, J.: Principles of Development. Oxford University Press (2002)

Computers from Plants We Never Made: Speculations

Andrew Adamatzky, Simon Harding, Victor Erokhin, Richard
Mayne, Nina Gizzie, Frantisek Baluška, Stefano Mancuso
and Georgios Ch. Sirakoulis

Abstract Plants are highly intelligent organisms. They continuously make distributed processing of sensory information, concurrent decision making and parallel
actuation. The plants are efficient green computers per se. Outside in nature, the
plants are programmed and hardwired to perform a narrow range of tasks aimed
to maximize the plants' ecological distribution, survival and reproduction. To 'per-
suade' plants to solve tasks outside their usual range of activities, we must either
choose problem domains which homomorphic to the plants natural domains or modify biophysical properties of plants to make them organic electronic devices. We
discuss possible designs and prototypes of computing systems that could be based

A. Adamatzky (✉) · S. Harding · R. Mayne · N. Gizzie
Unconventional Computing Centre, UWE Bristol, UK
e-mail: andrew.adamatzky@uwe.ac.uk

S. Harding
e-mail: slh@evolutioninmaterio.com

R. Mayne
e-mail: richard.mayne@uwe.ac.uk

N. Gizzie
e-mail: nina.gizzie@gmail.com

V. Erokhin
CNR-IMEM, Parma, Italy
e-mail: victor.erokhin@fis.unipr.it

F. Baluška
Institute of Cellular and Molecular Botany,
University of Bonn, Bonn, Germany
e-mail: baluska@uni-bonn.de

S. Mancuso
International Laboratory of Plant Neurobiology,
University of Florence, Firenze, Italy
e-mail: stefano.mancuso@unifi.it

G.Ch. Sirakoulis
Department of Electrical & Computer Engineering,
Democritus University of Thrace, Xanthi, Greece
e-mail: gsirak@ee.duth.gr

on morphological development of roots, interaction of roots, and analog electrical computation with plants, and plant-derived electronic components. In morphological plant processors data are represented by initial configuration of roots and configurations of sources of attractants and repellents; results of computation are represented by topology of the roots' network. Computation is implemented by the roots following gradients of attractants and repellents, as well as interacting with each other. Problems solvable by plant roots, in principle, include shortest-path, minimum spanning tree, Voronoi diagram, α-shapes, convex subdivision of concave polygons. Electrical properties of plants can be modified by loading the plants with functional nanoparticles or coating parts of plants of conductive polymers. Thus, we are in position to make living variable resistors, capacitors, operational amplifiers, multipliers, potentiometers and fixed-function generators. The electrically modified plants can implement summation, integration with respect to time, inversion, multiplication, exponentiation, logarithm, division. Mathematical and engineering problems to be solved can be represented in plant root networks of resistive or reaction elements. Developments in plant-based computing architectures will trigger emergence of a unique community of biologists, electronic engineering and computer scientists working together to produce living electronic devices which future green computers will be made of.

1 Introduction

Plants perform neuronal-like computation not just for rapid and effective adaptation to an ever-changing physical environment but also for the sharing of information with other plants of the same species and for communication with bacteria and fungi [17–19, 21–25, 33]. In fact, plants emerge as social organisms. Roots are well known for their ability to avoid dangerous places by actively growing away from hostile soil patches. By using a vast diversity of volatiles, plants are able to attract or repel diverse insects and animals, as well as to shape their biotic niche. The number of volatile compounds released and received for plant communication is immense, requiring complex signal-release machinery, and 'neuronal-like' decoding apparatus for correct interpretation of received signals. The plant screens continuously and updates diverse information about its surroundings, combines this with the internal information about its internal state and makes adaptive decisions that reconcile its well-being with the environment [97–99]. Until now, detection and contextual filtering of at least 20 biological, physical, chemical and electrical signals was documented. Plants discriminate lengths, directions and intensities of the signal. These signals induce a memory that can last, depending on the signal, for hours or days, or even up to years. Once learnt, these plant memories usually ensure a much quicker and more forceful response to subsequent signalling [20, 98]. Via feedback crosstalks, memories become associative, inducing specific cellular changes. Thus plant behaviour is active, purpose-driven and intentional.

In last decade we manufactured a range of high-impact experimental prototypes from unusual substrates. These include experimental implementations of logical gates, circuits and binary adders employing interaction of wave-fragments in light-sensitive Belousov-Zhabotinsky media [9, 40, 41], swarms of soldier crabs [63], growing lamellypodia of slime mould *P*hysarum polycephalum [6], crystallisation patterns in 'hot ice' [2], jet streams in fluidic devices [83]. After constructing over 40 computing, sensing and actuating devices with slime mould [6] we turned our prying eyes to plants. We thought that plant roots could be ideal candidates to make unconventional computing because of the following features. Roots interact with each of their proximal neighbours and can be influenced by their neighbours to induce a tendency to align the directions of their growth [23, 39, 78]. Roots, as already mentioned, communicate their sensory information within and between adjacent roots, and other organisms like bacteria and fungi [14, 95]. This signalling and communicative nature of plant roots is very important for our major task to generate root-based computing devices. Outside in Nature, almost all roots are interconnected into huge underground root-fungal information networks, resembling the Internet. Roots can be grown in wet air chambers, as well as within nutrition solutes which allow accomplishment of relevant experiments. Roots can be isolated from seedlings using root vitro cultures which can be maintained for several years. Each root apex acts both as a sensory organ and as a brain-like command centre to generate each unique plant/root-specific cognition and behaviour [18, 22, 25]. It is easy to manipulate root behaviour using diverse physical and chemical cues and signals. Roots efficiently outsource external computation by the environment via the sensing and suppression of nutrient gradients. The induction of pattern type of roots behaviour is determined by the environment, specifically nutrient quality and substrate hardness, dryness etc. Moreover, roots are sensitive to illumination and show negative phototropism behaviour when exposed to light [35, 109], electric fields, temperature gradients, diverse molecules including volatiles and, therefore, allow for parallel and non-destructive input of information.

In 2012 Adamatzky, Baluska, Mancuso and Erokhin submitted ERC grant proposal "Rhizomes". They proposed to experimentally realize computing schemes via interaction of plant roots and to produce electronic devices from living roots by coating and loading the roots with functional materials, especially conductive polymers. This particular grant proposal was not funded however some ERC evaluators 'borrowed' methods and techniques from the proposal and implemented them their own laboratories. Most notable example is plant electronics ideas. They were originally outlined in "Rhizomes" proposal yet borrowed from the proposal by some Nordic fellows, who evaluated for ERC, and published without crediting original authors. Realising that most ideas will be 'stolen' in a similar fashion sooner or later anyway, we decided to disclose them in full.

2   Morphological Computation

In morphological phyto-computers, data are represented by initial configuration of roots and configurations of sources of attractants and repellents. Results of computation are represented by the topology of the roots' network. Computation is implemented by the roots following gradients of attractants and repellents and interacting with each other. Gradients of fields generated by configurations and concentrations of attractants are an important prerequisite for successful programming of plant-based morphological processors. Chemical control and programming of root behaviour can be implemented via application of attractants and repellents in the roots' growth substrate and/or in the surrounding air. In particular, chemo-attractants are carbo-hydrates, peptones, amino-acids phenylalanine, leucine, serine, asparagine, glycine, alanine, aspartate, glutamate, threonine, and fructose, while chemo-repellents are sucrose, tryptophan and salt. Plant roots perform complex computations by the following general mechanisms: root tropisms stimulated via attracting and repelling spatially extended stimuli; morphological adaptation of root system architecture to attracting and repelling spatially extended stimuli; wave-like propagation of information along the root bodies via plant-synapse networks; helical nutation of growing root apexes, patterns of which correlate with their environmental stimuli; competition and entrainment of oscillations in their bodies.

2.1   Shortest Path

Shortest path is a typical problem that every rookie in computer science solves, typically by coding the treasured Knuth algorithm. In a less formal definition, given a space with obstacles, source and destination points, the problem is to calculate a collision-free path from the source to the destination which is the shortest possible relative to a predefined distance metric. A root is positioned at the source, the destination is labelled with chemo-attractants, obstacles are represented by source of chemo-repellents and/or localised super-threshold illumination. The root-based search could be implemented without chemicals but the source and destination loci are represented by poles of AC/DC voltage/current and the obstacles by domains of super-threshold illumination or sub-threshold humidity. The collision-free shortest path is represented by the physical body of the root grown from the source to the destination. Implementation of this task could require suppression of the roots branching.

An experimental laboratory demonstration for showing feasibility of solving the shortest path with roots was demonstrated in Baluska lab and published in [108]. Their experiments have been done in Y-junction, turned upside down because roots are guided by gravity. Seeds were placed in the end of the vertical channel and attracting or neutral substances in the slanted channels. When just a distilled water is placed in both slanted channels, the roots from several seeds split arbitrarily between the channels (Fig. 1a). When a chemo-attractant, e.g. diethyl ether, is placed in one of the slanted channels all roots move into the channel with the chemoattractant (Fig. 1b).

Experiments on path finding by roots in mazes with more complex geometry than Y-junction are proved to be inconclusive so far. In [5] we presented results of few scoping experiments, see two illustrations in Fig. 2, yet reached no conclusive results. When seeds are placed in or near a central chamber of a labyrinth their routes somewhat grow towards exit of the labyrinth. However, they often become stuck midway and do not actually reach the exit. These findings have been confirmed also with roots of Arabidopsis (unpublished data). The inconclusive results are possibly due to the fact that we used gravity as the only guiding force to navigate the routes.

Collision-free path finding of plant roots was tested in further experiments in the Unconventional Computing Centre (UWE, Bristol). We have 3D printed with nylon

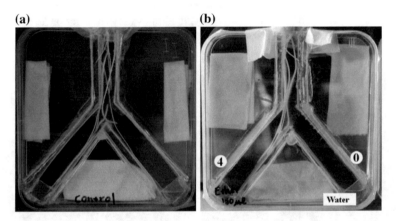

Fig. 1 Plant roots select a path towards attractant. Reprinted from [108]. **a** Control experiment. The maize roots are growing Y-maze with 1 ml of distilled water in both slanted channels. **b** Diethyl ether (100 μl ether and 900 μl distilled water is added to left slanted channel

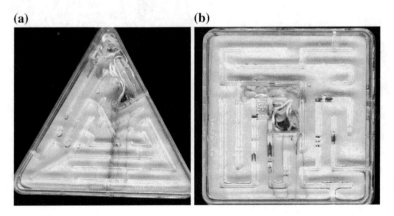

Fig. 2 Scoping experiments on routing plant roots in mazes. From [5]

templates of several countries, with elevation. Afterwards, we have been placing seeds either at bare templates or templates coated by 2% phytogel. Templates were kept in containers with very high humidity for up to several weeks. The templates rested horizontally: there was no preferentially directions of root growth, neither directed by gravity or light nor by chemo-attractants. Therefore roots explored the space. A single maize seed placed at the position roughly corresponding to Minneapolis, produced several roots. The roots propagated in different from each other directions thus optimising space explored (Fig. 3a). It seems that while navigating tips of roots avoid elevations: sometimes they go around localised elevation, sometimes they are reflected from the extended elevation. In another series of experiments, we scattered lettuce seeds on the template coated by phytogel. While scattering the seeds we were aiming to roughly approximate density of USA population: regions with high population density received a higher number of the seeds (Fig. 3b). We observed that roots tend to cluster in groups (Fig. 3c). This is a well known phenomenon [39] and groups of routes propagate along valleys or river beds, definitely

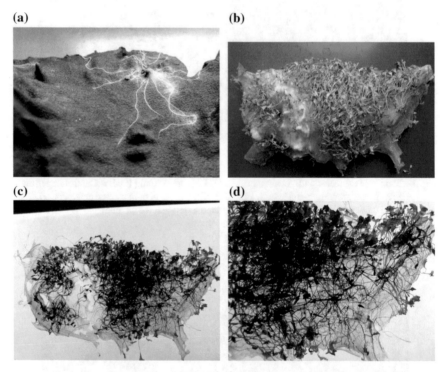

Fig. 3 Plant roots on 3D templates of the USA. **a** Maize seeds was placed at the location of the template corresponding to Minneapolis (the location was chosen for no apparent reason). Photo is made c. 7–10 days after inoculation. (**bcd**) Lettuce seeds were scatted in the USA template. **b** Dried template with lettuce seedlings. **c** Agar film removed from the template. **d** Zoomed part of the agar film representing eastern part of USA

avoiding any elevations (Fig. 3d). See also video of the experiment where lettuce seedlings grow on the 3D template of Russia: https://drive.google.com/open?id= 0BzPSgPF_2eyUYlNlWEVwenVoLUU.

2.2 Spanning Trees

Minimum spanning tree is a skeleton architecture of communication, transport and sensor networks. The Euclidean minimum spanning tree is a connected acyclic graph which has minimum possible sum of edges' lengths. To construct a spanning tree we can represent points of a data set with sources of nutrients and place a plant seed at one of the data points. The root will grow following gradients of the chemo-attractants, branch in the sites corresponding to the planar data set and, eventually, span the set with its physical body. In [7] we used live slime mould *Physarum polycephalum* to imitate exploration of planets and to analyse potential scenarios of developing transport networks on Moon and Mars. In that experiments we have been inoculating the slime mould on 3D templates of the planets, at the positions matching sites of Apollo or Soyuz landings; then we allowed the slime mould to develop a network of protoplasmic tubes. We have repeated similar experiments but used plant seeds instead of slime mould (Fig. 4). We found that roots could be as good as the slime mould in imitating exploratory propagation, or scouting, in an unknown terrains. The basic traits include maximisation of geographical distance between neighbouring roots and avoidance of elevations.

Fig. 4 Plant roots explore surface of 3D template of Moon. **a** Lettuce. **b** Basil

2.3 Crowd Dynamics

Can we study crowd dynamics using roots growing in a geometrically constrained environment? We printed a 3D nylon template of a part of Bristol city (UK), roughly a rectangular domain 2 miles × 2 miles, centered at Temple Mead Train station. In this template, blocks of houses were represented by elevations and streets as indentations. To imitate crowds we placed seeds of lettuce in large open spaces, mainly gardens and squares (Fig. 5a). The templates were kept in a horizontal position in closed

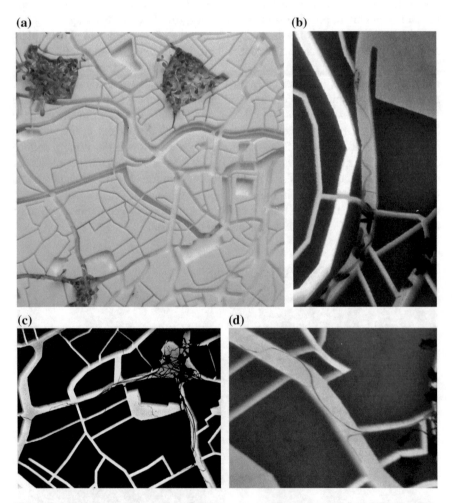

Fig. 5 Imitation of crowd propagation with plant roots on a 3D template of Bristol, UK. **a** Seedlings are growing in three open spaces, including Temple Gardens and Queen Square. **b** Example of bouncing movement of root apex. **c** Root apexes form two swarms separated by an obstacle. **d** Two apexes repel each other and choose different directions of propagation

transparent containers with very high moisture contents. Morphology of growing roots were recorded in 7–14 days after start of experiments. We have made the following observations so far.

Root apexes prefer wider streets, they rarely (if ever) enter side streets, and narrowing branches of main streets (Fig. 5b–d). This may be explained by the fact that plants emit ultrasound [86] and root apexes can sense ultrasound waves [57]. Chances are high that root apexes navigate in their constrained environment similarly to bats. Therefore entries to narrow streets are not detected.

In absence of attractants and repellents root apexes propagate ballistically: after entering a room a root grows along its original 'velocity' vector until it collides with an obstacle. The apex reflects on collision. This is well illustrated in Fig. 5b.

Root apexes swarm when propagating along wide streets (Fig. 5c), their growth is coherent, they often propagate in parallel, forming arrays of roots. Roots swarming is a known fact [39] yet still a valuable observation in the context of imitating crowds.

Rays of apexes often dissipate on entering the wider space as illustrated in Fig. 5d. As shown, two roots propagate westward along a narrow street. Then they enter main street and collide with its west side. On 'impact', one root deflects north-west another south-west.

2.4 Voronoi Diagram

Let P be a non-empty finite set of planar points. A planar Voronoi diagram of the set P is a partition of the plane into such regions that, for any element of P, a region corresponding to a unique point p contains all those points of the plane which are closer to p than to any other node of P. The planar diagram is combinatorially equivalent to the lower envelope of unit paraboloids centred at points of P. Approximation of the following Voronoi diagrams—planar diagram of point set, generalised of arbitrary geometrical shapes, Bregman diagrams on anisotropic and inhomogeneous spaces, multiplicative and furthest point diagrams—could be produced when implemented in experimental laboratory conditions with roots.

The Voronoi diagram can be approximated by growing roots similarly to the approximation of the diagram with slime mould [3] or precipitating reaction -diffusion fronts [8]. We represent every point of a data set P by a seed. The branching root system grows omnidirectionally. When wave-fronts of growing root systems, originated from different seeds, approach each other they stop further propagation. Thus loci of space not covered by roots represent segments of the Voronoi cells. The approach can be illustrated using a model of growing and branching pattern, developed by us originally to imitate computation with liquid crystal fingers [10]. Seeds are places in data points (Fig. 6a), root growth fronts propagate (Fig. 6b, c), collide with each other (Fig. 6d, e). The Voronoi diagram is approximated when the system becomes stationary and no more growth occurs (Fig. 6f).

(a)   (b)   (c)

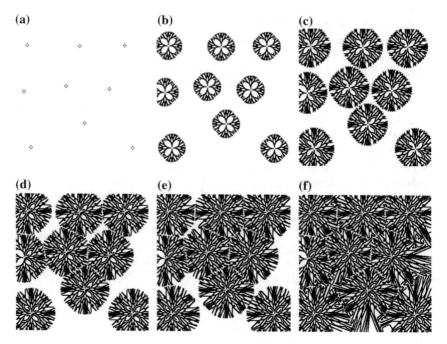

(d)   (e)   (f)

Fig. 6 Approximation of Voronoi diagram with growing and branching roots. See details of computer model in [10]. Snapshots are taken at **a** 160, **b** 216, **c** 234, **d** 246, **e** 256 and **f** 331 step of simulation

2.5 Planar Hulls

Computing a polygon defining a set of planar points is a classical problem of computational geometry. α-hull of a planar set P is an intersection of the complement of all closed discs of radius $1/\alpha$ that includes no points of P. α-shape is a convex hull when $\alpha \to \infty$. We can represent planar points P with sources of long-distance attractants and short-distance repellents and place a root outside the data set. The roots propagate towards the data and envelop the data set with their physical bodies. We can represent value of α by attractants/repellents with various diffusion constants and thus enable roots to calculate a wide range the shapes including concave and convex hulls. This approach worked well in experiments on approximation of concave hull with slime mould [4].

2.6 Subdivision of Concave Polygons

A concave polygon is a shape comprised of straight lines with at least one indentation, or angle pointing inward. The problem is to subdivide the given concave shape

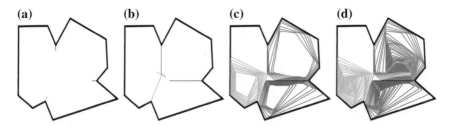

Fig. 7 Illustration on how a polygon can be subdivided by plant roots. The original model refers to liquid crystal fingers [10] but mechanisms of interaction could be the same. Snapshots are taken at different stages of simulation, see details in [10]

into convex shapes. The problem can be solved with growing roots as follows. Roots are initiated at the singular points of indentations (of the data polygon) and the roots propagation vectors are co-aligned with medians of the corresponding inward angles. Given a concave polygon, every indentation initiates one propagating root apex. By applying an external electro-magnetic field we can make root apexes turning only left (relatively to their vector of propagation). By following this "turn-left-if-there-is-no-place-to-go" routine and also competing for the available space with each other, the roots fill $n - 1$ convex domains. At least one convex domain will remain unfilled. See an example of subdivision of a concave polygon in Fig. 7.

2.7   Logical Gates from Plant Roots

A collision-based computation, emerged from Fredkin-Toffoli conservative logic [52], employs mobile compact finite patterns, which implement computation while interacting with each other [1]. Information values (e.g. truth values of logical variables) are given by either absence or presence of the localisations or other parameters of the localisations. The localisations travel in space and perform computation when they collide with each other. Almost any part of the medium space can be used as a wire. The localisations undergo transformations, they change velocities, form bound states and annihilate or fuse when they interact with other localisations. Information values of localisations are transformed as a result of collision and thus a computation is implemented. In [11] we proposed theoretical constructs of logical gates implemented with plant roots as morphological computing asynchronous devices. Values of Boolean variables are represented by plant roots. A presence of a plant root at a given site symbolises the logical TRUE, an absence the logical FALSE. Logical functions are calculated via interaction between roots. Two types of two-inputs-two-outputs gates are proposed [11]: a gate $\langle x, y \rangle \rightarrow \langle xy, x + y \rangle$ where root apexes are guided by gravity and a gate $\langle x, y \rangle \rightarrow \langle \overline{x}y, x \rangle$ where root apexes are guided by humidity. Let us show how a logical gate based on attraction of roots can be implemented.

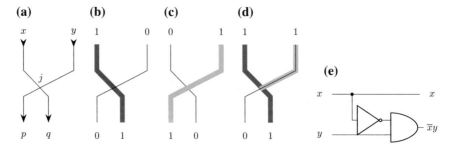

Fig. 8 **a** Scheme of humidity gate with two inputs x and y and two outputs p send q: $p = \bar{x}y$ and $q = x$. **b** $x = 1$ and $y = 0$. **c** $x = 0$ and $y = 1$. **d** $x = 1$ and $y = 1$. **e** Equivalent logic scheme. From [11]

Root apexes are attracted to humidity [22] and a range of chemical compounds [15, 62, 89, 92, 107, 108]. A root apexes grow towards the domain with highest concentration of attractants. The root minimises energy during its growth: it does not change its velocity vector if environmental conditions stay the same. This is a distant analog of inertia.

Assume attractants are applied at the exits of channels p and q (Fig. 8a). When an apex of the root, growing along channel x, reaches a junction between channels, the apex continues (due to energy minimisation) its growth into the channel q if this channel is not occupied by other root (Fig. 8b). A root in input channel y grows through the junction into the output channel p (Fig. 8c).

The gate Fig. 8a has such a geometry that a path along channel x to junction j is shorter than a path along channel y to the junction j. Therefore, a root growing in channel x propagates through the junction into channel q before root starting in channel y reaches the junction. When both roots are initiated in the input channels the x-root appears in the output q but the y-root is blocked by the x-root from propagating into the channel p: no signal appears at the output p (Fig. 8d). This gate realises functions $p = \bar{x}y$ and $q = x$ (Fig. 8e). If y is always 1 the gate produces a signal and its negation at the same time.

Two attraction gates Fig. 8a can be cascaded into a circuit to implement a one-bit half-adder, with additional output, as shown in Fig. 9a. We assume the planar gate is lying flat and sources of attractants are provided near exits of output channels p, q and r. The half-adder is realised on inputs $p = x \oplus y$ (sum) and $r = xy$ (carry); the circuit has also a 'bonus' output $q = x + y$ (Fig. 9e). The circuits work as follows:

- Inputs $x = 1$ and $y = 0$: Two roots are initiated in channels marked x in Fig. 9a; one root propagates to junction j_1 to junction j_3 and exits in channel q; another root propagates to junction j_4 to junction j_2 and into channel p (Fig. 9b).

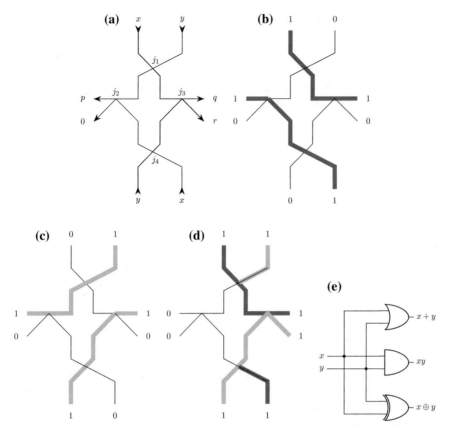

Fig. 9 A half-adder made of two humidity gates. **a** Scheme of the circuit, $p = x \oplus y$, $q = x + y$, $r = xy$. **b** $x = 1$ and $y = 0$. **c** $x = 0$ and $y = 1$. **d** $x = 1$ and $y = 1$. **e** Equivalent logic gates design. From [11]

- Inputs $x = 0$ and $y = 1$: Two roots are initiated in channels marked y in Fig. 9a; one root propagates to junction j_1 then to junction j_2 and exits at channel p; another root propagates to junction j_4 then to junction j_3 then into channel q (Fig. 9c).
- Inputs $x = 1$ and $y = 1$: Roots are initiated in all four input channels. The root initiated in the northern channel x propagates towards exit q. This root blocks propagation of the root initiated in the southern channel y, therefore the root from the southern channel x exits the circuit via the channel r. The root growing in the northern channel x blocks propagation of the root initiated in the norther channel y, therefore no roots appear in the output p (Fig. 9d).

3 Plant Electronics

In living electronic processors parts of a plant are functionalized via coating with polymers and loading with nano-particles, thus local modulations of the plant's electrical properties is achieved. A whole plant is transformed into an electronic circuit, where functionalized parts of the plant are basic electronic elements and 'raw' parts are conductors.

3.1 Plant Wire

In [5] we have exhibited that plants can function as wires, albeit slightly noisy ones. Namely, in laboratory experiments with lettuce seedlings we found that the seedlings implement a linear transfer function of input potential to output potential. Roughly an output potential is 1.5–2 V less than an input potential, thus e.g. by applying 12 V potential, we get 10 V output potential. Resistance of 3–4 day lettuce seedling is about 3 M Ω on average. This is much higher than resistance of conventional conductors yet relatively low compared to the resistance of other living creatures [60]. In our experiments [5] we measured resistance by bridging two aluminium electrodes with a seedling. If we did insert Ag/AgCl needle electrodes inside the seedling, we would expect to record much lower resistance, as has been shown in [76].

Resistance of plant wires can be affected by temperature [76], illumination and chemical substances. For example, when a lettuce seedling is exposed to vapour of chloroform (one 1 μL in a standard 90 mm Petri dish) the seedling exhibits high amplitude irregular oscillations of its resistance (Fig. 10).

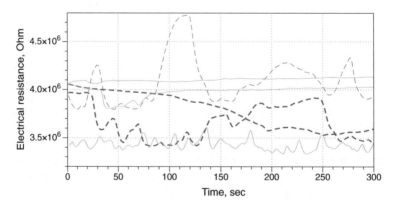

Fig. 10 Effect of chloroform on lettuce electrical resistance. Solid green lines show the resistance of an intact lettuce seedling and dashed red lines the resistance of a seedling in presence of chloroform

3.2 Functionalizing Plants

Electrical properties of plants can be changed by synthesis of inorganic materials by plants, coating roots with metal nano-particles and conductive polymers, growth of alloy networks. Here we overview general principles, some of them might not be applicable for plants, subject to further studies.

Many organisms, both unicellular and multicellular, produce inorganic materials either intra- or extra-cellular, e.g. include magneto-tactic bacteria (which synthesize magnetite nano-particles), diatoms (which synthesize siliceous materials) and S-layer bacteria (which produce gypsum and calcium carbonate layers). Biomimetic inorganic materials have been recently achieved by morpho-synthesis of biological templates such as viruses, bacteria, diatoms, biopolymers, eggshells, sea urchins, spider silks, insects, wood and leaves. We can employ recent results in biomorphic mineralization to produce reusable components of plants with conductive and magnetic substrates.

To coat roots with gold nano-particles (intake, transport and excretion) we can proceed as follows. Gold nano-particles (20–500 nm) pure or coated with bio-affine substances can be saturated in the feeding substrate of roots and/or applied as a liquid directly to growing roots. The gold particles will be in-taken by the roots, transported along the roots, distributed in the root network and eventually excreted onto the outer cell wall of root cells. When roots cease to function, the gold coating stays in place, providing passive conducting pathway.

Cellular synthesizes can be implemented via in plant growth of (semi-)conductive crystals from metal ions present in substrate. The growing roots will recover silver particle extracellularly from solutions containing Ag^+ ions, e.g. by saturating roots in 1 mM soluble silver in the log phase of growth. We can expose roots to aqueous $AuCl^{4-}$ ions; the exposure will result in reduction of the metal ions and formation of gold nano-particles of around 20 nm.

To growth alloy networks of plants we can use two approaches. First, co-growing of single metal networks. One root network is coated gold, and then another root network is grown on top of it and coated with silver. Another approach could be to synthesise nano-materials with programmed morphology. Some preliminary studies with biological substrates, were successful. Alloy nano-particles exhibit unique electronic, optical, and catalytic properties that are different from those of the corresponding individual metal particles. We can expose plants to equi-molar solutions of $HAuCl_4$ and $AgNO_3$ so that the formation of highly stable Au-Ag alloy nano-particles of varying mole fractions could be achieved.

Polyaniline (PANI) is a conducting polymer that can reach very high level of the conductivity (about 50–100 S/cm in our experiments), several forms of PANI can be deposited using layer-by-layer technique. Next layer is electro-statically attracted and, again, its thickness growth is blocked when the previous charge is compensated and new charge prevents further absorption due to the electrostatic repulsion. We can coat surface of roots with PANI, with a control of the thickness of about 1 nm. The approach has been successfully tested on several biological objects, notable slime mould [28, 38, 44]. When implementing passive electrical components from plants we can take into account the geometry and morphology of the roots network coated with PANI. The electrical conductivity of the signal pathway will depend only on the length and level of branching of the connection according to the Ohm's law. Of course, we will be able to set the basic level of the conductivity varying the thickness of the deposited layer. With regards to active electrical behaviour, conductivity state of PANI dependents on its redox state. Oxidized state is conducting and reduced state is insulating. The difference in conductivity is about 8 orders of magnitude. Thus, the conductivity of the individual zone of PANI will depend also on its actual potential (the use of plant roots implies the presence of the electrolyte, that will act as a medium for the redox reactions).The conductivity map will depend not only on the morphology, but also on the potential distribution map, that is connected to the previous function of the network. An attempt could be made to produce a simple bipolar junction transistor, with possible extension to an operational amplifier.

Another important feature of PANI layers, that can be useful for the plant computers, is its capability to vary the color according to the conductivity state [27]. This property will allow to register the conductivity map of the whole formed network, while usually we have the possibility to measure only between fixed points where electrodes are attached. If we are thinking about the system with learning properties [42], it can be not enough we must register the variation of the connections between all elements of the network. In the case of a double-layer perceptron, for example, it required a realization of rather complicated external electronic circuit, providing the temporal detachment of individual elements from the network for the registering of its conductivity [46]. Instead, the application of the spectroscopic technique allows monitoring of the conductivity state of all elements of the network in a real time for rather large area (up to half a meter), what will simplify, for example, the application of back propagation learning algorithms (Fig. 11).

Fig. 11 Lettuce functionalised with nanomaterials. **a** Scanning electron micrograph and EDX spectrum of lettuce seedlings treated with aluminium oxide: low magnification of a snapped lettuce seedling stem showing bright regions. **b** Light micrograph of 4 μm sections of lettuce seedlings treated with graphene in transverse orientation, haemotoxylin and eosin staining. From [61]

3.3   Case Study. *Modifying Lettuce with Nanomaterials*

In [61] we hybridised lettuce seedlings with a variety of metallic and non-metallic nanomaterials; carbon nanotubes, graphene oxide, aluminium oxide and calcium phosphate.   Toxic effects and the following electrical properties were monitored; mean potential, resistance and capacitance. Macroscopic observations revealed only slight deleterious health effects after administration with one variety of particle, aluminium oxide.

Mean potential in calcium phosphate-hybridised seedlings showed a considerable increase when compared with the control, whereas those administered with graphene oxide showed a small decrease; there were no notable variations across the remaining treatments. Electrical resistance decreased substantially in graphene oxide-treated seedlings whereas slight increases were shown following calcium phosphate and carbon nanotubes applications. Capacitance showed no considerable variation across treated seedlings. These results demonstrate that use of some nanomaterials, specifically graphene oxide and calcium phosphate may be used towards biohybridisation purposes including the generation of living 'wires'.

Graphene oxide and calcium phosphate were found to be, by a margin of at least 25%, the strongest modulators of the natural electrical properties of lettuce seedlings in this study, as is summarised in Fig. 12 and Table 1; although statistical significance was not achieved between the control and the nanomaterials, statistical significance was achieved between the resistance changes of graphene oxide and calcium phosphate.

We also provided evidence that nanomaterials are able to enter different histological layers of the plant, through demonstrating that graphene oxide and latex spheres become lodged in the epidermis whereas aluminium oxide and some latex spheres travel into the stem. With pores of plants being up to 8 nm in diameter, the size of the nanoparticles must be important when hybridising with lettuce seedlings. However, large nanomaterials may be able to embed and coat their surface, which provides extra benefits as any toxic effects caused by the nanomaterials would be reduced as there is less interference with proteins and intracellular mechanisms. Toxicity may

Fig. 12 Modifying lettuce electrical properties: resistance versus electrical potential plot. Nanomaterials used are graphene, carnon nanotybes (CNTs), calcium phosphate (CaPh) and aluminimum oxide (AO). From [61]

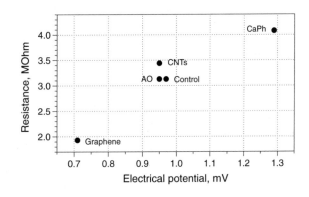

Table 1 Effects of selected nanomaterials on lettuce seedlings. If intake of the nanomaterial increases the measured parameters we indicate ↑, if it decreases ↓ and if parameter is within 5% of the control then the value is recorded as unchanged 0. If parameter is altered 25% above or below the control the arrow signs are encircled. Nanomaterials used are graphene, carnon nanotybes (CNTs), calcium phosphate (CaPh) and aluminimum oxide (AO). From [61]

Material	Potential	Resistance	Capacitance
Graphene	↓	ⓘ	↑
CNTs	0	↑	0
CaPh	ⓘ	↑	↓
AO	0	0	↑

also vary dependant on when the application of nanomaterials take place i.e. pre- or post- germination as well as different nanomaterials having different effects on plant species and organisms. So these factors i.e. size of nanoparticle, type of nanoparticle, when dispensed and plant species, all need to be considered when choosing a biological 'wire'.

3.4 Implementation of Logical Circuits Using Plant-Based Memristors

Memristor (memory resistor) is a device whose resistance changes depending on the polarity and magnitude of a voltage applied to the device's terminals and the duration of this voltage application. The memristor is a non-volatile memory because the specific resistance is retained until the application of another voltage [36, 37, 94]. A memristor implements a material implication of Boolean logic and thus any logical circuit can be constructed from memristors [32].

In 2013, we discovered memristive behaviour of a living substrate, namely slime mould *Physarum polycephalum* [58] while Volkov and colleagues reported that some plants, like Venus flytrap, *Mimosa pudica* and *Aloe vera*, exhibit characteristics analogous to memristors pitch curves in the electrical current versus voltage profiles [102]. A strong hypothesis is that more likely all living and unmodified plants are 'memristors' as well as living substrates, including slime mould [58], skin [77] and blood [68].

In our case, the basic device is an organic memristive system-element, composed of conducting polymer PANI, as described earlier, with a solid electrolyte heterojunction. Device conductivity is a function of ionic charge which is transferred through the heterojunction. Its application for the realization of adaptive circuits and systems, imitating synaptic learning, has been already demonstrated [30, 47–50]. By employing PANI coated plant roots as memristive devices, novel memristors-based electronics will be investigated and designed aimed at exploiting the potential advantages of this device in advanced information processing circuits. Original circuit design methodologies could be addressed to exploit the memristor non-linear behaviour and its memory properties in novel computational networks and especially when aiming at the design of beyond von Neumann computing systems.

In particular, memristor-based logic circuits open new pathways for the exploration of advanced computing architectures as promising alternatives to conventional integrated circuit technologies which are facing serious challenges related to continuous scaling [90], [72], [87]. However, up to now no standard logic circuit design methodology exists [104]. So, it is not immediately clear what kind of computing architectures would in practice benefit the most from the computing capabilities of memristors [32, 59, 69–71, 85, 103–105].

Even if we wish to apply some hybrid designs like 1 transistor—1 memristor (1T1M) to further explore the computing paradigms of such structures, while still lying in the living substrate level, the plants can be modified using a similar approach developed in [96]. Tarabella et al. implemented a transistor, a three-terminal active device that power amplifies an input signal, with slime mould. An organic electrochemical transistor is a semiconducting polymer channel in contact with an electrolyte. Its functioning is based on the reversible doping of the polymer channel. A hybrid Physarum bio-organic electrochemical transistor was made by interfacing an organic semiconductor, poly-3, 4-ethylenedioxythiophene doped with poly-styrene sulfonate, with the Physarum [96]. The slime mould played a role of electrolyte. Electrical measurements in three-terminal mode uncover characteristics similar to transistor operations. The device operates in a depletion mode similarly to standard electrolyte-gated transistors. The Physarum transistor works well with platinum, golden and silver electrodes. If the drain electrode is removed and the device becomes two-terminal, it exhibits cyclic voltage-current characteristics similar to memristors [96]. We are aiming to apply similar confrontation in the plant devices and discover by experiments the limits of the proposed approach.

As a result, basic logic and analog functionalities will be scouted for the specific organic memristor plant devices and should be compared with standard implementations on silicon. Our focus would be on the inherent low voltage and low power consumption of the proposed devices and it would be a key issue of the work. This approach will hopefully allow us to construct all the necessary universal logical circuits either by using one of the possible already existing, beyond von Neumann, computing architectures like, for example, material implication or by introducing a novel suitable, beyond-Von Neumann, computing system architecture.

4 Analog Computation on Electrical Properties of Plant Roots

In tasks of collision-based computing and implementation of conservative logical gates, apexes of roots represented discrete quanta of signals while acting in continuous time. When implementing analog computing devices with roots we adopt continuous representation of space and time and continuous values.

The following components will be analysed: resistors, capacitors, operational amplifiers, multipliers, potentiometers and fixed-function generators. We will evaluate a potential towards implementation of the core mathematical operations that will be implemented in experimental laboratory conditions: summation, integration with respect to time, inversion, multiplication, exponentiation, logarithm, division. The mathematical and engineering problems to be solved will be represented in plant root networks of resistive elements or reaction elements, involving capacitance and inductance as well as resistance, to model spatial distribution of voltage, current, electrical potential in space, temperature, pressure [66, 88, 106]. Implementations considered will included active elements analog computers, where no amplification required and passive elements computers, where amplification of signal is necessary.

Typically data in analogue circuits can be represented by resistors. By selectively modifying properties some parts of plant's root system, we make the plant to implement basic analog computing [65, 91]. A feasibility of constructing plant circuits with heterogeneous functionality of parts is supported by results showing that cerium dioxide nanoparticles in-taken by maize plants are not transferred to the newly grown areas of plants [31]. Let us consider an example of an analog circuit which can be produced from plants with modified electrical properties. This example is borrowed from our paper [61].

If we selectively adjust resistance of two branches of the same root and interface upper part of the root with an amplifier the plant will simulate certain equations. The classical (see e.g. [65, 91]) scheme is shown in Fig. 13. Root branches with modified electrical properties represent input resistors R_1 and R_2, upper modified root represents feedback resistance R_0. The output voltage v_0 is equal to the negative of the algebraic sums of the modified input voltages: $v_0 = -(\frac{R_0}{R_1}v_1 + \frac{R_0}{R_2}v_2)$. Depending on the ratios of the resistances and voltage polarity in each branch the scheme implements addition, subtraction and multiplication. The equation $v_0 = -(\frac{R_0}{R_1}v_1 + \frac{R_0}{R_2}v_2)$ can be seen as analogous to $z = ax + by$, where the output voltage v_0 has a polarity corresponding to the sign of z, the sign inversion of the amplifier is compensated by negative relationships between v_1 and x and v_2 and y: $v_1 = -x$ and $v_2 = -y$. Thus, we have $a = \frac{R_0}{R_1}$ and $b = \frac{R_0}{R_2}$.

Then we managed to selectively modify root branches of the same root with graphene $R_1 = R_{graphene} \approx 2M\Omega$ and calcium phosphate $R_2 = R_{CaPh} \approx 4M\Omega$ [61]. Part of the root above branching site remains unmodified: $R_0 = R_{control} \approx 3M\Omega$. Thus, the modified root simulates the equation $z = 1.3x + 2y$. The input voltages

Fig. 13 Summing amplifier simulating equation $z = ax + by$, $a = \frac{R_0}{R_1}$, $b = \frac{R_0}{R_2}$, $v_1 = -x$ and $v_2 = -y$, $v_0 = z$; see discussion in [61]

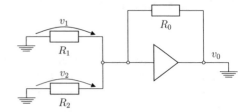

are multiplied by 1.3 and 2.0 and then added. If a and b are less than 1 then input voltage will be divided and then added: this can be achieved by, e.g. leaving one of the branches unmodified and loading part of the root above branching site with graphene. By making v_1 and v_2 with opposite polarity we simulate subtraction. To make a summing integration we need to substantially increase capacitance of some segments of a root [61].

The following tasks can be potentially implemented in plant-based analog computers:

- Hamilton circuit and Travelling Salesman Problem [29, 84]. Given cities, physically represented by current differences in the plant root network, find a shortest path through all cities, which visits each city once and ends in its source. Associated potential, or Lyapunov, function achieves its global minimum value at an equilibrium point of the network corresponding to the problem's solution.
- Satisfiability Problem [45, 67]. Values of variables of a Boolean logic formula are represented by voltage values of the root tree-like network and potential is indicated if the values can be assigned in a manner to make the Boolean formula true.
- Quadratic Diophantine Equation and partition problem [12]. Positive input numbers a, b, and c are represented by sources of electrical current to answer the question—are two positive integers x and y such that $(a \cdot x^2) + (b \cdot y) = c$.
- Majority-classification (earliest cellular automaton version is published in [54]). Given a configuration of input states a and b, represented by electrical characteristics of distant parts of plant root network, generate output value a if majority of inputs have value a and generated value b if the value b dominates in the inputs.
- Analog sorting of numbers [13]. Usually sorting of numbers assume discreteness of data. We can represent values of numbers to be sorted by electrical currents and apply principles of rational flow in a non-periodic Toda lattice [34] to undertake the smooth sorting.
- Implementation of Kirchhoff-Łukasiewicz machine [81, 82]. A Kirchhoff-Łukasiewicz machine was proposed by late Jonathan Mills [81, 82] to combine power and intuitive appeal of analog computers with conventional digital circuits. The machine is based on a sheet of conductive foam with array of probes interfaced with hardware implementation of Łukasiewcz logic arrays. The Łukasiewcz arrays are regular lattices of continuous state machines connected locally to each other. Arithmetic and logical functions are defined using implication and negated implication. Array inputs are differences between two electrical currents. Algebraic expressions can be converted to Łukasiewicz implications by tree-pattern matching and minimisation. The simplified expressions of implications can be further converted to layouts of living and mineralised/coated plant roots. The tasks to be implemented on plant-root based Kirchoff-Łukasiewicz machine are fuzzy controllers, tautology checkers, simulation of constraint propagation network with implications.

5 Evolution in Plants: Searching for Logical Gates

Given the success of implementing logical gates with the Mecobo evolvable hardware platform and Physarum [64], it was envisioned that plants may also be a suitable medium for 'evolution in materio' type experiments. The Mecobo was designed to provide a general purpose interface to allow for evolution in materio experiments and for probing the electrical properties of a substrate without understanding the underlying electrical properties of the substrate, and without having to develop new interface techniques for each material under investigation [73].

The same methodology was used to find gates as had previously been used for Physarum [64], and carbon nano-tubes and various polymers [74, 79, 80].

Eight electrodes were connected to the digital outputs of the Mecobo, via a current limiting 4.7 kΩ resistor, as shown in Fig. 14. The electrodes were then inserted through the stem of a plant. For these experiments, a common house hold plant, Schlumbergera, was used. Schlumbergera is a type of Brazilian cactus commonly known as a 'Christmas Cactus'. It has large, flat, stem segments that provide a good location to securely insert electrodes. All of the electrodes were inserted into a single segment. The positions chosen were essentially random, but with care taken to ensure that two electrodes were not very close together or touching as this would likely result in electrodes being shorted together, and reducing the chances of finding interesting behaviour. An example arrangement can be seen in Fig. 14a.

Similar to before, an exhaustive search was conducted by applying all possible binary combinations of various frequency pairs to 7 pins (which is a practical amount for time purposes). One pin was used as an output from the material, with the other 7 pins acting as inputs to the plant.

For each binary combination, each pair of frequencies was tried with one frequency representing a 'low' and the other representing a 'high' input. The frequency pairs were combinations of DC square waves of either (250 Hz, 500 Hz, 1 kHz, or 2.5 kHz). The amplitude of the square waves is 3.3 V.

(a)   **(b)**

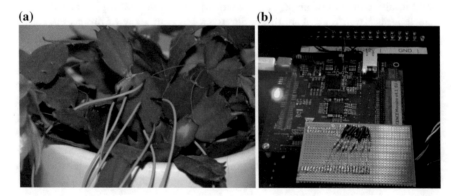

Fig. 14 Photos of experimental setup. **a** Electrodes in Schlumbergera cactus. **b** Mecobo board

The Mecobo measured the digital response from the plant for 32 ms. The digital threshold is 0.75 V for high, with voltages below this being classed as low. The sampling frequency was twice the highest input frequency applied.

Five different runs were completed. With different stem segments and electrode arrangements used each time.

Table 2, shows a summary for all of the runs. We see that all possible 2 input Boolean gates were implemented. As with previous work, we see that gates such as XOR and XNOR are found relatively infrequently. Looking at each run individually, Table 3 shows the same pattern. It is interesting to note that on each run all gates were found, and that in very similar proportions.

Table 4 shows how many XOR gates were found for each combination of frequencies used for representing the Boolean input states. We see that there is a bias towards the higher frequencies tested. We also see that not all combinations produce gates, and that the frequency pairs are not used asymmetrically. For example, true and false can be represented by either 2500 Hz or 1000 Hz, but representing false with 2500 Hz produces more viable gates. It appears that representing false by the higher frequency in the pair produces more solutions. More in depth analysis, and modelling of the results will be required to fully understand this behaviour, but it hints that expanding the search to use higher frequencies than 2500 Hz would result in more Boolean circuits being discovered.

Table 2 Number of gates mined from the frequency responses of the Schlumbergera

Cfg.	Inputs xy				Number of gates	Gate
	FF	FT	TF	TT		
1	F	F	F	F	95718	Constant false
2	T	F	F	F	366	x NOR y
3	F	T	F	F	304	NOT x AND y
4	T	T	F	F	430	NOT x
5	F	F	T	F	304	x AND NOT y
6	T	F	T	F	430	NOT y
7	F	T	T	F	74	x XOR y
8	T	T	T	F	314	x NAND y
9	F	F	F	T	510	x AND y
10	T	F	F	T	104	x XNOR y
11	F	T	F	T	863	y
12	T	T	F	T	307	NOT x AND NOT y OR y
13	F	F	T	T	863	x
14	T	F	T	T	307	x OR NOT y
15	F	T	T	T	512	x OR y
16	T	T	T	T	94564	Constant true

Table 3 Number of gates mined from the frequency responses of the Schlumbergera

Cfg.	Number of gates						Gate
	Run 1	Run 2	Run 3	Run 4	Run 5	Average	
1	19130	18870	19160	19198	19360	19144	Constant false
2	56	50	104	90	66	73	x NOR y
3	56	42	64	67	75	61	NOT x AND y
4	99	72	79	89	91	86	NOT x
5	56	42	64	67	75	61	x AND NOT y
6	99	72	79	89	91	86	NOT y
7	4	12	20	22	16	15	x XOR y
8	68	68	68	52	58	63	x NAND y
9	88	70	114	118	120	102	x AND y
10	8	8	38	32	18	21	x XNOR y
11	89	71	243	228	232	173	y
12	57	52	63	60	75	61	NOT x AND NOT y OR y
13	89	71	243	228	232	173	x
14	57	52	63	60	75	61	x OR NOT y
15	68	44	138	136	126	102	x OR y
16	19134	18844	18840	18842	18904	18913	Constant true

Table 4 Number of XOR gates mined from the frequency responses of the Schlumbergera for each frequency pair. Frequency A is used to represent False, Frequency B for True.

Frequency A	Frequency B	Count
2500	1000	28
1000	250	14
1000	2500	10
250	2500	8
500	250	8
2500	500	4
2500	250	2
1000	500	0
500	2500	0
500	1000	0
250	1000	0
250	500	0

Whilst the experiments with Schlumbergera were successful, experiments with other plants will be necessary to prove feasibility of the approach. Chances are high plants of different species will be showing differently shaped distributions of frequencies of logical gates discovered. Thus, we might construct a unique mapping between taxonomy of plants and geometries of logical gates distributions. Also, methodology wise, future experiments will investigate if it is possible to find solutions directly using evolution to find the pin configuration, rather than a time consuming exhaustive approach.

6 Brain Made of Plants

The survival of an organism depends on its ability to respond to the environment through its senses when neuronal systems translate sensory information into electrical impulses via neural code [43]. This enables multisensory integration that in turn leads to adaptive motor responses and active behavior. In plants, numerous physical environmental factors, especially light and gravity, are continuously monitored [21, 24, 25, 33, 55, 56, 75, 93]. Specialized plant cells have been optimized by evolution to translate sensory information obtained from the physical environment into motor responses known as tropisms, with root gravitropism representing one of the most intensively studied plant organ tropisms. Electrical signals are induced by all known physical factors in plants, suggesting that electricity mediates physical-biological communication in plants too [21, 51, 53, 100, 101]. Root gravitropism is a particularly instructive example in this respect. Sensory perception of gravity is accomplished at the very tip of the root apex, at the root cap [26], which includes a vestibular-like gravisensing organ composed of statocytes [17]. On the other hand, the motor responses, which begin almost immediately after root cap sensory events, are accomplished in relatively remote growth zones of the root apex [16]. Therefore, root gravitropism represents a nice example of a neuronal sensory-motor circuit in plants.

Quite possible future green computers will be using plants as elements of memory and neural network, followed by plant neuromorphic processors, will be developed. A rather artistic example of an attempt to make a neural-like network from plants is shown in Fig. 15. Seeds of lettuce were planted into the phytogel-made real-life sized model of a human brain cortex. When seedlings developed (Fig. 15a) recording electrodes were placed in the model brain (Fig. 15b); reference electrode was located medially between hemispheres. Electrical potential on each recording electrode was determined by electrical potential of c. 20 lettuce seedlings in the vicinity of the electrode.

To check if there is an interaction between seedlings in a response to stimulation we illuminated a group of seedlings. Electrical response of the lettuce population is shown in Fig. 15c. Indeed one experiment must not lead to any conclusion, however,

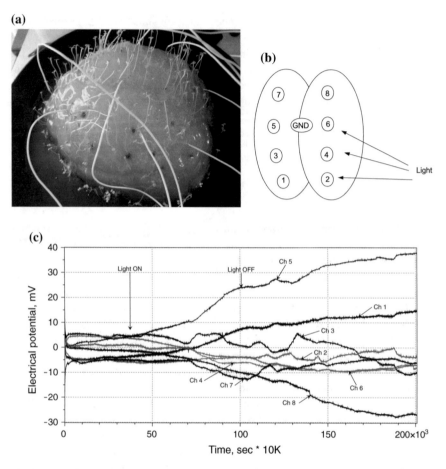

Fig. 15 Experimental setup towards a 'brain' made of lettuce seedlings. **a** A template of human cortex made from phytogel, lettuce seedlings and electrodes are visible. **b** Scheme of positions of electrodes and site of stimulation with light. **c** Electrical potential recorded over two days with sampling rate once per second

we can still speculate that light stimulus might lead to slight decrease in electrical potential of the illuminated population. The light stimulation might also lead to wider dispersion of electrical potential values in the population.

7 Discussion

We proposed designs of living and hybrid computing systems made of plants. The designs are original both in theory—collision-based computing, morphological computing, memristive circuits—and implementation—functional devices to be made of living or loaded with nanoparticles or coated with conductive polymers plant roots. When the proposed designs will be implemented they will contribute towards breakthroughs in computer science (algorithm/architectures of plant computing), computational geometry (plant based processors), graph-theoretic studies (proximity graphs by plant roots), biology (properties of plant at the interface with electronics), material science (functional elements), electronics (high density self-growing circuits), self-assembly and self-regenerative systems. The designs can be materialised in two type of phyto-chips: morphological chip and analog chips. The morphological chips are disposable computing devices based on computation with raw, unmodified, living roots; these chips represent result of computation by geometry of grown parts of plants. Computational geometry processors will be solving plane tessellations and generalised Voronoi diagrams, approximation of planar shapes (concave and convex hulls); these are classical problems of computational geometry. Almost all classical tasks of image processing, including dilation, erosion, opening and closing, expansion and shrinking, edge detection and completion can be solved using swarms of plant roots. Graph-optimisation processors will be solving tasks of parameterised spanning trees, Gabriel graph, relative neighbourhood graph, Delaunay triangulation, Steiner trees. Analog chips are based on biomorphic mineralisation of plant root networks and coating roots with metals and alloys; some architecture could be composed of living and artificially functionalised root networks. The analog phyto-chips are general purpose analog computers, capable for implementing multi-valued logic and arithmetic. What species of plants will be used in future phyto-chips? Best candidates are *Arabidopsis thaliana*, *Zea mays*, *Ocimum basilicum*, *Mentha genus*, *Raphanus sativus*, *Spinacia oleracea*, *Macleaya microcarpa*, *Helianthus annuus*. They could be chosen due to their growth characteristics, robustness, availability of mutant lines and research data on physiology and protocols of laboratory experiments.

References

1. Adamatzky, A.: Collision-Based Computing. Springer (2002)
2. Adamatzky, A.: Hot ice computer. Phys. Lett. A **374**(2), 264–271 (2009)
3. Adamatzky, A.: Physarum Machines: Computers from Slime Mould. World Scientific (2010)
4. Adamatzky, A.: Slime mould computes planar shapes. Int. J. Bio-Inspir. Comput. **4**(3), 149–154 (2012)
5. Adamatzky, A.: Towards plant wires. Biosystems **122**, 1–6 (2014)
6. Adamatzky, A. (ed.): Advances in Physarum Machines: Sensing and Computing with Slime Mould. Springer (2016)
7. Adamatzky, A., Armstrong, R., De Lacy Costello, B., Deng, Y., Jones, J., Mayne, R., Schubert, T., Sirakoulis, G.Ch., Zhang, X.: Slime mould analogue models of space exploration and planet colonisation. J. Br. Interplanet. Soc. **67**, 290–304 (2014)
8. Adamatzky, A., Costello, B.D.L., Asai, T.: Reaction-Diffusion Computers. Elsevier (2005)
9. Adamatzky, A., Holley, J., Bull, L., Costello, B.D.L.: On computing in fine-grained compartmentalised Belousov-Zhabotinsky medium. Chaos Solitons Fractals **44**(10), 779–790 (2011)
10. Adamatzky, A., Kitson, S., Costello, B.D.L., Matranga, M.A., Younger, D.: Computing with liquid crystal fingers: Models of geometric and logical computation. Phys. Rev. E **84**(6), 061,702 (2011)

11. Adamatzky, A., Sirakoulis, G.Ch., Martinez, G.J., Baluska, F., Mancuso, S.: On plant roots logical gates. arXiv preprint arXiv:1610.04602 (2016)
12. Adleman, L.M., McCurley, K.S.: Open problems in number theoretic complexity, ii. In: International Algorithmic Number Theory Symposium, pp. 291–322. Springer (1994)
13. Akl, S.G.: Parallel Sorting Algorithms, vol. 12. Academic press (2014)
14. Bais, H.P., Park, S.W., Weir, T.L., Callaway, R.M., Vivanco, J.M.: How plants communicate using the underground information superhighway. Trends Plant Sci. 9(1), 26–32 (2004)
15. Bais, H.P., Weir, T.L., Perry, L.G., Gilroy, S., Vivanco, J.M.: The role of root exudates in rhizosphere interactions with plants and other organisms. Annu. Rev. Plant Biol. 57, 233–266 (2006)
16. Baluška, F., Mancuso, S.: Plant neurobiology as a paradigm shift not only in the plant sciences. Plant Signal. Behav. 2(4), 205–207 (2007)
17. Baluška, F., Mancuso, S.: Deep evolutionary origins of neurobiology: turning the essence of'neural'upside-down. Commun. Integr. Biol. 2(1), 60–65 (2009)
18. Baluška, F., Mancuso, S.: Plant neurobiology: from sensory biology, via plant communication, to social plant behavior. Cognit. Process. 10(1), 3–7 (2009)
19. Baluška, F., Mancuso, S.: Vision in plants via plant-specific ocelli? Trends Plant Sci. 21(9), 727–730 (2016)
20. Baluška, F., Mancuso, S., Volkmann, D. (eds.): Communication in Plants: Neuronal Aspects of Plant Life. Springer (2007)
21. Baluška, F., Mancuso, S., Volkmann, D.: Communication in plants. In: Neuronal Aspect of Plant Life. Spriger, Heidelberg (2006)
22. Baluška, F., Mancuso, S., Volkmann, D., Barlow, P.: Root apices as plant command centres: the unique brain-like status of the root apex transition zone. Biologia (Bratisl.) 59(Suppl. 13), 1–13 (2004)
23. Baluška, F., Mancuso, S., Volkmann, D., Barlow, P.W.: Root apex transition zone: a signalling-response nexus in the root. Trends Plant Sci. 15(7), 402–408 (2010)
24. Baluška, F., Volkmann, D., Hlavacka, A., Mancuso, S., Barlow, P.W.: Neurobiological view of plants and their body plan. In: Communication in Plants, pp. 19–35. Springer (2006)
25. Baluška, F., Volkmann, D., Menzel, D.: Plant synapses: actin-based domains for cell-to-cell communication. Trends Plant Sci. 10(3), 106–111 (2005)
26. Barlow, P.W.: The response of roots and root systems to their environmentan interpretation derived from an analysis of the hierarchical organization of plant life. Environ. Exp. Bot. 33(1), 1–10 (1993)
27. Battistoni, S., Dimonte, A., Erokhin, V.: Spectrophotometric characterization of organic memristive devices. Org. Electron. 38, 79–83 (2016)
28. Battistoni, S., Dimonte, A., Erokhin, V.: Organic memristor based elements for bio-inspired computing. In: Advances in Unconventional Computing, pp. 469–496. Springer (2017)
29. Bellman, R.: Dynamic programming treatment of the travelling salesman problem. J. ACM (JACM) 9(1), 61–63 (1962)
30. Berzina, T., Erokhin, V., Fontana, M.: Spectroscopic investigation of an electrochemically controlled conducting polymer-solid electrolyte junction. J. Appl. Phys. 101(2), 024,501 (2007)
31. Birbaum, K., Brogioli, R., Schellenberg, M., Martinoia, E., Stark, W.J., Günther, D., Limbach, L.K.: No evidence for cerium dioxide nanoparticle translocation in maize plants. Environ. Sci. Technol. 44(22), 8718–8723 (2010)
32. Borghetti, J., Snidera, G.S., Kuekes, P.J., Yang, J.J., Stewart, D.R., Williams, R.S.: Memristive switches enable stateful logic operations via material implication. Nature 464(7290), 873–876 (2010)
33. Brenner, E.D., Stahlberg, R., Mancuso, S., Vivanco, J., Baluška, F., Van Volkenburgh, E.: Plant neurobiology: an integrated view of plant signaling. Trends Plant Sci. 11(8), 413–419 (2006)
34. Brockett, R.W.: A rational flow for the Toda lattice equations. In: Operators, Systems and Linear Algebra, pp. 33–44. Springer (1997)

35. Burbach, C., Markus, K., Zhang, Y., Schlicht, M., Baluška, F.: Photophobic behavior of maize roots. Plant Signal. Behav. **7**(7), 874–878 (2012)
36. Chua, L.: Memristor—the missing circuit element. IEEE Trans. Circuit Theory **18**(5), 507–519 (1971)
37. Chua, L.O., Tseng, C.W.: A memristive circuit model for p-n junction diodes. Int. J. Circuit Theory Appl. **2**(4), 367–389 (1974)
38. Cifarelli, A., Berzina, T., Erokhin, V.: Bio-organic memristive device: polyaniline–physarum polycephalum interface. Phys. Status Solidi (c) **12**(1-2), 218–221 (2015)
39. Ciszak, M., Comparini, D., Mazzolai, B., Baluska, F., Arecchi, F.T., Vicsek, T., Mancuso, S.: Swarming behavior in plant roots. PLoS One **7**(1), e29,759 (2012)
40. Costello, B.D.L., Adamatzky, A.: Experimental implementation of collision-based gates in Belousov-Zhabotinsky medium. Chaos Solitons Fractals **25**(3), 535–544 (2005)
41. Costello, B.D.L., Adamatzky, A., Jahan, I., Zhang, L.: Towards constructing one-bit binary adder in excitable chemical medium. Chem. Phys. **381**(1), 88–99 (2011)
42. Demin, V., Erokhin, V., Emelyanov, A., Battistoni, S., Baldi, G., Iannotta, S., Kashkarov, P., Kovalchuk, M.: Hardware elementary perceptron based on polyaniline memristive devices. Org. Electron. **25**, 16–20 (2015)
43. DeWeese, M.R., Zador, A.: Neurobiology: efficiency measures. Nature **439**(7079), 920–921 (2006)
44. Dimonte, A., Battistoni, S., Erokhin, V.: Physarum in hybrid electronic devices. In: Advances in Physarum Machines, pp. 91–107. Springer (2016)
45. Dowling, W.F., Gallier, J.H.: Linear-time algorithms for testing the satisfiability of propositional horn formulae. J. Log. Program. **1**(3), 267–284 (1984)
46. Emelyanov, A., Lapkin, D., Demin, V., Erokhin, V., Battistoni, S., Baldi, G., Dimonte, A., Korovin, A., Iannotta, S., Kashkarov, P., et al.: First steps towards the realization of a double layer perceptron based on organic memristive devices. AIP Adv. **6**(11), 111,301 (2016)
47. Erokhin, V., Berzina, T., Camorani, P., Fontana, M.P.: Non-equilibrium electrical behaviour of polymeric electrochemical junctions. J. Phys. Condens. Matter **19**(20), 205,111 (2007)
48. Erokhin, V., Berzina, T., Camorani, P., Smerieri, A., Vavoulis, D., Feng, J., Fontana, M.P.: Material memristive device circuits with synaptic plasticity: learning and memory. Bio-NanoScience **1**(1–2), 24–30 (2011)
49. Erokhin, V., Fontana, M.P.: Electrochemically controlled polymeric device: a memristor (and more) found two years ago. arXiv preprint arXiv:0807.0333 (2008)
50. Erokhin, V., Howard, G.D., Adamatzky, A.: Organic memristor devices for logic elements with memory. Int. J. Bifurcat. Chaos **22**(11), 1250,283 (2012)
51. Felle, H.H., Zimmermann, M.R.: Systemic signalling in barley through action potentials. Planta **226**(1), 203–214 (2007)
52. Fredkin, E., Toffoli, T.: Conservative logic. In: A. Adamatzky (ed.) Collision-Based Computing. Springer (2002)
53. Fromm, J., Lautner, S.: Electrical signals and their physiological significance in plants. Plant Cell Environ. **30**(3), 249–257 (2007)
54. Gács, P., Kurdyumov, G.L., Levin, L.A.: One-dimensional uniform arrays that wash out finite islands. Probl. Peredachi Informatsii **14**(3), 92–96 (1978)
55. Gagliano, M., Mancuso, S., Robert, D.: Towards understanding plant bioacoustics. Trends Plant Sci **17**(6), 323–325 (2012)
56. Gagliano, M., Renton, M., Depczynski, M., Mancuso, S.: Experience teaches plants to learn faster and forget slower in environments where it matters. Oecologia **175**(1), 63–72 (2014)
57. Gagliano, M., Renton, M., Duvdevani, N., Timmins, M., Mancuso, S.: Acoustic and magnetic communication in plants: is it possible? Plant Signal Behav **7**(10), 1346–1348 (2012)
58. Gale, E., Adamatzky, A., de Lacy Costello, B.: Slime mould memristors. BioNanoScience **5**(1), 1–8 (2015)
59. Gao, L., Alibart, F., Strukov, D.B.: Programmable cmos/memristor threshold logic. IEEE Trans. Nanotechnol. **12**(2), 115–119 (2013)

60. Geddes, L., Baker, L.: The specific resistance of biological materiala compendium of data for the biomedical engineer and physiologist. Med. Biol. Eng. **5**(3), 271–293 (1967)
61. Gizzie, N., Mayne, R., Patton, D., Kendrick, P., Adamatzky, A.: On hybridising lettuce seedlings with nanoparticles and the resultant effects on the organisms electrical characteristics. Biosystems **147**, 28–34 (2016)
62. Graham, T.L.: Flavonoid and isoflavonoid distribution in developing soybean seedling tissues and in seed and root exudates. Plant Physiol. **95**(2), 594–603 (1991)
63. Gunji, Y.P., Nishiyama, Y., Adamatzky, A., Simos, T.E., Psihoyios, G., Tsitouras, C., Anastassi, Z.: Robust soldier crab ball gate. Complex systems **20**(2), 93 (2011)
64. Harding, S., Koutnik, J., Greff, K., Schmidhuber, J., Adamatzky, A.: Discovering Boolean gates in slime mould. arXiv preprint arXiv:1607.02168 (2016)
65. James, M.L., Smith, G.M., Wolford, J.C.: Analog computer simulation of engineering systems. International Textbook Company (1966)
66. Johnson, C.L.: Analog Computer Techniques. McGraw-Hill Book Company, Incorporated (1963)
67. Kalmar, L., Suranyi, J.: On the reduction of the decision problem. J. Symb. Log. **12**(03), 65–73 (1947)
68. Kosta, S.P., Kosta, Y., Bhatele, M., Dubey, Y., Gaur, A., Kosta, S., Gupta, J., Patel, A., Patel, B.: Human blood liquid memristor. Int. J. Med. Eng. Inform. **3**(1), 16–29 (2011)
69. Kvatinsky, S., Belousov, D., Liman, S., Satat, G., Wald, N., Friedman, E.G., Kolodny, A., Weiser, U.C.: MAGIC - memristor-aided logic. IEEE Trans. Circuits Syst. **61-II**(11), 895–899 (2014)
70. Kvatinsky, S., Wald, N., Satat, G., Kolodny, A., Weiser, U.C., Friedman, E.G.: Mrl—memristor ratioed logic. In: 2012 13th International Workshop on Cellular Nanoscale Networks and their Applications, pp. 1–6 (2012)
71. Lehtonen, E., Tissari, J., Poikonen, J.H., Laiho, M., Koskinen, L.: A cellular computing architecture for parallel memristive stateful logic. Microelectron. J. **45**(11), 1438–1449 (2014)
72. Linn, E., Rosezin, R., Tappertzhofen, S., Bttger, U., Waser, R.: Beyond von Neumann logic operations in passive crossbar arrays alongside memory operations. Nanotechnology **23**(30), 305,205 (2012)
73. Lykkebø, O.R., Harding, S., Tufte, G., Miller, J.F.: MECOBO: A hardware and software platform for in materio evolution. In: International Conference on Unconventional Computation and Natural Computation, pp. 267–279. Springer (2014)
74. Lykkebø, O.R., Nichele, S., Tufte, G.: An investigation of square waves for evolution in carbon nanotubes material. In: 13th European Conference on Artificial Life (2015)
75. Mancuso, S.: Hydraulic and electrical transmission of wound-induced signals in vitis vinifera. Funct. Plant Biol. **26**(1), 55–61 (1999)
76. Mancuso, S.: Seasonal dynamics of electrical impedance parameters in shoots and leaves related to rooting ability of olive (Olea europea) cuttings. Tree Physiol. **19**(2), 95–101 (1999)
77. Martinsen, Ø.G., Grimnes, S., Lütken, C., Johnsen, G.: Memristance in human skin. In: Journal of Physics: Conference Series, vol. 224, p. 012071. IOP Publishing (2010)
78. Masi, E., Ciszak, M., Stefano, G., Renna, L., Azzarello, E., Pandolfi, C., Mugnai, S., Baluška, F., Arecchi, F., Mancuso, S.: Spatiotemporal dynamics of the electrical network activity in the root apex. Proc. Natl. Acad. Sci. **106**(10), 4048–4053 (2009)
79. Massey, M., Kotsialos, A., Qaiser, F., Zeze, D., Pearson, C., Volpati, D., Bowen, L., Petty, M.: Computing with carbon nanotubes: Optimization of threshold logic gates using disordered nanotube/polymer composites. J. Appl. Phys. **117**(13), 134,903 (2015)
80. Miller, J.F., Harding, S.L., Tufte, G.: Evolution-in-materio: evolving computation in materials. Evol. Intell. **7**(1), 49–67 (2014)
81. Mills, J.: Kirchhoff-Lukasiewicz Machines. Indiana University Web Sites Collection. (1995)
82. Mills, J.W.: The nature of the extended analog computer. Phys. D. **237**(9), 1235–1256 (2008)
83. Morgan, A.J., Barrow, D.A., Adamatzky, A., Hanczyc, M.M.: Simple fluidic digital half-adder. arXiv preprint arXiv:1602.01084 (2016)
84. Ore, O.: Note on Hamilton circuits. Am. Math. Mon. **67**(1), 55–55 (1960)

85. Papandroulidakis, G., Vourkas, I., Vasileiadis, N., Sirakoulis, G.Ch.: Boolean logic operations and computing circuits based on memristors. IEEE Trans. Circuits Syst. II Express Br. **61**(12), 972–976 (2014)
86. Perel'man, M.E., Rubinstein, G.M.: Ultrasound vibrations of plant cells membranes: water lift in trees, electrical phenomena. arXiv:preprint physics/0611133 (2006)
87. Pershin, Y.V., Ventra, M.D.: Neuromorphic, digital, and quantum computation with memory circuit elements. Proc. IEEE **100**(6), 2071–2080 (2012)
88. Peterson, G.R.: Basic Analog Computation. Macmillan (1967)
89. Schlicht, M., Ludwig-Müller, J., Burbach, C., Volkmann, D., Baluska, F.: Indole-3-butyric acid induces lateral root formation via peroxisome-derived indole-3-acetic acid and nitric oxide. New Phytol. **200**(2), 473–482 (2013)
90. Semiconductor Industry Association: International Technology Roadmap for Semiconductors (ITRS). Semiconductor Industry Association (2007). http://www.itrs2.net
91. Soroka, W.W.: Analog Methods in Computation and Simulation. McGraw-Hill (1954)
92. Steinkellner, S., Lendzemo, V., Langer, I., Schweiger, P., Khaosaad, T., Toussaint, J.P., Vierheilig, H.: Flavonoids and strigolactones in root exudates as signals in symbiotic and pathogenic plant-fungus interactions. Molecules **12**(7), 1290–1306 (2007)
93. Stone, B.B., Esmon, C.A., Liscum, E.: Phototropins, other photoreceptors, and associated signaling: the lead and supporting cast in the control of plant movement responses. Curr. Top. Dev. Biol. **66**, 215–238 (2005)
94. Strukov, D.B., Snider, G.S., Stewart, D.R., Williams, R.S.: The missing memristor found. Nature **453**(7191), 80–83 (2008)
95. Sugiyama, A., Yazaki, K.: Root exudates of legume plants and their involvement in interactions with soil microbes. In: Secretions and exudates in biological systems, pp. 27–48. Springer (2012)
96. Tarabella, G., D'Angelo, P., Cifarelli, A., Dimonte, A., Romeo, A., Berzina, T., Erokhin, V., Iannotta, S.: A hybrid living/organic electrochemical transistor based on the Physarum polycephalum cell endowed with both sensing and memristive properties. Chem. Sci. **6**(5), 2859–2868 (2015)
97. Trewavas, A.: Green plants as intelligent organisms. Trends Plant Sci. **10**(9), 413–419 (2005)
98. Trewavas, A.: What is plant behaviour? Plant Cell Environ. **32**(6), 606–616 (2009)
99. Trewavas, A.J., Baluška, F.: The ubiquity of consciousness. EMBO Rep. **12**(12), 1221–1225 (2011)
100. Volkov, A.G.: Electrophysiology and phototropism. In: Communication in Plants, pp. 351–367. Springer (2006)
101. Volkov, A.G., Ranatunga, D.R.A.: Plants as environmental biosensors. Plant Signal. Behav. **1**(3), 105–115 (2006)
102. Volkov, A.G., Tucket, C., Reedus, J., Volkova, M.I., Markin, V.S., Chua, L.: Memristors in plants. Plant Signal. Behav. **9**(3), e28,152 (2014)
103. Vourkas, I., Sirakoulis, G.Ch.: Memristor-based combinational circuits: a design methodology for encoders/decoders. Microelectron. J. **45**(1), 59–70 (2014)
104. Vourkas, I., Sirakoulis, G.Ch.: Emerging memristor-based logic circuit design approaches: a review. IEEE Circuits Syst. Mag. **16**(3), 15–30 (2016)
105. Vourkas, I., Sirakoulis, G.Ch.: Memristor-based nanoelectronic computing circuits and architectures. In: Emergence Complexity and Computation. Springer, Cham (2016)
106. Weyrick, R.C.: Fundamentals of Analog Computers. Prentice Hall (1969)
107. Xu, W., Ding, G., Yokawa, K., Baluška, F., Li, Q.F., Liu, Y., Shi, W., Liang, J., Zhang, J.: An improved agar-plate method for studying root growth and response of Arabidopsis thaliana. Sci. Rep. **3**, 1273 (2013)
108. Yokawa, K., Baluska, F.: Binary decisions in maize root behavior: Y-maze system as tool for unconventional computation in plants. IJUC **10**(5–6), 381–390 (2014)
109. Yokawa, K., Kagenishi, T., Kawano, T., Mancuso, S., Baluška, F.: Illumination of arabidopsis roots induces immediate burst of ros production. Plant Signal. Behav. **6**(10), 1460–1464 (2011)

Printed in the United States
By Bookmasters